DNA Engineering

DNA Engineering

Properties and Applications

edited by

Kenji Mizoguchi

Hirokazu Sakamoto

PAN STANFORD PUBLISHING

Published by

Pan Stanford Publishing Pte. Ltd.
Penthouse Level, Suntec Tower 3
8 Temasek Boulevard
Singapore 038988

Email: editorial@panstanford.com
Web: www.panstanford.com

British Library Cataloguing-in-Publication Data
A catalogue record for this book is available from the British Library.

DNA Engineering: Properties and Applications

ISBN 978-981-4669-46-7 (Hardcover)
ISBN 978-981-4669-47-4 (eBook)

Printed in the USA

To Keiko

Contents

PART IV APPLICATIONS OF DNA

Preface

I (K. M.) have been interested in the development of conducting polymers since 1980s, especially by making use of magnetic resonance techniques such as nuclear magnetic resonance (NMR) and electron spin resonance (ESR). In particular, measurement of ESR by varying the frequency from 10 MHz to 24,000 MHz is highly useful to study the electronic carrier dynamics in polycrystalline samples typical of conducting polymers, through which information about the anisotropic dynamics of the electronic carriers within the local crystalline regions can be obtained. In 2001, I had an opportunity to introduce the field of conducting polymers in butsuri (meaning physics in Japanese) in praise of Japanese Nobel laureates in chemistry in 2000, Drs. Hideki Shirakawa, A. MacDiarmid, and A. J. Heeger. After that, Dr. Kazumoto Iguchi let me know the DNA research field from the viewpoint of physics, which was a start of my research on DNA in materials science.

In early 2000s, many reports on the electrical properties of DNA making use of nanotechnologies appeared in the named journals, such as *Nature*, *Science*, and *Physics Review Letters*, with controversial conclusions, which strongly motivated us to make clear the basic physical properties of DNA. To avoid any physical contacts with DNA samples causing the controversial results, magnetic, optical, and AC measurements were mainly utilized for the purpose. This book is intended to provide the basic physical properties of DNA and the possible engineering and applications of DNA for the readers who are interested in DNA as one of the nanomaterials. Introduction to the book appears in Part I. Part II is devoted to describe the basic physical properties of DNA both experimentally and theoretically, which reconfirms that DNA has a semiconducting nature. In Part III, divalent metal ion M inserted

M-DNAs are described to introduce the charge carriers into the semiconducting DNA. Three examples of the applications of DNA to materials science, electronic devices, and fuel cell battery are demonstrated in Part IV.

The double-helical struture of the synthetic oligomer DNA is expressed in two ways in this book, poly(dA)-poly(dT) in Chapter 3 and Chapter 8 and poly(dA-dT) in Chapter 4 and Chapter 9, which is according to the already published presentations.

The chapters presented by us (K. M. and H. S.) are largely based on the contributions by many of the graduate and undergraduate students of the Tokyo Metropolitan University from 2002 to 2013. We would like to express our heartful thanks to them. We would also like to express our sincere thanks to all the contributors of this book.

Finally, we would like to express our sincere thanks to Stanford Chong, the director of Pan Stanford Publishing Pte. Ltd., for inviting us to edit this book.

Kenji Mizoguchi
Hirokazu Sakamoto
Tokyo, Japan
Summer 2016

Part I

Introduction

Chapter 1

Introduction

Kenji Mizoguchi
Department of Physics, Tokyo Metropolitan University, Hachioji, Tokyo, 192-0397 Japan
mizoguchi@phys.se.tmu.ac.jp

Genetic engineering or genetic modification of DNA with biotechnology in physiological and medical fields has been developing since the 1970s. An example is the production of human insulin through genetic engineering in 1978 [1], which was commercialized in 1982. Research in genetic engineering has expanded rapidly, and it is out of the focus of this book.

Deoxyribonucleic acid (DNA) is not only a genome but also a material with fascinating characteristics for a variety of applications. One of them is the self-organization ability through the complementarity of the base pairs, adenine–thymine and guanine–cytosine, with plannability of the base sequence. The double helical structure discovered by Watson and Crick [2] with the π–π stacking of the flat base molecules (Fig. 1.1) along the helical axis is also another characteristic property of DNA, which provides intercalation of flat aromatic molecules. Physical properties of DNA are strongly sensitive to the humidity of circumstances. The summary of DNA characteristics is as follows:

DNA Engineering: Properties and Applications
Edited by Kenji Mizoguchi and Hirokazu Sakamoto

ISBN 978-981-4669-46-7 (Hardcover), 978-981-4669-47-4 (eBook)
www.panstanford.com

- Self-organization based on the complementarity of base pairing
- Flexibility of base sequences
- π electron band formation with base stacking
- Flexibility of intercalation with a variety of flat molecules
- Physical properties controlled by the content of water molecules
- Unique optical properties

This book focuses on several approaches to new materials based on DNA with additional functions, such as electronic, optical, and other physical properties.

In the early 2000, controversial results on the electrical properties of natural DNA studied by emerging nanotechnologies with direct contacts to samples aiming at the future DNA engineering came into spotlight and created confusion [3–9]. This phenomenon has strongly suggested the importance of basic and scientific knowledge of DNA as a material for further development in the DNA engineering field. Thus, the main part of this book is devoted to the establishment of the basic knowledge of DNA, both experimentally and theoretically.

In the applications of semiconducting DNA as an electro-active material, charge carrier doping is inevitably required. Several approaches have been proposed so far: chemical doping [10–15], including divalent metal ion insertion into DNA [16–28], direct charge doping with FET structure [7, 9, 29, 30], and natural DNA itself with quantum current loops [31–34].

In this chapter, brief introductions of the basic information of DNA and DNA engineering are given. In this book, the term DNA engineering is used for all usages of DNA other than genetic engineering or genetic modification of DNA with biotechnology in physiological and medical fields. As DNA engineering, DNA origami and the first report on the metal ion insertion study are introduced. DNA origami utilizes the fascinating characteristics of DNA to form arbitrary nanostructures in 2D and 3D in a computer-aided style, which opens new developments in nanotechnology and nanomachines, such as drug delivery systems [35].

In Part II, the basic physical properties of DNA are summarized: optical properties, magnetic properties, electrical properties, structural properties through infrared (IR) absorption, and a basic theoretical approach with Hückel theory.

In Part III, the recent development of DNA engineering with divalent metal ion insertion into DNA aiming a modification of the physical properties of DNA is described in the viewpoint of scientific understanding and a new material. Optical, magnetic, IR properties; a special case of freeze-dried Zn-DNA (FD-Zn-DNA) with carrier doping; and a unique approach of the structural study in *M*-DNA with a scanning tunneling microscope (STM) are presented.

In Part IV, three examples of applications in DNA engineering are presented, which widely extend over optical (fiber, switch, and laser) and electrical (memory, LED, and fuel cell) materials based on natural, especially marine, resources with molecular weight control, together with physiological and medical materials.

1.1 Natural DNA

1.1.1 *Structure of DNA*

DNA has been known as the container of life information, which is carried by the base codon made of three sequential nucleobases from four bases: guanine (G), adenine (A) (purine), cytosine (C), and thymine (T) (pyrimidine). A purine base connects to a pyrimidine base with two hydrogen bonds for adenine–thymine (AT) pair and three for guanine–cytosine (GC) pair, which leads to the complementarity of base pairs, as described in Fig. 1.1. The complementarity has been experimentally suggested by Chargaff's rule that DNA contains approximately the same number of adenine and thymine bases or guanine and cytosine bases [36]. The function "R" represents DNA backbone made of periodically repeating unit of a deoxyribose and phosphate, as shown in Fig. 1.2. In 1953, Watson and Crick discovered the famous double helical structure of DNA in Fig. 1.3 [2]. Their *Nature* paper has been followed by the two important papers on the X-ray diffraction diagrams of DNA by

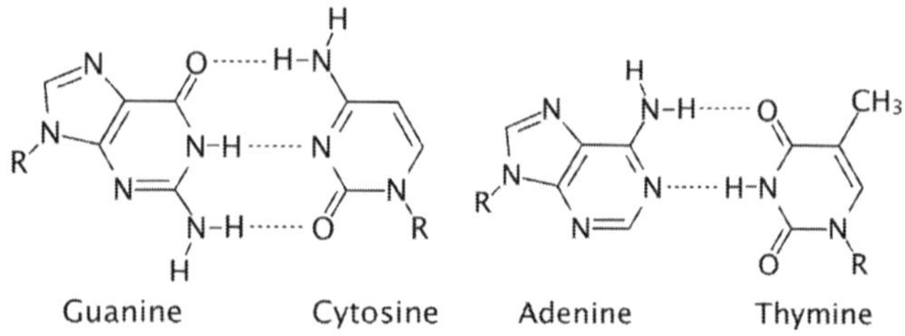

Figure 1.1 Two complementary sets of base pairs connected by hydrogen bonds: guanine–cytosine (GC) and adenine–thymine (AT) pairs. No hydrogen bonds can be formed in the other pairs of purine and pyrimidine, such as GT and AC pairs. This confinement to the base pair formation is known as complementarity.

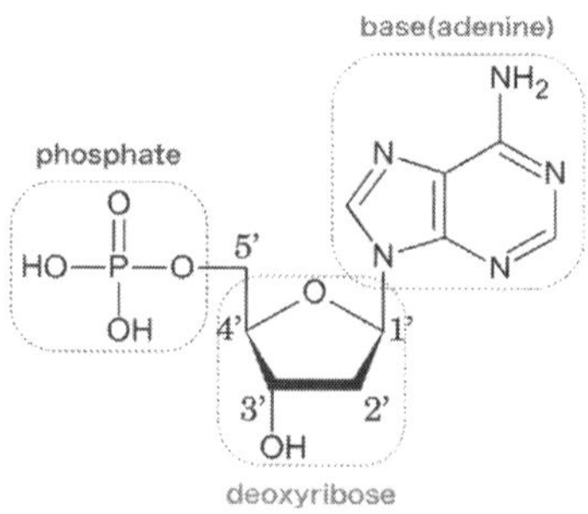

Figure 1.2 An example of nucleotide with adenine. A group of deoxyribose sugar and phosphate is a repeating unit of DNA backbone. The 3′-OH of deoxyribose and the 5′-phosphate-OH of another nucleotide make the strong covalent bonds, phosphodiester bonds, to form a sugar-phosphate backbone polymer of DNA.

Wilkins and colleagues [37] and Franklin and Gosling [38]. Watson, Crick, and Wilkins received the 1962 Nobel Prize in physiology or medicine for their contributions to the discovery of the DNA structure. Franklin contributed to the discovery with the exceptional X-ray diffraction data, Photo 51 in Fig. 1.3, but she was ineligible for the Nobel Prize because she died in 1958.

Figure 1.4 shows the schematic helical structures of B-form DNA with a twisted ladder form and A-form DNA with a coil form of a ladder. Table 1.1 summarizes several average parameters of the three major helical structures of DNA: A-DNA, B-DNA, and

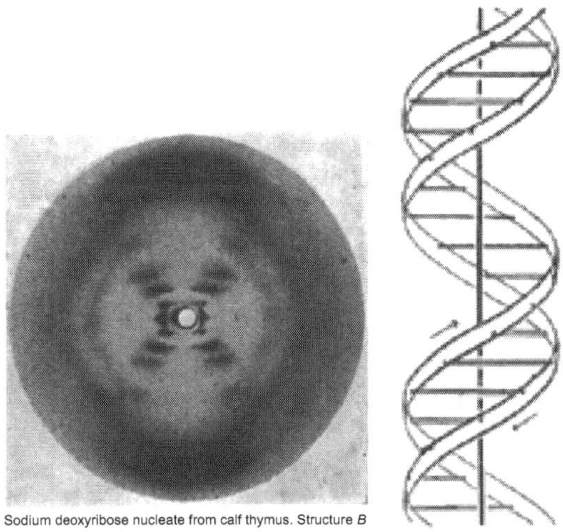

Figure 1.3 (Left) Exceptional X-ray diffraction diagram of sodium deoxyribose nucleate from calf thymus with B-form in a fine fiber DNA sample measured by Raymond Gosling in May 1952 in the lab of Rosalind E. Franklin [38], which is labeled as "Photo 51." The "X"-form diagram corresponds to a helical structure of DNA. The DNA sample showed two distinct types of diagram, structure A (A-form) with a crystalline form and structure B (B-form) with a lower degree of order, depending on humidity. B-form appears at higher humidities and persists over a wide range of ambient humidity. The change from A to B is reversible. B-form is derived from A-form when DNA takes up more than 14–17 water molecules in a unit structure with one base pair, accompanied by 30% of elongation in the DNA fiber length. Reprinted by permission from Macmillan Publishers Ltd: *Nature*, Ref. [38], copyright 1953. (Right) Double helix structure of DNA proposed by Watson and Crick in 1953, in which the ribbons and the horizontal bars represent the sugar-phosphate backbones and the base pairs, respectively. The arrow at each ribbon represents the direction of DNA backbone, 3′ to 5′ (upstream) or 5′ to 3′ (downstream: DNA sequences are described along downstream), that is, each backbone directs opposite direction to each other. Reprinted by permission from Macmillan Publishers Ltd: *Nature*, Ref. [2], copyright 1953.

Z-DNA. Franklin and Gosling have found that B-form structure shows a lower degree of order and is stable in humid circumstances, like in living cells, more than 75% of relative humidity at room temperature [38] because of the hydrophobic nature of the bases, which stabilizes the base stacking in B-form DNA. The hydrophilic

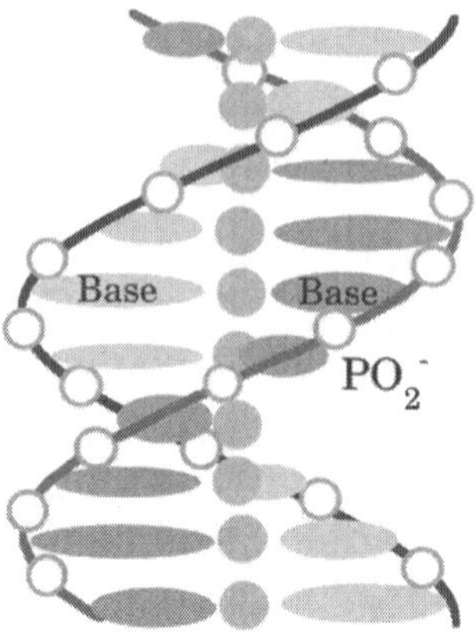

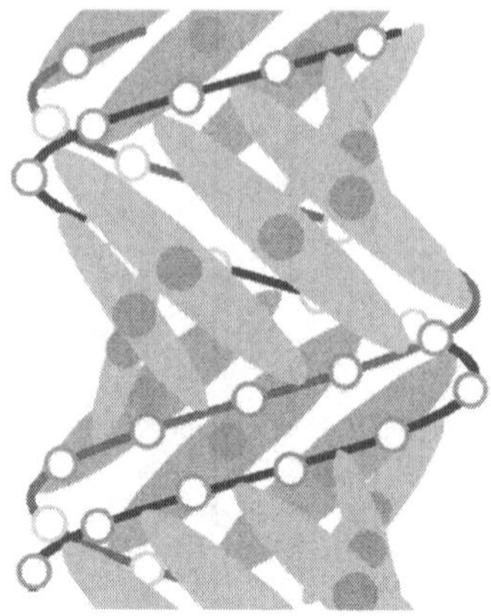

Figure 1.4 (Left) Schematic structure of B-form DNA. (Right) A-form structure. The open circles represent the phosphate groups. The ellipses do the bases for the left and the base pairs for the right. The circles symbolize the hydrogen bonds.

Table 1.1 Structural parameters of the three major helical forms of DNA [39–42]

Helix sense	A-form Right-handed	B-form Right-handed	Z-form Left-handed
Base pairs per turn	11	10.5	12
Rise per base pair (nm)	0.26	0.34	0.37
Major groove width (nm)	0.27	1.17	0.88
Minor groove width (nm)	1.1	0.57	0.2
Diameter (nm)	2.3	2.0	1.8

sugar-phosphate DNA backbones surround the base pairs, as shown in Fig. 1.4. By removing water molecules from B-DNA, B-form is transformed into A-form reversibly, at which the water content in B-form DNA spreads over from 40%–50% (14–17 water molecules per base pair (bp)) to several hundred percent of the dry weight ($\approx$610 g/mol-bp) of DNA. Matsui *et al*. reported that DNA contains 6–12 water molecules per base pair (3–6 wpn) at the relative humidity of $\approx$60% and 30°C [43], as summarized in Table 3.2.

A-form DNA is stable in dried circumstances, like in vacuum, and has a crystalline structure. In film samples, the bases of a double helix of B-form interact with each other, but the bases of the neighboring helices are isolated from each other because DNA backbones surround them. In contrast, the bases of A-form DNA interact with those of the neighboring DNA because of the bases in the surface of the A-form helix. This feature will be reflected in physical properties of DNA, especially metal ion–incorporated DNA.

Z-form of DNA is a left-handed helical structure with a zig-zag pattern of two base pairs as a repeat unit. Z-form is one of the three biologically active structures shown in Table 1.1. This form is regarded as a transient form in biological activities under certain conditions, such as negative DNA supercoiling, high salt and some cations at physiological temperature, 37°C, and pH = 7.3–7.4 [44].

1.1.2 *Electronic States of DNA*

Electronic states of natural DNA are semiconducting with the optical energy gap of $\approx$4 eV [27, 45, 46]. Thus, it is natural to think that the electric conduction in DNA is semiconducting and approaches insulation with decreasing temperature. In 1962, 9 years from the discovery of DNA structure by Watson and Crick [2], Eley and Spivey suggested that the π-orbital stacking of the base pairs along a double helical axis of B-DNA could lead to conduction [47], as the aromatic organic conductors, such as Bechgaard salts, suggest to us [48].

At the beginning of the 21st century, many researches on conductivity in a single molecule of double helical DNA or a bundle with nanotechnology have been reported with diverse conclusions, such as insulation [8], semiconduction with a small activation gap [4, 5], FET formation [7], conduction [3], and metallic down to cryogenic temperatures with a proximity effect of superconductivity [6]. This diversity of conclusions was discussed and analyzed by Endres *et al.* on the basis of the theoretical point of view in details [31].

1.2 Examples of DNA Engineering

1.2.1 *DNA Origami*

An interesting application of the self-organization and flexibility of base sequences is found in 3D nanostructures of DNA with a

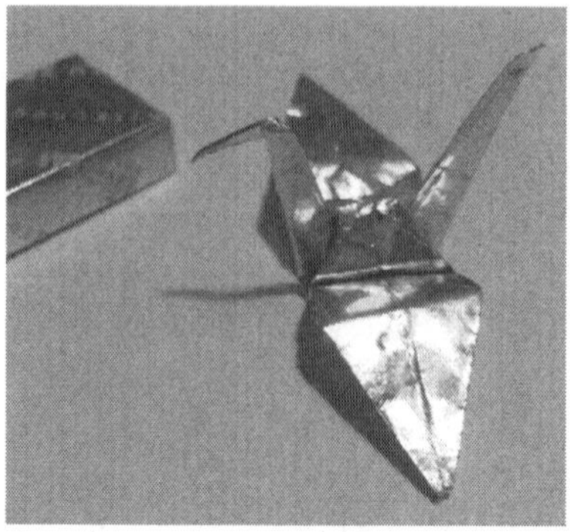

Figure 1.5 Origami is a handicraft with a square paper, which is folded many times along a designed way. An example of the origami crane made of a wrapping paper of chocolate.

cube [49], an octahedron [50], a truncated octahedron [51], and a 4×4 array of 16 addressable pixels [52]. A recent advancement of the nanostructure formation with DNA is a computer-aided programmable 2D pattern named scaffolded "DNA origami" [35]. Origami is a Japanese handicraft of square papers, as demonstrated in Fig. 1.5 by an example of the origami crane produced by sophisticated folding of the paper with 3D structure. DNA origami is a generalized method to form systematically computer designable 2D nanoscale patterning with a single-stranded (SS) DNA of a common virus M13mp18, 7249 base-long, ring-form genome (≈2.6 μm long), as the scaffold strand, which can be bought commercially for $30. An arbitrary 2D pattern describable in a unicursal way (raster fill method), as shown in Fig. 1.6, can be produced with "single-stranded" one long "scaffold strand" and numerous short "helper strands," which form double helices with each other based on the programmed sequences of helper strands complementary to the corresponding segments of the scaffold strand. Helper strands join neighboring parts of the scaffolded strand with double helices crossed between them, thus, working as "staples" to fix unicursal 2D patterns. One pot reaction with these strands produces self-organized 2D patterns under some thermal treatment for a few hours, as demonstrated by AFM image in Fig. 1.6.

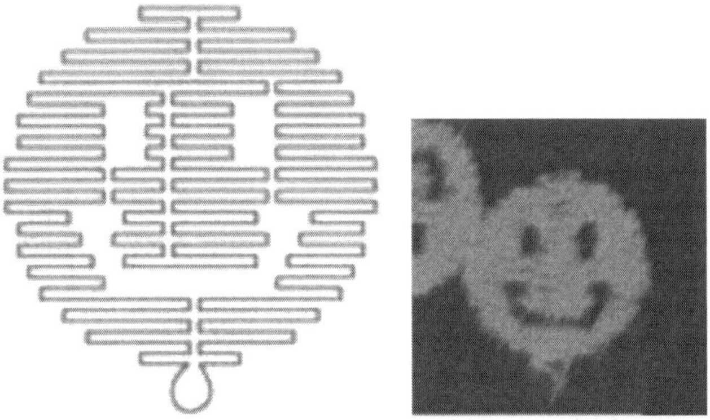

Figure 1.6 Strategy of DNA origami. (Left) Single closed ring of a scaffold strand, which fills a smiley in a unicursal way (raster fill method). Each neighboring line of the scaffold strand is connected by short helper strand(s), which form double helical structure with the scaffold strand. Refer the original report in details. (Right) AFM image of the smiley self-organized by DNA origami. Reprinted by permission from Macmillan Publishers Ltd: *Nature*, Ref. [35], copyright 2006.

With some modifications of the staple strands, one can implement information into the 2D pattern, such as "0" and "1" arbitrary sequences with the theoretically expected pixel size of 5.4 nm × 6 nm. Furthermore, attachments of proteins can provide nano-breadboard for biological experiments, and by attaching nanowires, carbon nanotubes, and gold nanoparticles, electric nano-breadboards might be created. Two-dimensional DNA origami was further extended to 3D origami, with 3D arrangement of staple strands [53], and to twisted and curved DNA origami, with a block of honeycomb lattice of antiparallel scaffold helices, by adding and subtracting bases in staple strands [54]. These developments of DNA origami provided a new research field to design and produce 3D molecular nanomachines with functional staple strands. From 2011, a biomolecular design competition (BIOMOD) has been held for providing undergraduates an opportunity to design nanomachines with self-assembling biological macromolecules (http://biomod.net/). A drug delivery machine with DNA origami has attracted great interest for nanotherapeutic applications [55]. A 3D design application of

DNA origami, "CaDNAno," can be downloaded free from the cadnano site "http://cadnano.org/."

1.2.2 *Metal Ion Incorporation Into DNA*

In 1993, Lee and colleagues found that divalent metal ions, Zn^{2+}, Co^{2+}, and Ni^{2+}, bind to duplex DNA at pHs above 8 and cause a conformational change, i.e., a metal DNA complex *M*-DNA [16]. They studied a fluorescence assay of ethidium, which binds duplex DNA. Only when the aromatic part of ethidium molecules intercalates between the base stacking of DNA, ethidium emits strong fluorescence at 600 nm with excitation at 525 nm. This is because water, as a highly efficient fluorescent quencher, is removed from the intercalated ethidium molecules. The point of their study is that the divalent metal ions bound between the bases of a base pair in duplex DNA prevent ethidium from binding DNA, resulting in the loss of ethidium fluorescent. Figure 1.7 reported by Lee and colleagues represents the kinetics of fluorescence assay with 300 μL of the ethidium buffer solution containing 75 μM DNA, 15 mM NaCl, 10 mM tris-HCl (pH 8), and, if not specified, 1 mM Zn^{2+} at pH 8.5 and 20°C. Figure 1.7A implies a threshold concentration of $\approx$1 mM Zn^{2+} to realize the Zn ion insertion into the most of the base pairs under the above conditions. Lee and colleagues suggested a cooperative nature of this reaction. In this buffer solution, pH and temperature are also crucial conditions to attain the Zn^{2+} insertion into DNA, as demonstrated in Fig. 1.7B and C. Figure 1.7D suggests that ion species that form *M*-DNA are limited to Zn, Ni, and Co under these conditions, but not Mn, Mg, and Cu. This restriction on the ion species substantially limits the magnetic studies to investigate the nature of *M*-DNA, such as electron spin resonance (ESR or EPR), magnetic susceptibility, etc., which utilize the electron spin and the nuclear spin of ions. Especially, Mn^{2+} ions are expected to work as an excellent magnetic probe, as will be demonstrated, including the revised preparation method of *M*-DNA in this book.

The structure of *M*-DNA is not well known yet. The aforementioned fluorescence assay of ethidium shows that when ethidium intercalates into DNA, the fluorescence of ethidium revives by escaping from the quencher of water molecules, but the metal ion

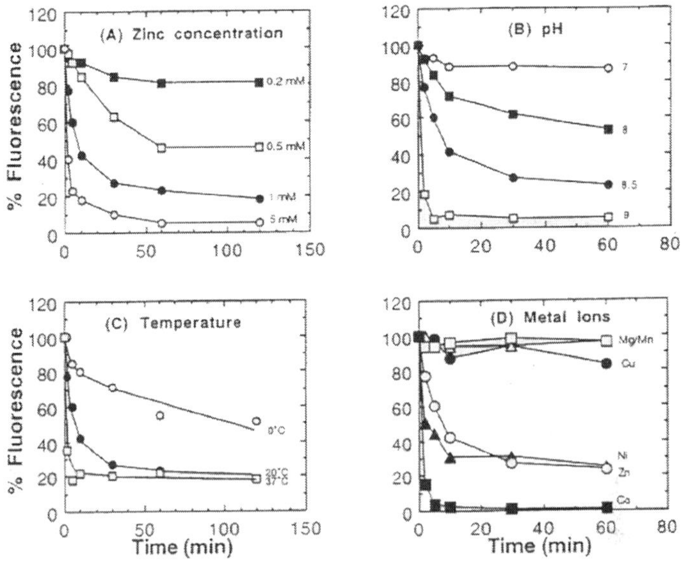

Figure 1.7 Kinetics of ethidium fluorescence assay with 300 μL of a solution containing 75 μM DNA, 15 mM NaCl, and, if not specified, 1 mM Zn^{2+} at pH 8.5 and 20°C. The loss of ethidium fluorescence reflects the insertion of divalent metal ion M^{2+} in place of the hydrogen bond between the bases of a base pair. (A) Shows the Zn ion concentration dependence, (B) demonstrates the effect of pH on the Zn ion insertion, (C) suggests a crucial role of temperature in the Zn ion insertion, and (D) represents the insertion rate of 1 mM of various metal ion species. Copyright 2008 Canadian Science Publishing or its licensors. Reproduced with permission from Ref. [16].

in *M*-DNA repels ethidium with positive charge from DNA, resulting in quenching of the fluorescence. This suggests that the location of Zn^{2+} should be near the base pair of DNA. Lee and colleagues also studied ^1H NMR of the imino protons of the bases [16]. The NMR intensity of the imino protons decreased with increasing Zn^{2+} insertion into DNA and finally disappeared at 100% insertion of metal ions. There are two possible reasons for the disappearance of the NMR signal: (1) replacement of the imino proton by the Zn^{2+} ion or (2) a rapid exchange of the imino protons with solvent being free from the hydrogen bonds by the inserted Zn^{2+} ions, which suggests that the Zn^{2+} ion is located between the bases of a base pair in place of the hydrogen bonds. Rakitin and colleagues reported the conductivity result on Zn-DNA on the basis of the first interpretation [17]. In contrast, recent studies on *M*-DNA prepared with a precipitation method from aqueous DNA + MCl_2 solution suggest that the most probable is the second case for the elimination of the NMR signal, a complex formation of the Zn^{2+} ion with water molecules [27, 56].

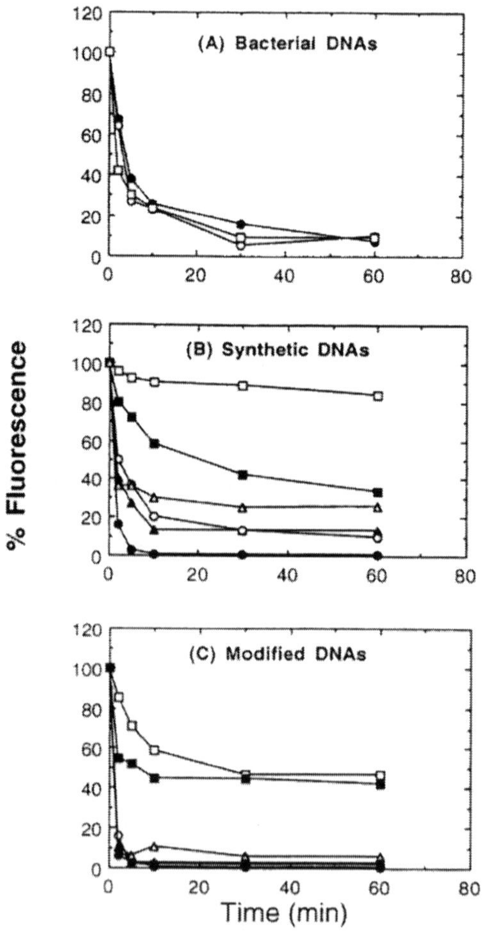

Figure 1.8 Effect of base composition and sequence on the rate of Zn ion insertion. Copyright 2008 Canadian Science Publishing or its licensors. Reproduced with permission from Ref. [16].

In connection with the location of the metal ions, Lee and colleagues reported the effect of base composition and sequence on the rate of Zn ion insertion as shown in Fig. 1.8. The insertion rate of Zn^{2+} into the three bacterial DNAs, *Clostridium perfrengens* DNA, *E. coli* DNA, and *Micrococcus luteus* DNA, showed similar kinetics with each other. In contrast, remarkable differences were observed in several synthetic DNAs. Poly[d(A-T)] (original notation, which is the same as poly(dA-dT) or poly(dA)-poly(dT) in this book) reacts very slowly, less than 10% , but 40% in poly[d(G-C)] and almost 100% in poly[d(T-G)]·poly[d(C-A)], at 10 min. A similar tendency has also been found in the kinetics of the Fe ion insertion into DNA, as will be discussed in Section 7.2 [27]. Fe^{2+} ions react with double-stranded (DS) poly(dG-dC) twice as fast as the case of DS poly(dA-dT), to form Fe^{3+}-poly(dG-dC) and Fe^{3+}-poly(dA-dT), respectively. The large difference reflecting the characteristic of the bases is another supporting evidence for the insertion of a Zn^{2+} ion between the bases of a base pair. Model structures for base pair and the divalent metal ion of Zn were proposed by Aich and colleagues [17, 21, 22, 57].

The electrical conductivity measurement on Zn-DNA was reported by Rakitin and colleagues [17]. They directly measured the conductivity of a bundle of B-DNA and Zn-DNA and found that the I–V curve showed a small conductivity plateau between 0 and 0.25 eV for B-DNA, but no plateau like a metal for Zn-DNA. Thus, they concluded that the engineering DNA had been successfully discovered and expected that it would be applied in a variety of fields such as nanoelectronics.

Part II

Natural DNA

Chapter 2

Basic Physical Properties of Natural DNA

Kenji Mizoguchi and Hirokazu Sakamoto

Department of Physics, Tokyo Metropolitan University, Hachioji, Tokyo, 192-0397 Japan

mizoguchi@phys.se.tmu.ac.jp, sakamoto@phys.se.tmu.ac.jp

2.1 Introduction

In this section, we describe the basic physical properties of natural DNA: (1) UV/Vis absorption spectra and (2) magnetic properties. Absorption spectra provide basic information about the electronic states of DNA: Is there energy gap? If yes, then how much is the energy gap and how large are the interactions between the π electrons of bases of intra- and inter-strand? If the DNA is semiconducting with the energy gap, magnetism based on the π electrons will not be expected. On the other hand, it has been reported that natural DNA is expected to show sizable electrical conductance on the basis of orbital paramagnetism [32, 34]. A lot of magnetic properties have been reported so far, and the origin and mechanism of the observed magnetism have been discussed. With these viewpoints, it is valuable to reexamine precisely the physical properties of natural DNA.

DNA Engineering: Properties and Applications
Edited by Kenji Mizoguchi and Hirokazu Sakamoto

ISBN 978-981-4669-46-7 (Hardcover), 978-981-4669-47-4 (eBook)
www.panstanford.com

2.2 UV/Vis Absorption of DNA

2.2.1 *Introduction*

Optical or visible light gives us basic information on the electronic states of materials. The product of wavelength λ (in meters) and frequency ν (in hertz or 1/s) is equal to the constant velocity of light, $c = 2.99792 \times 10^8$ m/s in a vacuum,

$$\nu \times \lambda = c, \tag{2.1}$$

and the energy of light (in joules) is proportional to its frequency with Plank's constant $h = 6.62607 \times 10^{-34}$ J·sec,

$$E = h\nu = \frac{hc}{\lambda}, \tag{2.2}$$

as described in Fig. 2.1. The wavelength of visible light is $\approx$400 nm ($\approx 8 \times 10^{14}$ Hz) for the violet color and $\approx$750 nm ($\approx 4 \times 10^{14}$ Hz) for the red color, whose energy corresponds to $\approx 5.3 \times 10^{-19}$ J and $\approx 2.6 \times 10^{-19}$ J, respectively. In solid state physics, one frequently uses electron volt (eV) to measure the energy of electrons with the elementary charge of $e = 1.60217 \times 10^{-19}$ C, that is, 1 eV = 1.60217×10^{-19} J, which corresponds to the energy that the electron obtains by the acceleration of 1 V. Thus, the red light corresponds to $\approx$1.6 eV, and the violet light corresponds to $\approx$3.5 eV. To investigate various materials with a variety of energy gaps, such as $\approx$0.7 eV

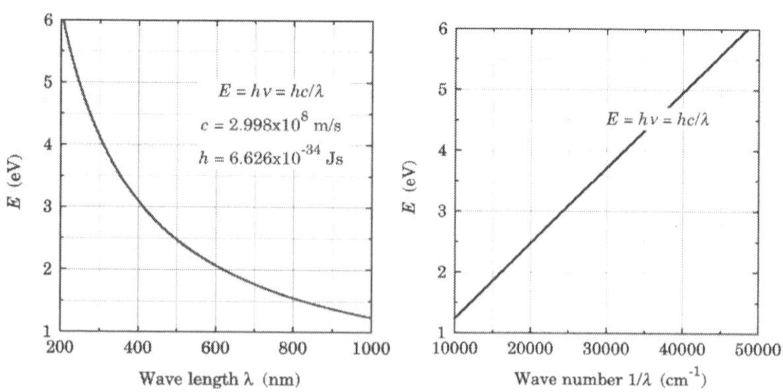

Figure 2.1 Relation of photon energy E (eV) and wavelength λ (nm) [left] or wave number $1/\lambda$ (cm^{-1}) [right].

for Ge and ≈5 eV for Diamond, UV/Vis absorption spectroscopy is generally applied over 2–3 eV of visible to 3.5–6 eV of ultraviolet region. Throughout this chapter, we use eV as the unit of energy.

2.2.2 *Experimental*

DNA samples used in this chapter are (1) double-stranded (DS) salmon sperm DNA purchased from Wako Pure Chemical Ind. Ltd., and (2) single-stranded (SS) 30mer-DNAs, poly(dA), poly(dT), poly(dG), and poly(dC), purchased from Hokkaido System Science Co. Ltd. A sample of aqueous solution was prepared by dissolving DNA into pure water by 0.05 mM/L as a standard condition. A transparent film sample was made by casting and drying DNA aqueous solution of several mM/L in a plastic case. UV/Vis absorption spectra were measured in a quartz sample cell for solutions with Shimadzu UV-1700 spectrophotometer. The background of aqueous solution was corrected by subtracting the absorption spectrum of pure water in a quartz sample cell located in a reference optical arm. Since no subtraction of background was made for the film samples, absorption tail by the dispersion of light with unexpected submicron structures will sometimes appear.

2.2.3 *Nucleotide Solution*

Historically, a lot of measurements on the base molecules of vapor and/or solution forms [58–65] were reported to unveil the mechanism of the UV absorption damage of DNA, and it has been a long way to reach an answer with the help of femtosecond technology [66]. A related topic will be described in Chapter 5. The characteristic absorption spectra of four nucleosides (nucleobase + ribose), adenosine, cytidine, guanosine, and thymidine, in aqueous solution measured by Bouvier and colleagues are reprinted in Fig. 2.2. The dotted lines represent the analyzed absorption bands with half-width at the half-maximum of ≈0.25 eV. Interestingly, the absorption spectra of the vapor phase did not show narrower spectral width compared to that of the solution spectra [59]. Thus, high resolution studies with magnetic circular dichroism (MCD) [62, 63] and polarized absorption in crystal [67–71] suggested a

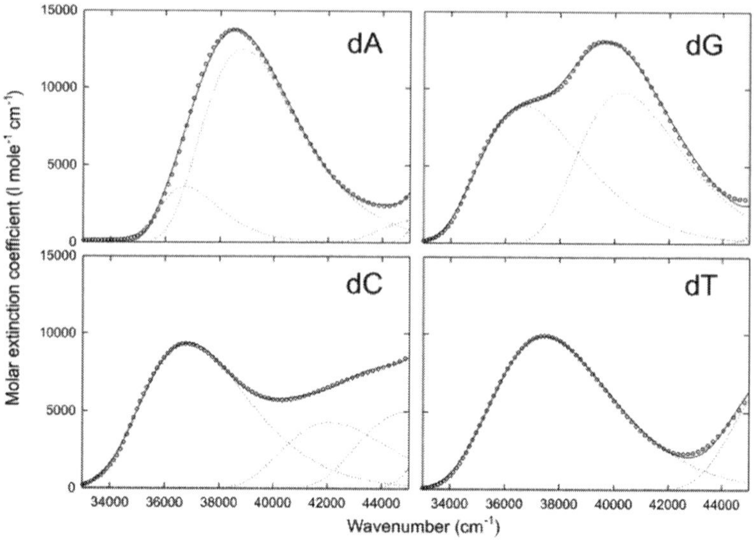

Figure 2.2 Absorption spectra (circles) of the adenosine, cytidine, guanosine, and thymidine molecules in ultrapure water (10^{-5} M) fitted with a sum (solid lines) of log-normal curves (dotted lines), as a function of cm^{-1}, corresponding to the energy range from 4.1 to 5.6 eV (refer to Fig. 2.1). Reprinted from Ref. 65, Copyright 2002, with permission from Elsevier.

fine structure in the lowest excitation peak of nucleobases. On the basis of experimental and theoretical considerations, it has been concluded that the observed spectra below 6 eV are dominated by the optical transition from π to π^* states in the bases of DNA [64, 72–77].

2.2.4 *SS and DS 30mer-DNA Solution*

With the 30mer-DNAs, poly(dG), poly(dA), poly(dC), and poly(dT) instead of the nucleosides in Fig. 2.2, further analysis of the electronic states of DNA is expected. Figure 2.3 demonstrates the absorption spectra of the four SS 30mer-DNAs in solution, which is compared with the spectra of nucleosides in Fig. 2.2 [27]. The absorption spectra are compared with the gap energies between the lowest unoccupied molecular orbital (LUMO) and the highest

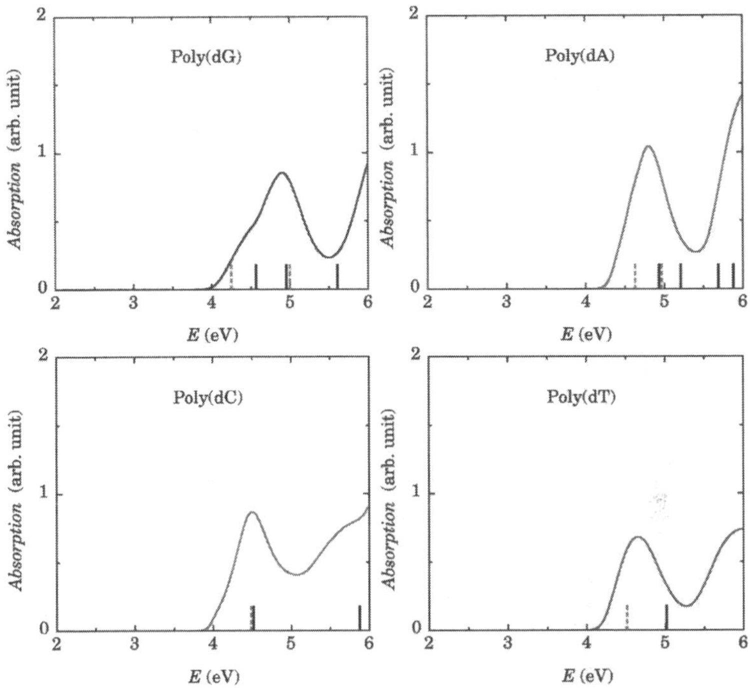

Figure 2.3 Absorption spectra in aqueous solutions of the four SS 30mer-DNAs, poly(dG), poly(dA), poly(dC), and poly(dT). The bars in the bottom of each panel represent the Hückel estimation [77]. The leftmost bar represents the energy difference between LUMO and HOMO. The second bar represents the difference of 2nd LUMO and HOMO for poly(dG) and poly(dA), and LUMO and 2nd HOMO for poly(dC). The third bar shows that of 3rd LUMO and HOMO for poly(dG) and LUMO and 2nd HOMO for poly(dA). The fourth bar for poly(dA) corresponds to the difference of 3rd LUMO and HOMO. The broken bars represent the CS-INDO estimation of the first one or two absorption bands [65].

occupied molecular orbital (HOMO) predicted by the CS-INDO estimation [65] and by the Hückel estimation (see Chapter 6 for details) [77], which are indicated with short bars at the bottom of Fig. 2.3.

The spectrum of poly(dG) shows a strong absorption at 4.9 eV with a weak shoulder around 4.4 eV, which is similar to that of the guanosine in Fig. 2.2, but the intensity and the peak position of band

Table 2.1 Excited energies of nucleobases, nucleosides, and 30mer-DNAs. The first row represents the experimental results: poly(dA), poly(dG), poly(dT), and poly(dC) in solution. The second is the experimental data for the nucleosides in solution, and the next two rows show the nucleobases in solution and vapor. The fifth and below are the theoretical estimations with different approaches and approximations. Purine bases both of band I (LUMO–HOMO) and band II (2nd LUMO–HOMO) are indicated

		adenine	guanine	thymine	cytosine
	eV in unit	band I/band II		band I	
water	SS 30mer-DNA[a]	≈4.40/4.81	≈4.37/4.90	4.65	4.49
	nucleoside[b]	4.56/4.82	4.56/5.01	4.65	4.57
	nucleobase[c]	4.77	4.51/4.96		
vapor	nucleobase[c]	4.92			
	CS-INDO[b]	4.62/4.95	4.25/4.99	4.51	4.48
	INDO/S[d]	4.44/4.74	3.94/4.51	4.87	3.94
theory	*Ab initio* MRCI[e]	4.52/4.80	4.50/5.30		
	CASPT2	4.73/4.75[f]	4.45/4.58[g]	4.89[h]	4.50[i]
	CNDO/OPTIC-2[e]			4.87	
	Hückel for SS DNA[j]	4.93/5.21	4.56/4.95	5.01	4.51

[a] Fig. 2.3 [b] [65] [c] [59] [d] [78] [e] [79] [f] [64] [g] [80] [h] [74] [i] [75] [j] [77]

I look different from those of guanosine, weak and redshifted, as summarized in Table 2.1. On the poly(dA) spectrum, it looks like a single peak, but two peaks closely located at ≈4.6 and ≈4.8 eV (as shown by the dotted curves for adenosine in Fig. 2.2) have been reported experimentally, by the high resolution MCD studies [62, 63] and polarized absorption studies [67–71], and theoretically [64, 65, 76–81]. The typical experimental data of the nucleosides, nucleobases, 30mer-DNAs, and a variety of theoretical estimations are summarized in Table 2.1.

It is interesting to note that 4.77 eV in the adenine solution substantially redshifted from 4.92 eV in the adenosine vapor, which suggests that water molecules cause a redshift by 0.15 eV. A small redshift of the peak positions of band I in the SS poly(dA) from the adenosine, similar to that of the SS poly(dG) from the guanosine, suggests the possibility of additional interaction caused by base stacking in the SS 30mer-DNA.

CS-INDO estimation reproduces fairly well the spectrum of the 30mer-DNAs [65]. The Hückel estimation of the energy level diagram (see Chapter 6) demonstrates a characteristic electronic structure: the LUMO level of the purine (guanine and adenine) base is accompanied by the closely located next level, which causes the first absorption band to split, as discussed earlier. In contrast, the spectra of pyrimidine bases in poly(dC) and poly(dT) are rather simple compared with that of the purine bases, which is demonstrated by the isolated LUMO levels of the pyrimidine bases in poly(dC) and poly(dT) [77]. The Hückel estimation reproduces well the peak positions of poly(dC) but fails in poly(dT). This is due to the omission of the methyl group of thymine in the calculation.

2.2.5 *DNA Solution*

Figure 2.4 shows the absorption spectrum of a DNA aqueous solution [27]. The absorption starts at 4 eV and has the peak at

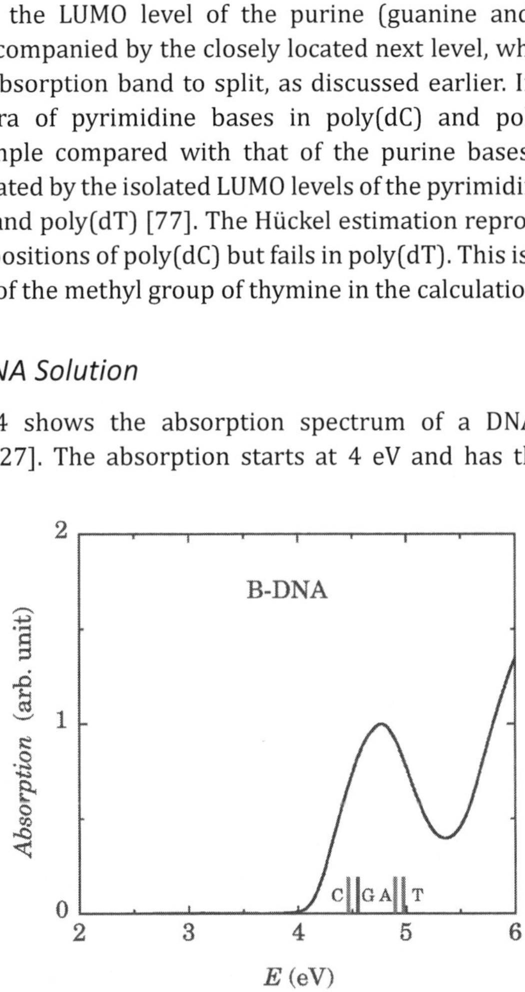

Figure 2.4 Absorption spectrum in an aqueous solution of DNA. The small bars in the bottom represent the HOMO–LUMO gaps for the four bases—cytosine (C), guanine (G), adenine (A), and thymine (T)—from the left to the right, calculated by Iguchi with the Hückel approximation (see Chapter 6 for details) [77]. Reprinted with permission from Ref. [27], Copyright 2014 by the American Physical Society.

4.78 eV. These features are consistent with the reported data in solution and film forms so far [45, 46, 82]. The small vertical bars at the bottom of Fig. 2.4, labeled C, G, A, and T, represent the energy difference between LUMO and HOMO for each base estimated by the Hückel calculation [77], which agree reasonably with the absorption band of DNA. The energy at the absorption edge of 4 eV corresponds approximately to the band gap of B-form DNA (B-DNA), and the peak energy of 4.78 eV represents approximately the energy difference between the HOMO and LUMO energy levels.

The half-width at the half-maximum of the absorption band of $\approx 1/2$ eV suggests the maximum size of the energy bandwidth in B-DNA, as will be discussed in Chapter 6. The source of the bandwidth is the interactions between π electrons of the nucleobases by (1) forming the base stacking of SS DNA and by (2) making double helix with hydrogen bondings. First, we can estimate it by comparing the absorption bandwidth of the nucleosides in Fig. 2.2 and that of the SS 30mer-DNAs in Fig. 2.3. By comparing these by converting the unit of the photon energy from cm^{-1} to eV, we find that the half-width at the half-maximum of the SS 30mer-DNA is very similar to that of the nucleoside, suggesting that the stacking interaction between the bases along the DNA backbone is weak enough compared with the 0.25 eV of their absorption bandwidth. Second, we can compare the spectra of DS 30mer poly(dG-dC) and DS poly(dA-dT) with the sum of the corresponding SS 30mer-DNAs represented by the dotted curves in Fig. 2.5. Curves of poly(dG-dC) and poly(dA-dT) are reproduced very well with the sums of the SS 30mer-DNAs, suggesting the weakness of the inter-strand interaction of the π electrons.

To study the electronic states of the photo-excited states through the π–π interactions in DNA double helix, Bouvier and colleagues have discussed theoretically the electric dipole interaction within the short double helices of $(dA)_{20}(dT)_{20}$ and $(dAdT)_{10}(dAdT)_{10}$ and concluded that optically excited states are delocalized over the whole length of the double helix and that the extent of exciton states over the two strands depends on the base sequence [65]. They estimated that the excitation energy of some chromophore states in these double helices blue-shifted by ≈ 0.1 eV from that of the nucleoside molecules. However, the similarity of the absorption

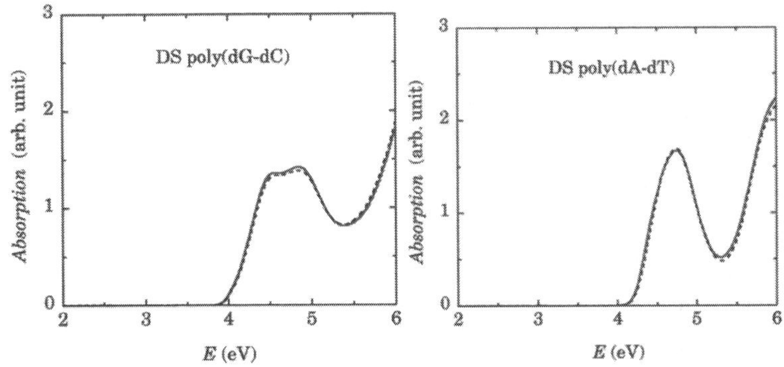

Figure 2.5 Absorption spectra in aqueous solutions of DS 30mer poly(dG-dC) [left] and DS poly(dA-dT) [right] are shown together with the sum of the corresponding spectra of SS 30mer-DNAs in Fig. 2.3 by dotted curves. Reprinted with permission from Ref. [27], Copyright 2014 by the American Physical Society.

peak energy of (1) the SS 30mer-DNA and the nucleosides in Table 2.1 and that of (2) the DS 30mer-DNAs and the sum of the corresponding SS 30mer-DNAs in Fig. 2.5 does not demonstrate such a large blue-shift of $\approx$0.1 eV in the DS DNA from the nucleosides. Concerning with this strong similarity, Eisinger *et al.* have discussed that the absorption band of DS DNA does not necessarily represent the excitation across the band gap of DNA in usual meaning, but would be superposition of the excitations between the occupied and unoccupied states of each base molecule—G, C, A, and T [82].

The absorption spectra of B-DNA in Fig. 2.4 can be compared with that of the sum of the DS 30mer-DNAs with G-C or A-T base pairs in Fig. 2.5. The composition ratio of the GC and AT base pairs for the salmon sperm DNA is known to be GC : AT = 41.2 : 58.8 [83]. Figure 2.6 demonstrates the reproduction of the DNA spectrum with a combination of the 30mer-DNAs in this ratio for the salmon sperm DNA [27]. There is small deviation between the B-DNA and the sum of the 30mer-DNAs. This might be caused by the interaction of the π electrons between the strands of double helix much longer than the 30mer-DNAs and/or by the deviation of the ratio of GC to AT in the present sample from the reported ratio because they are actually not the same portion from the salmon DNA.

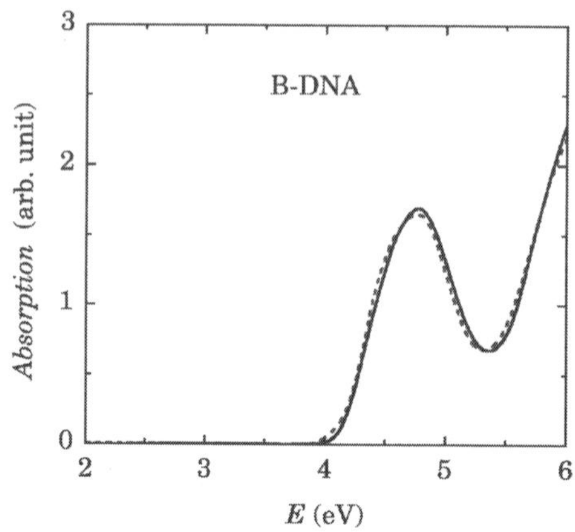

Figure 2.6 The spectrum of DNA is compared with the sum of the spectra of the DS 30mer poly(dG-dC) (Fig. 2.5 [left]) and the DS poly(dA-dT) (Fig. 2.5 [Right]) with the ratio of 0.37 : 0.63 by the dotted curve, which is in good agreement with the experimentally observed ratio of 0.4 (GC) : 0.6 (AT) in the salmon sperm DNA [83]. Reprinted with permission from Ref. [27], Copyright 2014 by the American Physical Society.

2.2.6 *DNA Film*

In a solution sample, DNA double strand is surrounded by water molecules and isolated from the other strands. In contrast, DNA strands form solids and interact strongly with each other in the film samples. Thus, it is expected that some changes from the solution spectra would be induced in the film spectra. The absorption spectra of B-form DNA film are compared with those of DNA solution in Fig. 2.7. Small but finite changes from the solution case are found: (1) a tiny redshift of the peak position by $\approx$0.01 eV in the first band and (2) a small change in the amplitude ratio of the peak at 4.78 eV to the minimum around 5.4 eV, which would be caused by the broadening of the π electron bandwidth.

Omerzu and colleagues found a similar redshift and the broadening of the absorption spectrum of the DNA film from the solution with EDTA in the calf thymus (sigma) DNA [46]. These changes are a kind of solid state effect. They interpreted that these changes were caused by the structural transformation from the more ordered B-form to the less ordered A-form. However, this interpretation is in contradiction with the report by Franklin and Gosling on the high

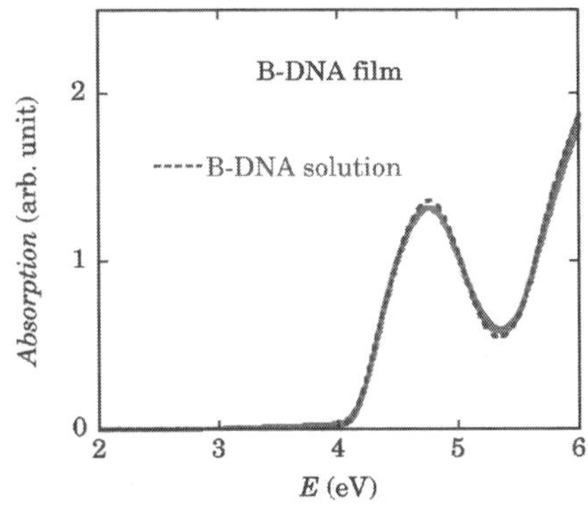

Figure 2.7 Absorption spectrum of DNA film (solid curve) along with that of DNA solution (dotted curve).

quality X-ray data, as shown in Fig. 1.3, that A-form is a crystalline form and B-form is a lower degree of ordered form, as reported in *Nature* along with the famous report by Watson and Crick [38].

In the crystalline form, the broadening of the absorption spectrum is a reasonable result of the increase in the interactions between the π electrons of the neighboring double-helices and in turn the energy bandwidth. The other possible factor to increase the bandwidth is the increase in the base stacking interaction, which accompanies a possible looseness of the torsion angle of the double helix caused by the Coulomb repulsion between the charges of Na^+ ions of the neighboring DNA strands. This repulsive interaction tends to expand the separation of the Na^+ ions between the neighboring double strands, which would, at the same time, slightly expand the double helix along the helical axis and thus loosen the helicity, that is, the torsion angle of the double strands, giving rise to tiny increase in the base stacking interaction by virtue of the reduced torsion angle. In contrast to the DNA case, in which the Coulomb repulsion is partly cancelled by the counterion of PO_4^-, the metal ion inserted DNA complexes, *M*-DNA, should show more pronounced changes in the film spectra from the solution spectra, because the PO_4^- ions are not accompanied by the nearby Na^+ counterions. Thus, the PO_4^- ions, with the hydrating water molecules in the solution, directly repel each other in the film (see Section 7.2.4 for details).

2.2.7 *Summary of UV/Vis Absorption in DNA*

With the UV/Vis absorption study, the semiconducting nature of natural DNA has been reconfirmed. The HOMO–LUMO energy difference is approximately 4.78 eV, and the energy gap is approximately 4.0 eV [27]. However, the spectrum of B-DNA can be almost reproduced with the sum of the four SS 30mer-DNAs, poly(dA), poly(dG), poly(dT), and poly(dC) in the known relative fraction of GC : AT = 41.2 : 58.8 for the salmon sperm DNA [83], which means the absorption spectrum is composed of the four absorption spectra for the four bases G, C, A, and T. Thus, the absorption band of DS DNA does not necessarily represent the excitation across the band gap of B-DNA in usual meaning but would be a superposition of the excitations between the occupied and unoccupied states of each base molecule, G, C, A, and T [82]. The half-width at the half-maximum of the spectra of the bases and SS 30mer-DNAs is approximately 0.25 eV. The size of the inter-strand interaction in a double helical DNA is bound by the half-width of the absorption spectrum but is much smaller than the width, for example, <0.01 eV. In contrast, theoretical estimation suggests that the optically excited states delocalized over the whole length of the double helix of $(dA)_{20}(dT)_{20}$ system and the extent of exciton states over the two strands depend on the base sequence [65]. Such a nature of excited states extended over the whole length of the double helix is assumed to be concerned with lethal or mutagenic effects on the cells by UV irradiation.

2.3 Magnetic Property of DNA

2.3.1 *Introduction*

Does natural DNA possess paramagnetic spin? This question was asked in 1959. Blois and colleagues described the detailed story of this issue as follows (Fig. 2.8) [84]. Shields and Gordy [85] and Shen and colleagues [86] reported the pioneering electron spin resonance (ESR) studies on nucleic acid samples with high energy irradiation. Subsequently, Bliumenfeld and colleagues [87] reported unexpected ESR in un-irradiated nucleic acid. Some nucleic acid they

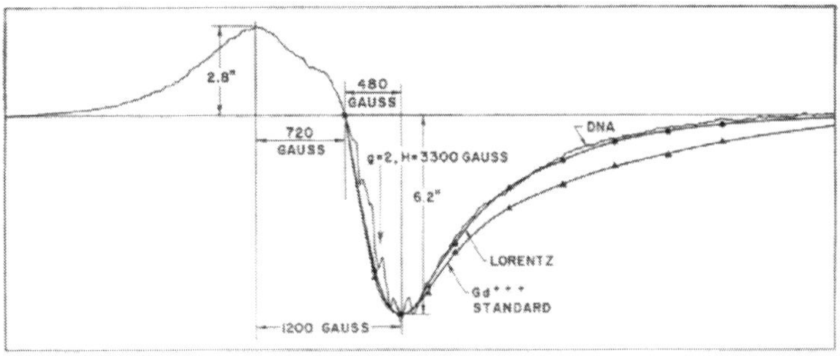

Figure 2.8 An example of the "anomalous" ESR of nucleic acid, seen in the commercial preparation of salmon sperm DNA. The high field tail (to the right) fits a Lorentzian function. The high field portion of the resonance of an S-state rare earth ion, Gd^{3+}, in glycerin solution is included for a line shape comparison. The Gd^{3+}–glycerin system is used as a standard to determine the spin concentration in the DNA sample. The Gd^{3+} line is 200 gauss wide and has been scaled up to fit the high field portion of the line in width and amplitude. Reprinted from Ref. [27], Copyright 1963, with permission from Elsevier.

prepared gave broad (120–500 gauss) asymmetric ESR signals with considerable intensity even at zero magnetic field. They estimated the unpaired spin concentration of up to 10^{21} spins/g ($\approx$1 spin per base pair) from the ESR intensity. Interestingly, they did not find such strong signals in denatured DNA, in protein alone, or in the purine and pyrimidine bases. In addition, the spin magnetization of some samples was found to saturate at higher fields. Following their work, many studies on this issue were reported [84, 88–92]. Müller and colleagues found similar signals in highly purified and lyophilized (freeze-dried) samples of T1 and T2 phage with 10^{18} to 10^{19} spins per gram and four different resonances: one near zero field; one near $g = 4$, 100–200 gauss broad; the third at $g = 2.2$, about 1200 gauss broad; and the fourth, a narrow line, at $g = 2$ [88].

Blois and Maling found 10^{19} to 10^{20} spins per gram in several DNA and RNA samples, and they suggested iron contribution to ESR on the basis of trace analysis of the samples [89]. However, the quantity of iron impurity in the samples was not enough to elucidate the ESR intensities, and in addition, a large static susceptibility was observed, which was too large to be accounted for by the ESR intensities. Shulman and colleagues observed ESR with a temperature-independent intensity down to liquid helium temperature and supposed that iron in the ferromagnetic state provided ESR signals observed in DNA [90]. Walsh and colleagues supported this hypothesis by demonstrating the presence of 100–1000 Å diameter inorganic crystalline particles bound tightly to nucleic acid [91].

Malino *et al.* and Blois *et al.* studied this issue in details using a variety of lyophilized samples prepared by different ways and treatments, along with comparative standard materials, and concluded that the signals near $g = 4$ and $g = 2.2$ were originated by iron nanoparticles of diameter less than 0.1 μm contaminated by Waring Blendor used in the DNA sample preparation [84, 92]. Such nanoparticles show superparamagnetic behavior without remanent magnetization. The DNA samples used in the present chapter do not give, like before, strong ESR signals caused by nanoparticle contamination but still present ESR signals similar to the above in the commercially supplied DNA samples as demonstrated in the following sections.

2.3.2 *ESR Study*

DNA samples used in this chapter are (1) DS salmon sperm DNA purchased from Wako Pure Chemical Ind. Ltd., and Sigma-Aldrich, and kindly supplied by Ogata Materials Science Lab., and (2) DS 30mer-DNAs, poly(dG-dC), purchased from Hokkaido System Science Co. Ltd. Figure 2.9 demonstrates ESR spectra of well-evacuated salmon sperm DNA ([left] Wako Pure Chemical, [right] Sigma-Aldrich) [93]. The backgrounds of ESR cavity and quartz sample tube were subtracted from the data. Note that the quartz tube frequently provides background signals, for example, broad signal centered around 0.15 T and long tail up to 0.5 T, as sometimes reported as the ESR signal of DNA [34, 87, 88]. In both DNAs, weak but clear ESR signals are found around $g = 2$. This broad signal can be simulated with the characteristic spectrum of Fe(III)-DNA, Fe^{3+} incorporated DNA complex (refer Section 7.2 in detail) [25, 27, 93–95]. Magnetism of Fe-DNA is paramagnetic with five d-electrons, and thus three ESR signals appear at $g = 2$ corresponding to $S = \frac{5}{2}$ of the

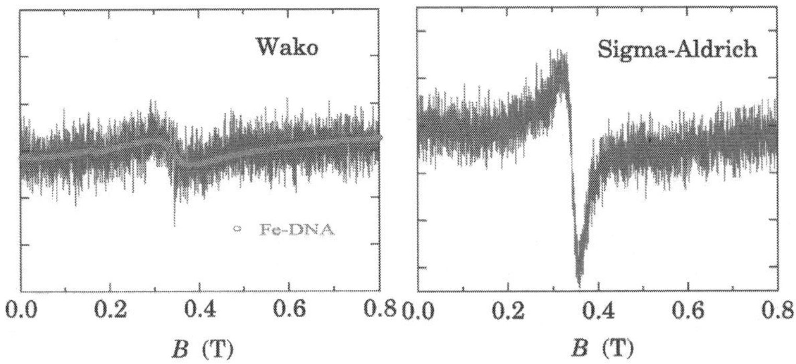

Figure 2.9 [Left] ESR derivative spectrum of the 13 mg salmon sperm DNA under the dry condition provided from Wako Pure Chemicals, taken with a Jeol FE-30 spectrometer: microwave power = 1 mW, frequency = 9.447 GHz, modulation amplitude = 0.5 mT, gain = 500, time constant = 0.01 s, average number = 25 times. Backgrounds of ESR cavity and quartz sample tube are subtracted from the raw data. The signal of DNA around $g = 2$ could be compared with that of Fe(III)-DNA, suggesting Fe contamination in the extraction processes from salmon sperm, resulting in the formation of Fe-DNA in part. The estimated number of spin $S = 5/2$ is around 100 ppm/bp = 0.01% /bp ($\approx$0.00001% /g = 0.1 ppm/g), which is low enough to consider that the origin would be contamination. [Right] ESR derivative spectra of the salmon DNA purchased from Sigma-Aldrich with the same sample quantity and the same vertical scale as the left panel. All the measurement conditions are the same as the left, suggesting three times larger signal intensity than the Wako case.

high spin state of Fe^{3+} with the width of $\approx$220 mT, $S = \frac{1}{2}$ of the low spin state of Fe^{3+} with 44 mT, and $S = \frac{1}{2}$ of π electron with 15 mT. The spin concentration of Fe^{3+} is estimated as 100 ppm for Wako and 300 ppm for Sigma-Aldrich, if one assumes the high spin state of Fe^{3+}. Actually, the low spin state of Fe^{3+} is the dominant species in Fe-DNA; thus, estimation of the spin concentration approaches 0.1% and 0.3% for Wako and Sigma-Aldrich, respectively. The magnetic susceptibility behaves approximately according to the Curie–Weiss law with several K of antiferromagnetic Curie–Weiss temperature. Figure 2.10 demonstrates the temperature dependence of ESR spectra in one of salmon sperm DNAs (Ogata Materials Science Lab.). The line shape and the resonance position of $g = 2$ is similar to that of Fe-DNA, and the temperature dependence of the intensity is

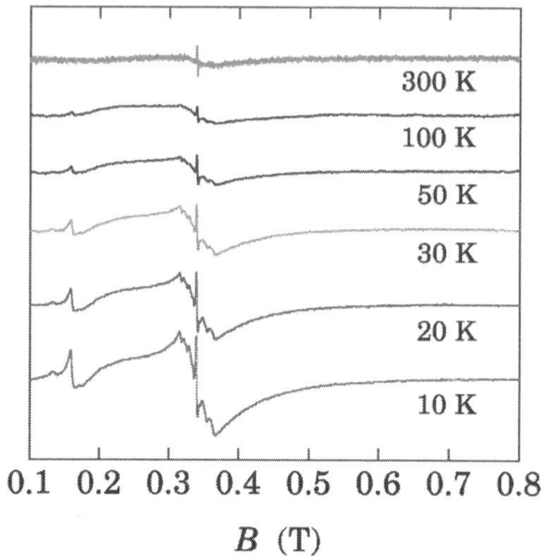

Figure 2.10 Temperature dependence of ESR derivative spectra in salmon sperm DNA supplied by Ogata Materials Science Lab. Three signals are found: at $g \approx 2$—(1) very sharp and (2) broad width $\approx$0.5 T, and (3) at $g \approx 4$ (0.16 T). Fine structure found at $g \approx 2$, especially at 10 K, would be caused by Mn ESR signal. The broad signal is similar to that of Fe-DNA in Fig. 2.9 (left), where Fe ion is located between the bases of a base pair. The temperature dependence is approximately assigned to the Curie–Weiss law with the antiferromagnetic Curie–Weiss temperature of several K, which suggests that the source material is not in a ferromagnetic state, consistent with Fe-DNA as the electronic states of Fe impurity. These features are almost independent of the humidity conditions, in air, in vacuo, and in freeze-drying. Unfortunately, the source of the sharp signal at $g = 2$ is not resolved yet.

consistent with the Curie–Weiss law, which suggests the presence of iron impurities incorporated between the bases of a base pair, but not magnetic particles in a ferromagnetic state. Lee and colleagues have reported that the humidity condition affects ESR signal [34], but all the data in Figs. 2.9 and 2.10 are almost independent of the humidity condition. Figure 2.11 describes an ESR derivative spectra taken in the synthesized 30mer-DNAs with poly(dG-dC) and poly(dA-dT) structures. No signal is found around $g = 2$, but a low field tail below 0.2 T appeared, at least in part, as the background of the quartz sample tube.

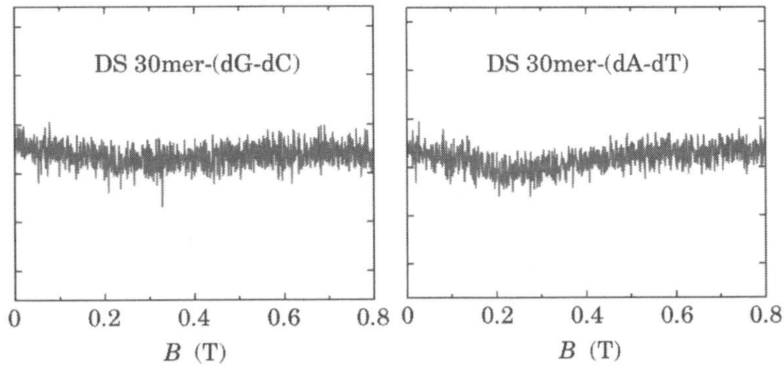

Figure 2.11 ESR derivative spectra of DS 30mer poly(dG-dC) [left] and DS 30mer poly(dA-dT) [right] evacuated for 50 h at 40°C, taken with the same parameters as Fig. 2.9, except for time constant = 0.1 s, average number = 10 times, purchased from Hokkaido System Science. Since the background of quartz sample tube was not subtracted, a low field upturn appears below 0.2 T at 300 K. There is no signal around $g = 2$, $H = 340$ mT, within the experimental uncertainty.

Thus, the aforementioned experimental results and considerations strongly suggest that the salmon sperm DNA provides no ESR signal other than the magnetic impurities of mainly Fe and Mn, which is consistent with the nature of semiconducting electronic states of DNA with the optical energy gap of 4 eV as was discussed in Section 2.2.

2.3.3 *Magnetic Property with SQUID*

Magnetic susceptibility of DNA was measured with a SQUID susceptometer, as shown in Figs. 2.12 and 2.13. The magnetization curve of B-DNA at 100 K in Fig. 2.12 behaves as a straight line with a negative slope, which corresponds purely to the diamagnetism of DNA in the order of -3.8×10^{-4} emu/mol-bp. Very interestingly, the freeze-dried DNA (FD-DNA) shows a huge paramagnetism saturating at less than 1 T without hysteresis, which is almost temperature independent up to 300 K, except for the oxygen contribution below 50 K, as shown in Fig. 2.13 [33]. It was reported that the long-term evacuation of DNA leads to an emergence of this kind of saturating magnetization, but as small as one-tenth of that of Fig. 2.12 [34].

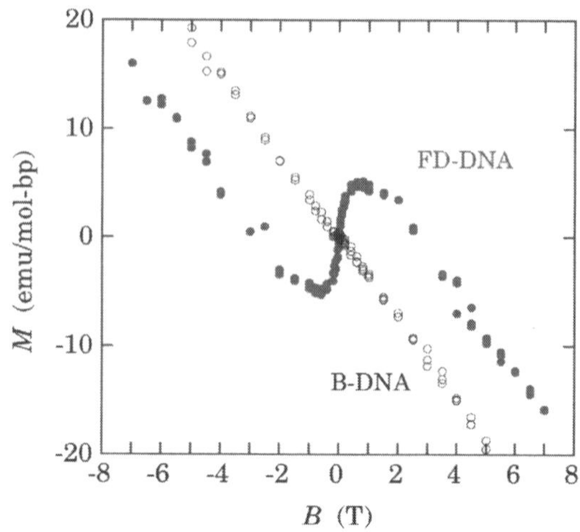

Figure 2.12 Magnetization at 100 K in B-DNA and freeze-dried DNA. B-DNA shows pure diamagnetism, but the freeze-drying (FD-DNA) treatment at -3°C induces the large nonlinear paramagnetism. Reprinted with permission from Ref. [28], Copyright 2014, The Physical Society of Japan.

The temperature dependence of the magnetic susceptibility of FD-DNA taken at the fixed magnetic field of 1 T is shown in Fig. 2.13. It looks like temperature-independent paramagnetism as large as 4×10^{-4} emu/mol-bp (to evaluate the net paramagnetism, subtract the DNA diamagnetism of -3.8×10^{-4} emu/mol-bp), which mimics Pauli paramagnetism. Thus, if one does not have any magnetization data, such as Fig. 2.12, the constant paramagnetism is not able to be assigned. The upturn below 50 K shows a typical temperature dependence of oxygen contamination [33].

What is the origin of the huge saturating magnetization? If one assumes that this is superparamagnetism without hysteresis, the number of spins can be estimated as 0.1% per base pair for $S = \frac{1}{2}$ spin or 0.01% for $S = \frac{5}{2}$ spin, which is consistent with the concentration of Fe impurity in the measured sample estimated by the X-ray fluorescence analysis. Thus, the ferromagnetic-type paramagnetism in A-form DNA is certainly due to the magnetic impurity of DNA on the basis of its concentration, but the mechanism has not been settled yet. Some discussion on possible mechanisms will be given in Chapter 9.

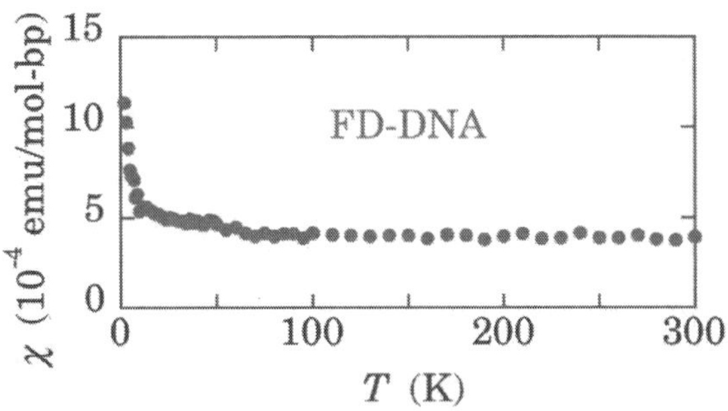

Figure 2.13 Temperature dependence of the susceptibility at 1 T in freeze-dried DNA (FD-DNA). The freeze-drying treatment leads to temperature-independent paramagnetism up to 4×10^{-4} emu/mol-bp (to obtain the paramagnetism, subtract DNA diamagnetism of -3.8×10^{-4} emu/mol-bp), which mimics Pauli paramagnetism. The upturn below 50 K is caused by oxygen contamination [33]. Reprinted with permission from Ref. [28], Copyright 2014, The Physical Society of Japan.

2.3.4 *Other Reports on Magnetic Property of DNA*

Nakamae and colleagues reported a magnetic study of λ phage DNA with two different counterion types, Na^+ and Mg^{2+}, in conjunction with the structural change between A-form and B-form at low temperature [32]. They measured magnetization of DNA (400 μg) with different quantities of water ($\approx$1.9 μl, $\approx$0.9 μl, and $<$ 0.1 μl) as a function of temperature at 5 T. They found the diamagnetic susceptibility of B-form DNA under a humid condition and that of A-form without water molecules are equal to $-(0.63\pm0.1)\times10^{-6}$ emu/g (-4.1×10^{-4} emu/mol-bp with the average molecular weight 660 of DNA), which is compared with the calculated atomic diamagnetic value of -0.52×10^{-6} emu/g (-3.4×10^{-4} emu/mol-bp).

They observed the water content dependence of susceptibility versus temperature. DNA with $<$ 0.1 μl has A-form structure typical of a dehydrated isomer of double helical DNA and showed almost temperature-independent, diamagnetic susceptibility down to 2 K. However, B-form DNA with increasing water content exhibited a paramagnetic enhancement at low temperatures below $\approx$50 K, like the Curie–Weiss law [32]. The magnetization curve at 2 K shows a weak saturation behavior as given by $m(H) = [\chi_L k_F\sqrt{S(T)}/[1 + (\frac{HS(T)}{\Phi_0})^2]$ [32]. They carefully checked the possibilities of impurities in origin for paramagnetism and concluded that paramagnetism was not originated from the magnetic moment of the electron spin, but was due to the orbital magnetic moment, thus an intrinsic property unique to B-form DNA. As an interesting possibility, they discussed the existence of persistent current loops along the DNA molecules on a mesoscopic micron scale. There is a theoretical model of the mesoscopic orbital current loop that the magnetism is paramagnetic and nonlinear when repulsive electron–electron interactions dominate over single particle effects [96, 97]. The maximum nonlinear magnetization of the persistent coherent loops occurs at $H_0 = \Phi_0/S(T)$, where $\Phi_0 = h/e = 4.14 \times 10^{-7}$ Gcm^2 is the magnetic flux quantum and $S(T)$ is the maximum surface area enclosed inside the coherent current loop. On the basis of the report that λ DNA would be metallic and shows proximity-induced superconductivity at $T < 1$ K [6], they estimated the coherence length of $\approx$1 μm in λ DNA with their experimental values, $H_0 = 1 \approx 2$ T, via $S(T) = d \times L(T)$, where d is the distance between bases of B-form DNA molecules and $L(T)$ is the electron coherent length. They assumed that the circulation of electrons enclosing a finite flux can be achieved through the combination of intra- and inter-strand transfer of π electrons on bases.

As another interpretation of the low-temperature magnetization of B-form DNA, oxygen contribution to the magnetization was proposed [33]. Oxygen molecules in a quartz tube with quartz wool provided a bump and Curie–Weiss magnetization below 100 K, as shown in Fig. 2.14. The Curie–Weiss law would be due to 2D layer of the adsorbed oxygen molecules on quartz wool and the bump would be due to 3D solid oxygen. Here, note that the temperature dependence of the magnetization in Fig. 2.14 along with the magnetization curve in Fig. 2.15 of the oxygen molecules is remarkably similar to that of the low-temperature magnetization of B-form DNA. Thus, we have to be careful in the contribution of oxygen molecules to the evolution of magnetization. B-form DNA contains a lot of water molecules probably along with oxygen molecules, which provides enhancement of the low-temperature paramagnetism. After evacuation of the water, the oxygen was also evacuated from the dried A-form DNA, which resulted in losing the enhancement of the magnetization.

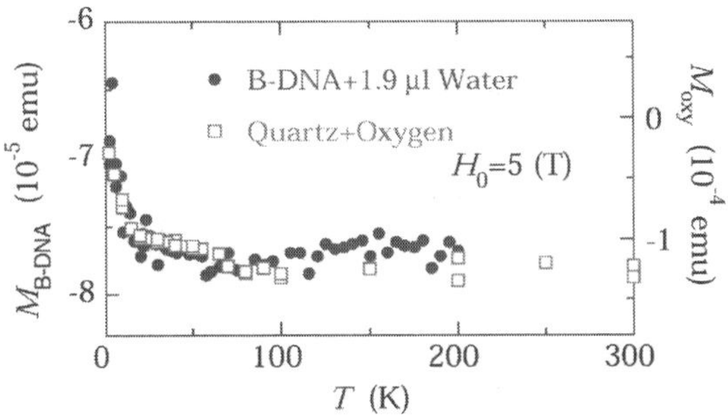

Figure 2.14 Comparison of the magnetization M_{oxy} of the oxygen adsorbed on quartz wool with the reported data of B-form of λ DNA with 1.9 μl of water [32]. Reprinted with permission from Ref. [33], Copyright 2006 by The American Physical Society.

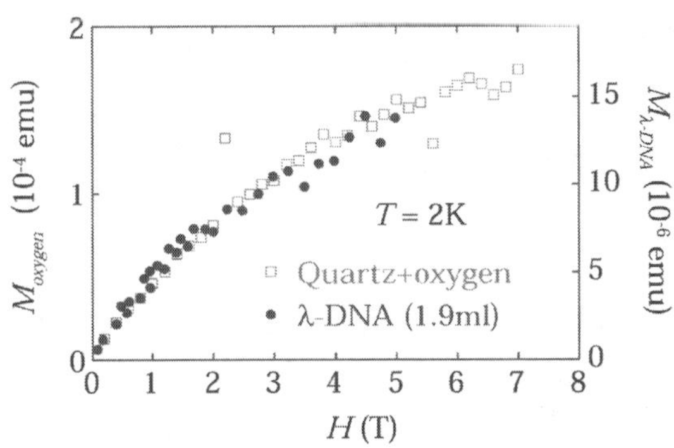

Figure 2.15 Comparison of M–H curves at 2 K between the oxygen molecules on quartz wool and the reported data of B-form of λ DNA with 1.9 μl of water [32]. Reprinted with permission from Ref. [33], Copyright 2006 by The American Physical Society.

Lee and colleagues reported ESR and SQUID magnetization of A-form DNA under an evacuated, dried condition and tried to interpret their data with the coherent current loops proposed by Nakamae and colleagues [34]. They measured magnetization of salmon sperm DNA purchased from Sigma-Aldrich as a function of magnetic field down to 4.5 K. They tried to reproduce the observed magnetization curve with nonlinear behavior in terms of (1) diamagnetism of DNA, (2) Pauli paramagnetism of the amorphous part of the sample, and

(3) orbital magnetism of the crystalline part due to the persistent current loop proposed by Nakamae and colleagues. However, the reproduction looks like being unsuccessful from several viewpoints.

First, DNA is well known as a semiconducting material with the energy gap of 4 eV, as demonstrated in Section 2.2. Thus, Pauli paramagnetism should not be observed and nobody has observed in DNA yet, within the author's knowledge. No experimental evidence for the Pauli paramagnetism is found in their report. Actually, we found no Pauli paramagnetism both in B-form and A-form DNA, as demonstrated in Fig. 2.12.

Second, the nonlinear magnetization interpreted as the orbital magnetization looks like usual ferromagnetic type saturation. Figure 2.12 also demonstrates that the large ferromagnetic type saturation is observed in A-form DNA, prepared with the freeze-drying procedure, up to 7 T (the large scattering of the data is found around $M \approx 0$ as a nature of the SQUID susceptometer). This ferromagnetic type saturation is temperature independent up to 300 K, and its magnitude is strongly dependent on the water content in the DNA, which is inconsistent with the orbital magnetization model proposed by Nakamae and colleagues, but is assigned to magnetic impurities such as Fe (see also Chapter 9).

Third, the isomeric structure of the DNA sample in which Nakamae and colleagues ascribed the observed paramagnetism to the orbital paramagnetism is B-form. In contrast, Lee and colleagues assigned the orbital paramagnetism to the nature of the A-form DNA in their report. Nakamae and colleagues used λ DNA, which shows no paramagnetism in the A-form structure down to 2 K, suggesting less magnetic impurity contamination like Fe than the commercial salmon sperm DNA.

Thus, the data analysis and the interpretation given in the report have to be carefully re-examined. Any paramagnetism is not found in B-form DNA, but, in contrast, the large ferromagnetic type paramagnetism emerges in A-form DNA prepared with the freeze-drying procedure, as found in Fig. 2.12. On the assumption that this is superparamagnetism, the number of spins can be estimated as 0.1% per base pair for $S = \frac{1}{2}$ spin or 0.01% for $S = \frac{5}{2}$ spin, which is consistent with the concentration of Fe impurities estimated by the X-ray fluorescence analysis. Thus, the ferromagnetic type paramagnetism in A-form DNA is certainly due to the magnetic impurity of DNA on the basis of its concentration, but the mechanism has not been settled yet.

2.3.5 *Summary of Magnetic Property in DNA*

It is reconfirmed that natural DNA is a substantially nonmagnetic semiconductor with the energy gap of 4.0 eV. Thus, the correspondingly expected magnetic property is diamagnetism of DNA. Actually, ESR spectra of DS 30mer-DNA show no distinguishable signal with sufficiently large intensity corresponding to the number of base pairs, if natural DNA would have the intrinsic magnetic spin in each base pair. SQUID magnetic susceptibility is consistent with the diamagnetism of DNA of the order of $\approx 4 \times 10^{-4}$ emu/mol-bp. Sometimes, abnormal paramagnetism has been reported, in which the magnetization saturates at $\ll 1$ T. Magnitude of such abnormal magnetism strongly depends on the quantity of water molecules in DNA double helix. In B-form DNA with a lot of water molecules, almost no abnormal magnetism appears, but in A-form DNA, especially a perfectly dried sample with the freeze-drying procedure shows remarkably large abnormal magnetism. However, the amplitude still corresponds to 0.1% per base pair for $S = \frac{1}{2}$ spin or 0.01% for $S = \frac{5}{2}$ spin, which is consistent with the concentration of Fe impurities. In conclusion, the nature of natural DNA is semiconducting and nonmagnetic states. Thus, natural DNA is not suitable for the direct application for nanowire in nanoelectronics.

Chapter 3

Infrared Spectral Studies on Structure and Hydration State of Dry DNA

Hiroshi Matsui

Department of Physics, Tohoku University, Sendai, Miyagi, 980-8578 Japan

matsui_ldp@m.tohoku.ac.jp

3.1 Introduction

By applying infrared spectroscopy for dry DNA, we can conveniently obtain information on conformation, content of hydrated water molecules, and hydration structure. In the regulated base-pair sequence of poly(dG)–poly(dC) double strand, the characteristic absorption band emerges at 1400 cm^{-1} in contrast to poly(dA)–poly(dT) and natural DNA. Through the spectral difference from Li-DNA and Na-DNA with monovalent counterions, the intercalation of divalent and trivalent metallic counterions into natural DNA yields a novel N–M bonding in the base molecules. Around the negatively charged phosphate group, disappearance of such counterions is found to induce anisotropy against the electric dipole and to drastically change the hydration structure.

DNA Engineering: Properties and Applications

Edited by Kenji Mizoguchi and Hirokazu Sakamoto

ISBN 978-981-4669-46-7 (Hardcover), 978-981-4669-47-4 (eBook)

www.panstanford.com

Infrared spectroscopy [98–100], which is employed in a wide variety of fields, is a powerful probe to investigate the structure and conformation of DNA, complementary to X-ray, and neutron diffraction experiments [101–107]. From the infrared spectra, we can identify molecular vibrations due to fundamental subgroups of DNA: heterocyclic base molecules, sugar groups, and phosphate groups [2, 101, 108, 109]. As for experimental advantage, infrared spectroscopy needs a tiny amount of sample in contrast to sizable single crystals required in X-ray and neutron diffraction experiments. Moreover, radiation power of infrared light is usually small, and then infrared spectroscopy is known to be a non-destructive method for DNA [110]. By using commercial Fourier-transformed infrared spectrometer (FT-IR), molecular vibrations of DNA can be easily measured under various conditions such as concentration of counterions, temperature, and relative humidity. Since dry DNA is important for application and engineering, infrared spectroscopy is a useful skill to obtain detailed information on structural property and hydration state at low water content.

Infrared-active molecular vibrations originate from a change of electric dipole moment inherent in molecule with respect to an electric field of infrared light. The larger the change, the more intense the infrared absorption. Molecular vibrations can be conveniently divided into two components associated with a variation in bond length (stretching mode) and bond angle (bending mode) [98–100]. Some stretching modes can vibrate in-phase (symmetric stretching mode) or out-of-phase (asymmetric stretching mode). Bending modes have different types of vibrations such as deformation, rocking, wagging, and twisting.

In conventional frequency region of $500 \sim 3500$ cm^{-1}, many absorption bands appear with different relative intensities, and we are available to assign those bands. The DNA structure is affected by various conditions such as concentration of counterions, base sequence, temperature, and relative humidity [111–117]. In particular, for dry DNA, the conformation is known to be strongly dependent on environmental relative humidity. Thanks to the systematic study combining infrared spectroscopy to X-ray diffraction experiments previously carried out, conformation of DNA can be determined with the set of marker bands observed in the

Table 3.1 Major marker bands for A-form and B-form

A-form (cm^{-1})	B-form (cm^{-1})	Assignment
1705	1715	Base molecule
1418	1425	Sugar group
1240	1225	Asymmetric stretching PO_2^-
882, 864	840	Sugar group

infrared spectra [110, 118–120]. For DNA studies, the marker band makes infrared experiment convenient to confirm the structural property influenced by environmental factors. The major marker bands for A-form and B-form are listed in Table 3.1.

Relative humidity varies water content corresponding to hydration strength in DNA double helix. Under high humid conditions, DNA fiber and film take B-form at room temperature. When relative humidity decreases below a certain threshold, the structure transforms from B-form to other conformations such as A-form or Z-form [101, 120]. In much lower humidity, a disordered structure appears, where the hydrogen bond connecting base pairs is partially broken and the base stacking damages. Other damage for DNA structure arises from denaturation, which is introduced by heating.

Infrared spectra provide hydration properties of DNA [121–124]. A phosphate group in a backbone is charged negatively, and hence water molecules with a dipole moment are strongly hydrated around it. It is well known that water molecules are connected to each other through hydrogen bonds [125–127], the strength of which is usually enhanced in the presence of ions or charged groups. The hydration structure in A-form and B-form was studied by X-ray and neutron diffraction experiments employing natural DNA samples so far [102–104, 128–131].

In computer simulations for water networks and various nanospaces involving water molecules, hydrogen bonds are expressed by the sum of Coulomb interaction and Lennard-Johns potential [126]. A charged site generates additional Coulomb interaction. Consequently, hydrogen bonds in a water network with such charged sites are enhanced in comparison with free water. In the hydrophilic nanochannel of molecular porous crystals $[Co^{III}(H_2bim)_3](TMA){\cdot}20H_2O$ and $[Co^{III}(H_2bim)_3](TATC){\cdot}7H_2O$

[132–134], the regulated water networks are strongly affected by the charged sites. Such interactions between nanochannel framework and water molecules are referred to as an interfacial interaction, which strongly depends on nanochannel size and number of water molecules embedded in the nanochannel [133, 134]. In general, biological materials contain many water molecules and various kinds of charged sites, so that a charge effect is significant not only for the hydration structure, but also for the biological functionality. Among them, for instance, a membrane nanochannel of aquaporin possesses a hydrophobic pore region and a charged site situating in the narrowest region of hourglass structure [135]. By the existence of charged site, the hydrogen bond between water molecules forming one-dimensional water network is broken, and proton transfer is blocked.

In the next section, we first mention the infrared spectra of four base molecules and nucleotides and identify the molecular vibrations of these molecules. Second, we deal with the infrared spectra of dry poly(dG)–poly(dC) and poly(dA)–poly(dT) with regulated base-pair sequence. Na counterions contact to the phosphate group in these double helices. We will show that the conformation strongly depends on relative humidity and temperature.

3.2 Infrared Spectra of Base Molecules and Nucleotides

Figures 3.1–3.4 demonstrate the infrared spectra of four base molecules (purine base: adenine (A) and guanine (G); pyrimidine base: cytosine (C) and thymine (T)) at room temperature under low relative humidity [136–141]. Several stretching modes due to NH, NH_2 and CH, CH_3 overlap in rather broad and large absorptions at 2500~3400 cm^{-1} [100, 141]. The bandwidth of thymine with CH_3 is narrower compared to the widths of other three bases with NH_2. For both CH_3 and NH_2, there are symmetric and asymmetric stretching vibrations. Two peaks of NH_2 stretching bands are observed at around 3200 and 3400 cm^{-1} in adenine and guanine, while both peaks of cytosine shift to higher wavenumber. According to the classical harmonic oscillator model [100], the force constants of NH_2 vibrations in cytosine are considered larger than those in adenine

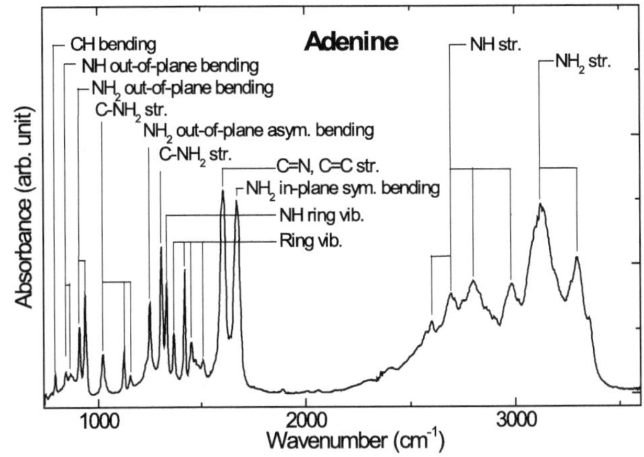

Figure 3.1 Infrared spectrum of adenine, and the assignment of each absorption bands.

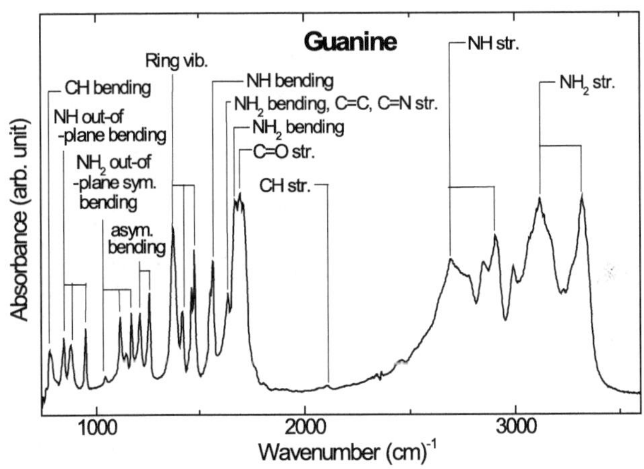

Figure 3.2 Infrared spectrum of guanine, and the assignment of each absorption bands.

and guanine. Such blue shift reflects the local structural difference between purine and pyrimidine.

As shown in Figs. 3.1–3.4, a number of sharp absorption bands are observed below 1700 cm^{-1}. The stretching modes (C=O, C=C, C=N) appear at 1600~1800 cm^{-1}. Some absorption bands at

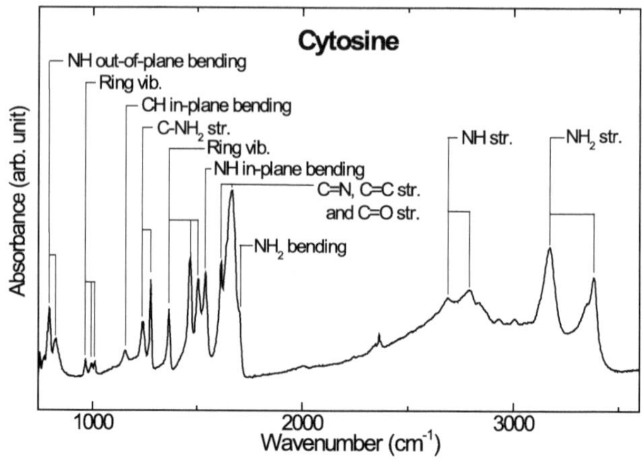

Figure 3.3 Infrared spectrum of cytosine, and the assignment of each absorption bands.

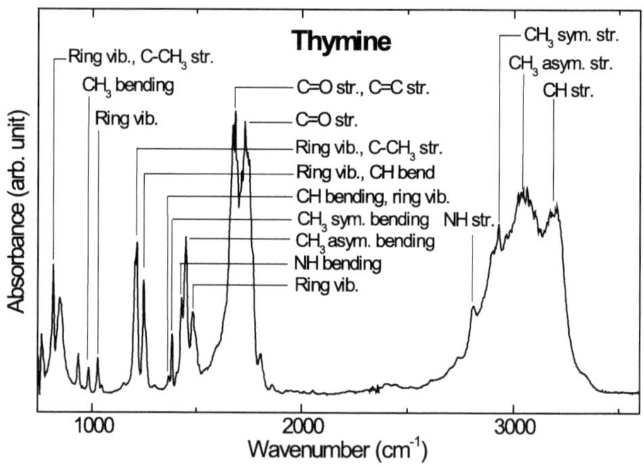

Figure 3.4 Infrared spectrum of thymine, and the assignment of each absorption bands.

1200~1500 cm^{-1} are assigned to a ring vibration induced on the six- and five-membered heterocyclic rings. The bending bands come from deformations of NH_2 and CH_3, and additionally vibrations of NH or CH deforming against the heterocyclic rings.

A nucleotide composed of a base molecule, a sugar, and a phosphate group can be synthesized by hydrolysis. The nucleotide is a fundamental monomer, which becomes a building block of double helix. There are four principle nucleotides: dAMP (deoxyadenosine monophosphate), dGMP (deoxyguanosine monophosphate), dCMP (deoxycytidine monophosphate), and dTMP (deoxythymidine monophosphate). A DNA chain is formed with those four types of nucleotides bonded together by a phosphodiester linkage.

Figure 3.5 depicts the infrared spectra of four nucleotides at room temperature under low relative humidity condition. Absorption bands around 3000 cm^{-1} are extremely broadened in contrast to Figs. 3.1–3.4. Several stretching absorptions due to base molecules, as mentioned previously, are hidden in those bands. Stretching vibrations due to sugar molecule are generated at around 3300 cm^{-1} (O-H), 2990 cm^{-1} (CH_2), and 3150 cm^{-1} (C-H). Since the valence of phosphate group is 2− (PO_3^{2-}), large amounts of water molecules are strongly hydrated in the vicinity of the charged

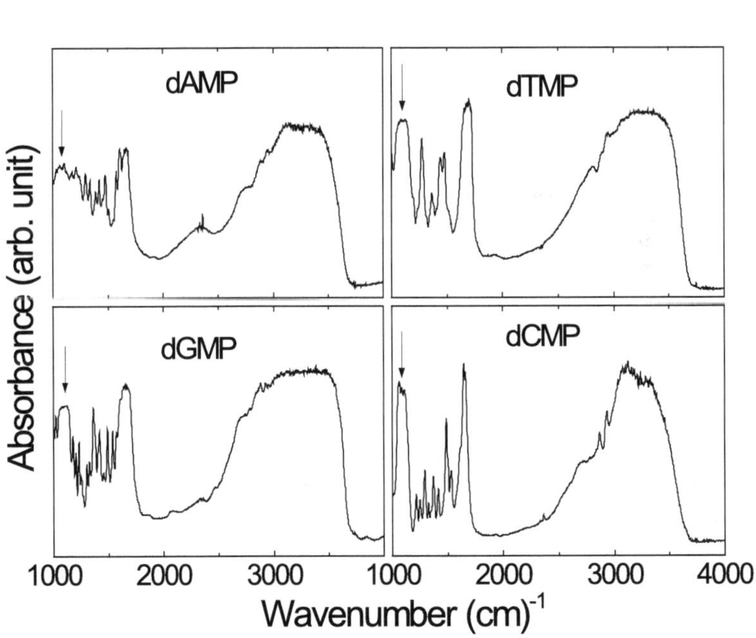

Figure 3.5 Infrared spectra of four kinds of nucleotides.

site to screen the charge. Such water molecules are absorbed from the atmosphere. For this reason, the band around 3000 cm^{-1} is hugely broadened by the OH stretching mode of water molecules [98, 100, 125].

From 1000 to 1700 cm^{-1}, many sharp bands emerge similar to the base molecules, as shown in Figs. 3.1–3.4. In particular, strong absorption bands, indicated by arrows in Fig. 3.5, which are never detected in base molecules, appear at 1100 cm^{-1} in all nucleotides. The absorption is attributed to a stretching mode of divalent PO_3^{2-}. In a DNA double helix, however, the band at 1100 cm^{-1} is never observed, because the phosphate group changes to monovalent PO_2^-. The absorption bands around 1600 cm^{-1} involves the contribution of HOH bending vibration of water molecules.

3.3 Hydration Structure of Dry Poly(dA)-Poly(dT) and Poly(dG)-Poly(dC)

Natural DNA holds a random base-pair sequence, which carries genetic codes and information of protein synthesis. In contrast, a regular base-pair sequence has much interest to clarify the structural property resulting from different types of sequences. The infrared spectra of dry poly(dA)–poly(dT) and poly(dG)–poly(dC) have been measured as changing relative humidity and temperature [142–144]. The double helical structures for each DNA are illustrated in Fig. 3.6.

To remove proteins and other impurities, the samples in solution were purified and collected by conventional phenol extraction and ethanol precipitation [14, 143]. The sample quality was checked by ultraviolet spectroscopy. We confirmed that the absorbance ratio at 280 nm to 260 nm was larger than 1.8, indicating a high-quality sample. In the case of dry DNA, counterions are necessary to neutralize negatively charged phosphate group (PO_2^-). To stabilize the structure, Na^+ ions were introduced in the present poly(dA)–poly(dT) and poly(dG)–poly(dC) samples. To exclude excess Na^+ in the solution, the sodium concentration was properly reduced to 1.2 $\times 10^3$ ppm by dialyzation with deionized water. The solution was

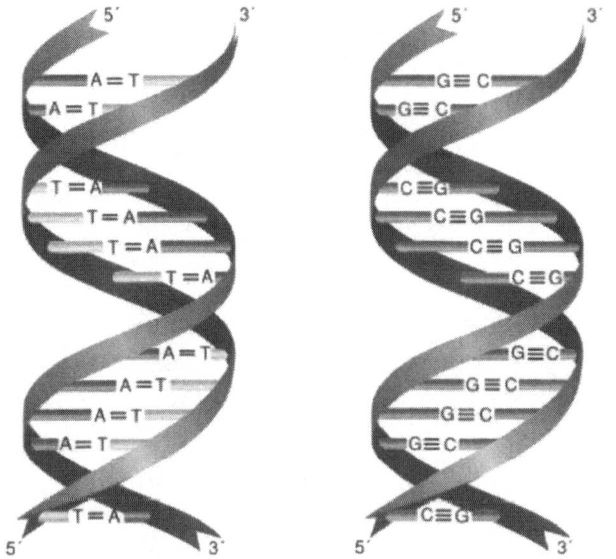

Figure 3.6 Schematic illustrations of poly(dA)–poly(dT) and poly(dG)–poly(dC).

rapidly frozen by liquid nitrogen and dried in vacuum, and finally dry fibrous samples were obtained.

Figure 3.7(a) shows the infrared spectra under relative humidity of about 60% at 30°C for poly(dG)–poly(dC) [142]. Two broad peaks at around 3200 and 3400 cm^{-1} are associated with the stretching bands due to NH and NH_2. By dry nitrogen gas exposed inside optical cryostat, the relative humidity around sample environment was controlled to be about 0%. The absorbance at 3400 cm^{-1} becomes half as large as that for 60% relative humidity. The absorbance is gradually reduced with increasing temperature up to 120°C, above which the absorbance becomes constant. Since the amount of DNA itself should be preserved, the decrease in integrated absorbance at 3000~3600 cm^{-1} is associated with the variation in OH stretching in water molecules. Here the integrated absorbance is associated with the area of absorption band.

Similar experiments were performed in poly(dA)–poly(dT) sample [144], but the process of temperature change differed from the aforementioned measurement in the poly(dG)–poly(dT) sample.

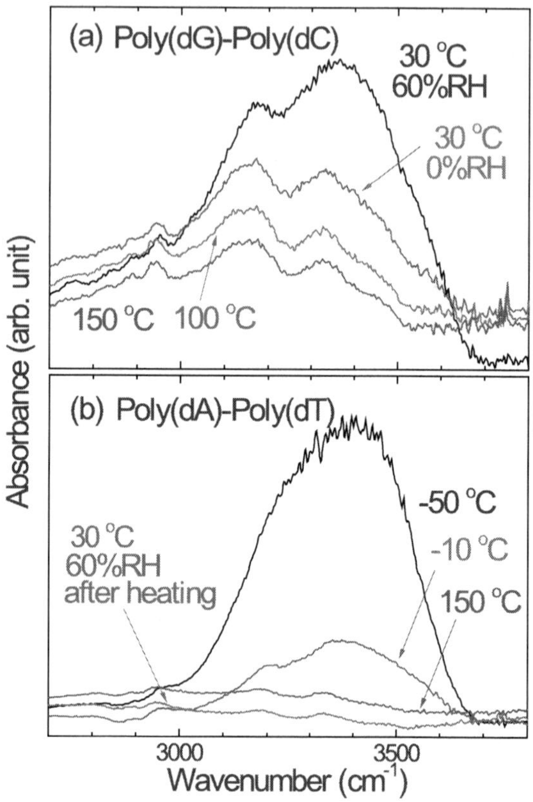

Figure 3.7 Infrared spectra at 2700∼3800 cm^{-1} for different temperatures and relative humidity in poly(dG)–poly(dC) (a) and poly(dA)–poly(dT) (b).

After cooling down to −50°C, the sample was heated up to 150°C, and then the sample temperature was returned to 30°C. As shown in Fig. 3.7(b), the maximum absorbance decreases up to 120°C, above which temperature dependence disappears as well as in poly(dG)–poly(dC) sample. Once the sample reaches 150°C, the broad maximum almost diminishes even at 30°C. The spectra above 120°C are dominated by the stretching modes of DNA molecules as well.

To lead the contribution of water molecules, each spectrum is subtracted from the spectra of dehydrated DNA samples at 150°C. The differential spectra for either DNA are shown in

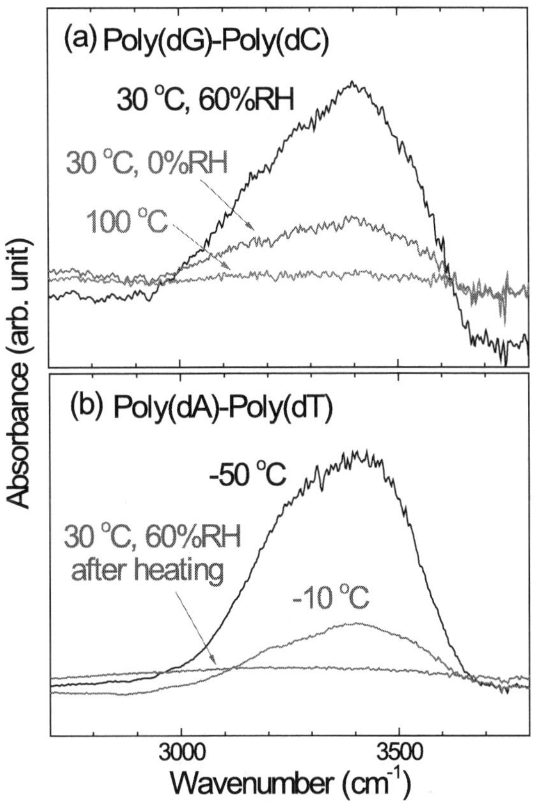

Figure 3.8 Differential spectra at 2700~3800 cm^{-1} reflecting the OH stretching vibration of hydrated water molecules in poly(dG)–poly(dC) (a) and poly(dA)–poly(dT) (b).

Fig. 3.8. A remarkable absorption band at 3300 cm^{-1} is attributed to the OH stretching band of hydrated water molecules. The integrated absorbance decreases with increasing temperature. Generally speaking, variations of total integrated absorbance reveal that the number of oscillators resonating to infrared electric fields is not conserved. As a consequence, the decrease in integrated absorbance of OH stretching band is caused by the reduction in water molecules from these samples. In other words, dehydration occurs with increasing temperature and at low humid conditions.

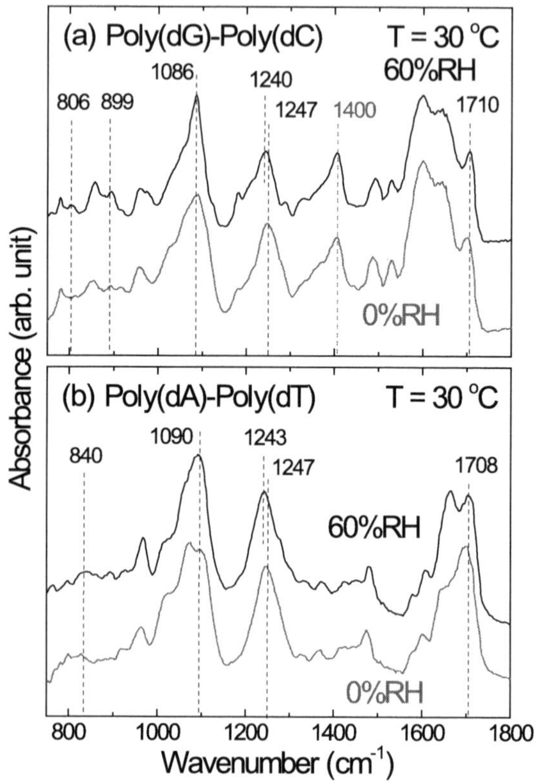

Figure 3.9 Infrared spectra at 750∼1800 cm^{-1} for 60% relative humidity and 0% relative humidity in poly(dG)–poly(dC) (a) and poly(dA)–poly(dT) (b).

Figure 3.9 shows the infrared spectra in the range of 750∼1800 cm^{-1} at 30°C. The poly(dG)–poly(dC) and poly(dA)–poly(dT) samples were placed in the atmosphere of 60% relative humidity and 0% relative humidity as well. In the infrared spectra of Figs. 3.1–3.4, broad bands around 1600 cm^{-1} are mainly composed of molecular vibrations due to base molecules. The C═N and C═C stretching modes appear at around 1710 cm^{-1}, which is close to the wavenumber of A-form marker band in Table 3.1. The peak wavenumber (1708 cm^{-1}) in poly(dA)–poly(dT) is slightly smaller than that in poly(dG)–poly(dC). The environment around C═N and C═C is not equivalent due to the different structure of base molecules. The absorption bands observed at 806, 890, and 1180 cm^{-1} reflect a puckering of the sugar group. According to Table 3.1, these bands are also identified to marker bands characterizing A-form.

In particular, distinct spectral difference between poly(dA)–poly(dT) and poly(dG)–poly(dC) samples emerges at 1400 cm^{-1} for both values of relative humidity. In the poly(dG)–poly(dC) sample (Fig. 3.9(a)), the absorption peak at 1400 cm^{-1} has comparable absorbance to that at 1240 cm^{-1}. It must be noticed that such absorption at 1400 cm^{-1} is never observed in the poly(dA)–poly(dT) sample (Fig. 3.9(b)) and also natural DNA with random base-pair sequence. Then the absorption at 1400 cm^{-1} is concluded to be a characteristic band in poly(dG)–poly(dC) double strand.

Absorption bands at around 1086 and 1240 (1243) cm^{-1} are assigned as symmetric and asymmetric stretching modes of PO_2^-, respectively. The former intensity slightly depends on relative humidity [145], and the latter wavenumber shifts to 1247 cm^{-1} for 0% relative humidity owing to dehydration.

A hydration structure is illustrated in Fig. 3.10, where the regions from 1 to 5, denoted by shaded circles, are the hydration sites of water molecules. On increasing the number, hydration

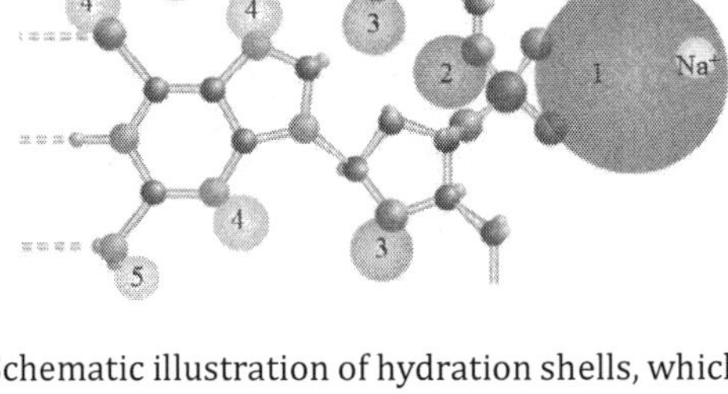

Figure 3.10 Schematic illustration of hydration shells, which are numbered from 1 to 5. Na^+ situates in the vicinity of phosphate group. Red, blue, black, and green circles stand for P, O, C, and N atoms, respectively. Small blue circles denote H atom.

strength weakens. The negatively charged phosphate group (PO_2^-) is compensated by counterions Na^+, and additionally it is surrounded by water molecules belonging to the primary hydration shell. The Coulomb repulsion force between the adjacent phosphate groups is increased by the reduction of water molecules. The electric interaction sensitively affects the resonating state of asymmetric stretching mode in PO_2^-, which becomes a significant marker band to determine the conformation of DNA. From Table 3.1, the present wavenumber of asymmetric stretching mode in PO_2^- indicates A-form as well.

Through the change in relative humidity and/or temperature, the integral absorbance reveals a variation in water content in hydration shells. As shown in Fig. 3.10, the strongest hydration occurs in the vicinity of phosphate group. Hydration at base molecules is weak, and dehydration is considered to start from base molecules. In the first hydration shell positioned around PO_2^-, the maximum water content is about 20 water molecules per nucleotide (wpn), as shown in Table 3.2. [121]. The higher hydration shells around sugar and base molecules can fulfill 10 wpn at most [123]. These hydrated water molecules are contacted through hydrogen bonds [101]. It is commented here that the melting temperature of hydration water molecule is lower than 0°C [143].

Water content can be evaluated from methods such as thermogravimetry or Karl–Fisher titration [103]. Employing the formula obtained by Falk *et al.* [122, 123], the water content n is empirically estimated by the ratio R of infrared absorption band at 3400 cm^{-1} to one at 1240 cm^{-1}; $n = 4.52(R - 0.86)$.

Table 3.2 Summary of hydration states in DNA

Relative humidity (% RH)	Number of water molecules per nucleotide (wpn)	Hydration state of phosphate group	Hydration state of sugar group and base molecules
100	~30	Completely filled	
90	~20	Almost filled	Slightly occupied
80	~10		
50	~4		
20	~2	Formation of water bridge	Empty
0	~1		

From Figs. 3.8 and 3.9, we can calculate the water content at different conditions. At 30°C in Fig. 3.8(a), 3∼6 wpn are hydrated for 60% relative humidity, but a single water molecule is connected to the phosphate group for 0% relative humidity. After heating up to 100°C, almost all water molecules are dehydrated, though there still remain 0.1 wpn [146].

Under 60% relative humidity in Fig. 3.8(b), the water content at −50°C is estimated to be about 16 wpn in the poly(dA)-poly(dT) sample. At low temperature, the sample absorbs water molecules from the atmosphere. With increasing temperature, the water content gradually decreases through dehydration: ∼7 wpn at −10°C and ∼4 wpn at 30°C. After heating up to 150°C, very less water molecules contact the sample.

From the previous thermogravimetry experiment on natural DNA at room temperature [121], the water content is determined as 26 wpn near 90% relative humidity, ∼6 wpn for 60% relative humidity, and ∼2 wpn for 10% relative humidity. The present water contents in the poly(dG)-poly(dC) and poly(dA)-poly(dT) samples with Na^+ counterions are almost comparable to those in natural DNA. Therefore, the hydration structure is not strongly dependent on the type of base-pair sequence.

For relative humidity lower than 60%, the first hydration shell is partially filled, while the higher hydration shells are empty. The present infrared response basically corresponds to the water molecules in the first hydration shell. According to the previous X-ray and neutron diffraction experiments on natural DNA [128], the hydrated water molecules are confirmed to construct a water network via hydrogen bonds. Such water molecules make a bridge between neighboring phosphate groups in the major groove. In order to form the water bridge, 1∼3 wpn are required (Table 3.2). The water bridge contributes to the stabilization of A-form double helical structure. Below 60% relative humidity in poly(dG)-poly(dC) and poly(dA)-poly(dT) samples, the water content estimated as 3∼6 wpn is enough to form a water bridge. It is noteworthy that the dielectric constant measured for the same sample with 3∼4 wpn takes the anomalous maximum ($\epsilon \sim 7$) at 7°C [143]. The characteristic dielectric response is responsible for a collective excitation due to the water dipole on the water bridge.

3.4 Summary of the Infrared Spectra in Poly(dG)–Poly(dC) and Poly(dA)–Poly(dT)

Employing infrared spectroscopy, the molecular vibrations of base molecules and nucleotides can be conveniently measured and assigned. Furthermore, we have clarified the fundamental physical properties of conformation, content of hydrated water, hydration structure in poly(dG)-poly(dC) and poly(dA)-poly(dT) samples with regulated base-pair sequence. In particular, the characteristic absorption band emerges at 1400 cm^{-1} in poly(dG)-poly(dC) in contrast to poly(dA)-poly(dT) and natural DNA.

Chapter 4

Proton Conduction in DNA

Yasumitsu Matsuo[a] **and Seiichiro Ikehata**[b]

[a]*Department of Life Sciences, Setsunan University, 17-8 Ikeda-Nakamachi, Neyagawashi, Osaka, 572-8508 Japan*
[b]*Department of Applied Physics, Tokyo University of Science, 1-3 Kagurazaka, Shinjyuku-ku, Tokyo, 162-8601 Japan*
ymatsuo@lif.setsunan.ac.jp

The present chapter explains the electrical properties of DNA from the viewpoint of ion transfer, which is an important feature of bio-tissue materials. The fuel cell application in Chapter 13 is developed on the basis of the characteristic features in the ionic conduction of DNA. Based on the investigation of ion transports using impedance analysis and nuclear magnetic resonance (NMR) measurement, the proton conductivity of DNA under humidified conditions is revealed. Impedance analysis indicates that water bridges play an important role in the appearance of proton conductivity. Moreover, ^{2}D-DNA and DNA-Na also become ionic conductors under humidified conditions. The chapter also describes the process of water bridge formation and the role of deuterons and Na ions in ionic conductivity.

DNA Engineering: Properties and Applications
Edited by Kenji Mizoguchi and Hirokazu Sakamoto

ISBN 978-981-4669-46-7 (Hardcover), 978-981-4669-47-4 (eBook)
www.panstanford.com

4.1 Introduction

4.1.1 *Proton Transfer in Bio-Tissue Materials*

The use of biomaterials in the medical and recombinant fields is well known. Recently, the detailed characteristics of biomaterials and their application to devices have been clarified through the development of computer technology and nanotechnology [147–152]. Therefore, since 2000, a lot of researches on devices using bio-tissue-derived materials have been carried out. Since bio-tissue materials have characteristic features such as ionic pumps, transporters, and ionic channels, they have the potential to be an ionic conductor. The schematic diagram of an ionic pump is shown in Fig. 4.1(a). A lipid bilayer translates M^+ ions (for example, H^+ ion) from inside the cell to outside the cell accompanied by the hydrolysis of ATP. In this case, the proton motive force (pmf) is expressed with the proton concentrations $[H^+]_i$ and $[H^+]_o$ inside and outside the cell, respectively, by the following equation:

$$\mathrm{pmf} = -(\psi_i - \psi_o) - \frac{k_B T}{e}(\log[\mathrm{H}^+]_\mathrm{i} - \log[\mathrm{H}^+]_\mathrm{o}). \tag{4.1}$$

Here, k_B is the Boltzmann constant, e is the elementary charge, and ψ_i and ψ_o are the potentials inside and outside the cell, respectively. Obeying this equation, the proton transport is realized in the bilayer.

The schematic diagram of ionic transport by the ionic channel is shown in Fig. 4.1(b). The lipid bilayer has a selective ionic transport for K^+, Na^+, and H^+ ions, in which characteristic features such as the

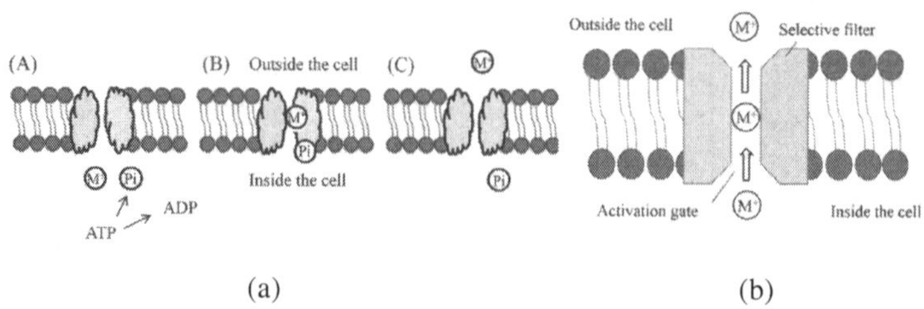

Figure 4.1 Schematic figure of ionic transports in a lipid bilayer with an ionic pump (a) and an ionic channel (b).

Table 4.1 Ionic transport in bio-tissue materials. The transfer velocity corresponds to the product of the mobility of mobile ions and the number of mobile ions

	Pump	Channel
Transfer velocity	10^2 ion/s	10^8 ion/s
Example	H^+ - ATPase Ca^{2+} - ATPase	K^+ channel Ca^{2+} channel Proton channel

number of transfer ions are different. Table 4.1 shows the difference of ionic transports for the ionic pump and the ionic channel. The transfer velocity of ionic transport in the ionic channel is larger than that in the ionic pump. Therefore, bio-tissue materials become good ionic conductors by applying ionic channels. Very recently, for example, it was found that collagen, which is a bio-tissue material, becomes a good proton conductor under humidified conditions and acts as electrolyte for fuel cells [153, 154]. Hence, bio-tissue materials such as DNA and collagen have the potential to develop a new type of ionic conductor, which can reduce environmental loads accompanying production and waste.

4.1.2 *DNA Conductor*

Since 2000, in the field of electric devices, many efforts have been devoted to the investigation of electrical conduction in DNA [3, 4, 8, 155–160]. For example, Fink and Schenenberger as well as Kasumov *et al.* reported that a DNA molecule behaves as a good conductor [3, 155], and Porath *et al.* and Kutnjak *et al.* reported that the electrical conductivity of DNA molecules is similar to that of semiconductors [4, 156]. On the other hand, Zhang *et al.* reported that DNA molecules are insulators and that contamination of DNA is crucial for electrical conductivity [8]. It is known that the electrical properties of DNA depend on the atmosphere. In particular, electrical conductivity is drastically changed by relative humidity (RH). Thus, hydration of DNA changes the electrical properties of DNA, and therefore its behavior is of great interest. Recently, we have investigated the electrical conductivity of DNA under humidified conditions and found that conductivity is caused by ionic conductivity. We will show the characteristics of ionic transport in DNA films in this section and the application of DNA films to fuel cells in Chapter 13.

4.2 Experimental

Impedance analysis and NMR measurement are described in this section. Sample preparation will be shown in each section with experimental results.

4.2.1 *Impedance Analysis*

Impedance analysis enables us to obtain information on dielectric and conductive behaviors of a sample. Especially, the existence of polarization and molecular motion can be obtained from the frequency dependence of impedance Z. A lot of materials are approximately described by the parallel circuit of ionic conductivity and dielectric polarization, as shown in Fig. 4.2. In this case, the admittance Y $(= 1/Z)$ is given by the following equation:

$$Y = G + j\omega C. \tag{4.2}$$

Here, ω is the angular frequency, j is the imaginary unit, C is the capacitance of dielectric polarization, and G is the ionic conductance.

When the equivalent circuit of a specimen corresponds to that of Fig. 4.2, the specimen is described by the frequency-independent

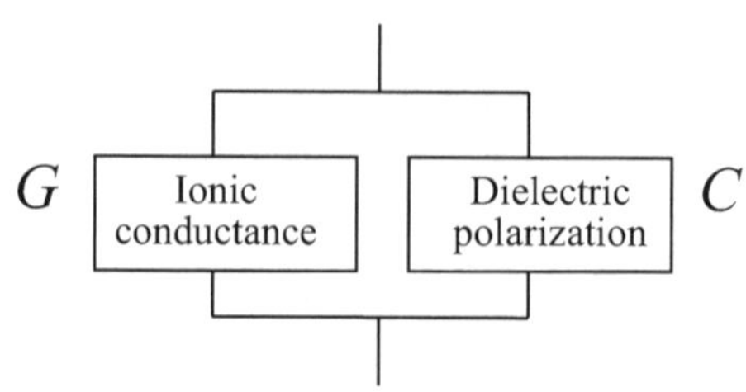

Figure 4.2 Parallel equivalent circuit with ionic conductivity and dielectric polarization for bio-tissue materials.

dielectric polarization C and the ionic conductance G. In some materials, however, C and G estimated with the equivalent circuit could show strong frequency dependence. In this case, we cannot describe the equivalent circuit only by the simple parallel circuit of capacitance C and conductance G, but the dispersion of dielectric polarization has to be taken into account. For example, it is well known that water molecules, which are small molecules, show dielectric dispersion around 22 GHz. In contrast, in large molecules such as bio-tissue materials, dielectric dispersion is observed at relatively low frequencies. In this way, from the frequency dependence of admittance, we can obtain information on molecular motions. The ionic conductivity including the dielectric dispersion is given by [161]

$$\begin{aligned}\sigma(\omega) &= \sigma_0 - \mathrm{Im}\left[\omega\epsilon_0\epsilon_\infty + \frac{\omega\epsilon_0(\epsilon_s - \epsilon_\infty)}{1 + (j\omega\tau)^\beta}\right] \\ &= \sigma_0 + \frac{\omega\epsilon_0(\epsilon_s - \epsilon_\infty)(\omega\tau)^\beta \sin\left(\frac{\pi}{2}\beta\right)}{\left(1 + (\omega\tau)^\beta \cos(\frac{\pi}{2}\beta)\right)^2 + \left((\omega\tau)^\beta \sin(\frac{\pi}{2}\beta)\right)^2},\end{aligned} \tag{4.3}$$

where ϵ_s is the static dielectric constant, ϵ_0 is the dielectric constant of vacuum, and ϵ_∞ is the unrelaxed (or optic frequency) dielectric constant. ω and τ show the angular frequency and the relaxation time of dielectric response, respectively. An empirical exponent of β shows the degree of mono-dispersion (distribution of τ, flat distribution for $\beta = 0$). Figure 4.3 shows the typical frequency dependence of $\sigma(\omega)$ calculated with Eq. (4.3). As shown in Fig. 4.3, by the existence of dielectric dispersion, the frequency dependence of $\sigma(\omega)$ strongly depends on the relaxation time τ, and $\sigma(\omega)$ exhibits the step-like anomaly below the frequency of $\approx 1/\tau$. With this equation, we can derive the DC conductivity σ_0 and the dielectric dispersion parameters from the frequency dependence of AC conductivity.

The impedance data were obtained with the precision LCR meter (HP4284A and Agilent E4980A), and the frequency dependence is analyzed in detail.

4.2.2 *Nuclear Magnetic Resonance*

NMR spectroscopy is a powerful and theoretically supported analytical tool. Especially, the motion of protons and ions in a

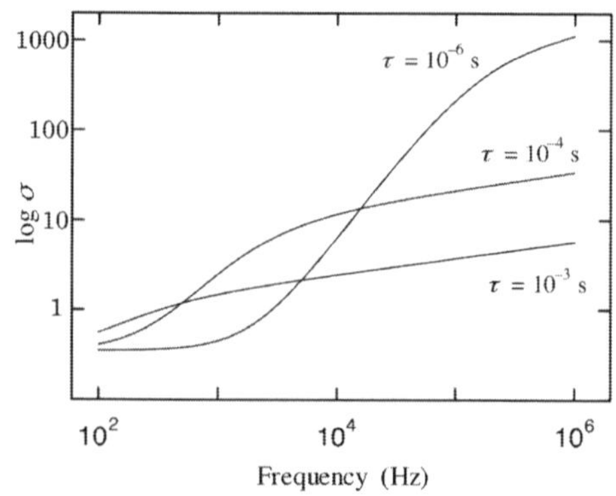

Figure 4.3 Qualitative frequency dependence of conductivity with dielectric dispersion in Eq. (4.3).

specimen is sensitively reflected in the NMR absorption line. In the present chapter, we will show the results of ^{1}H-NMR, deuterium-NMR (^{2}D-NMR), and ^{23}Na-NMR. The ^{1}H-NMR absorption lines were measured using the house-made continuous-wave NMR system, while the deuterium-NMR and ^{23}Na-NMR absorption lines were observed using Chemagnetics CMX infinity 300 spectrometer.

4.3 Proton Conductivity in a DNA-H Film

4.3.1 *Sample Preparation*

The present work used λ-DNA (BioDynamics Laboratory Inc.), which has 300–700 base pairs. DNA films were prepared by the casting method. First, purified DNA was dissolved in a DNA solution of distilled water with the concentration of 10 mg/ml, without any salt such as NaCl. Next, the DNA solution was dropped on a glass or a Teflon substrate and was then dried. Figure 4.4 shows a transparent DNA-H film. The thickness, which can be controlled by the concentration of DNA solution, was measured with an atomic force microscope (AFM) and a digital film thickness measurement system. The temperature of specimens was kept at 30°C, and the relative humidity was controlled by adjusting the gas humidity around the specimens.

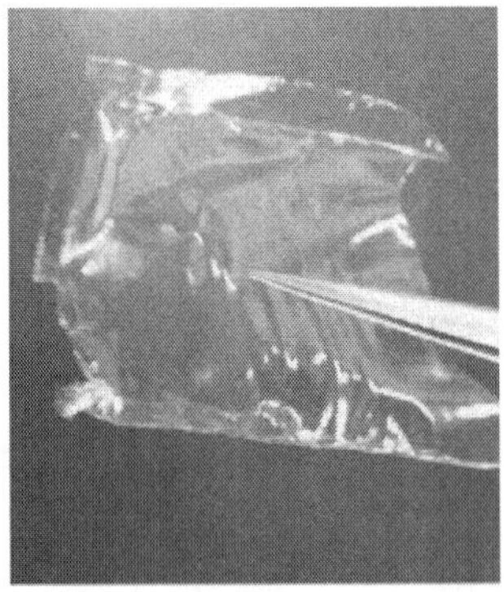

Figure 4.4 Example of a DNA-H film.

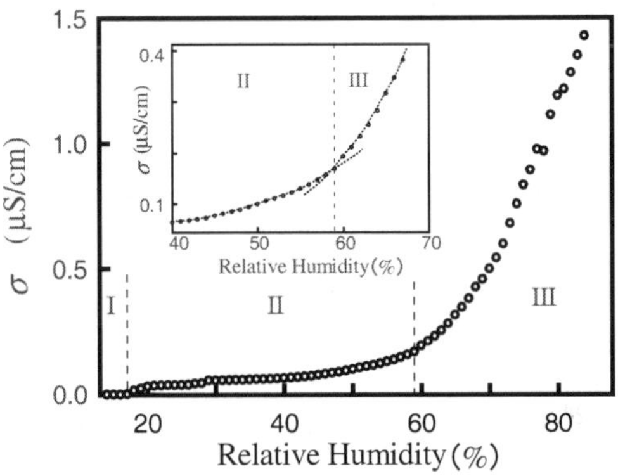

Figure 4.5 Relative humidity dependence of the conductivity in a DNA-H film. Reprinted with permision from Ref. [163], Copyright 2005, Elsevier.

4.3.2 *DC Conductivity and ^{1}H-NMR*

Figure 4.5 shows the humidity dependence of DC conductivity in the DNA-H film [163]. We can clearly see that DC conductivity begins around 17% RH, thereafter increases monotonically with the relative humidity and then drastically goes up above 60% RH. In this way, the DNA-H film becomes an electrical conductor under humidified conditions. Thus, it is suggested that the appearance of DC conductivity originates from the H^+ transfer, because DC conductivity strongly depends on humidity (that is, the amount

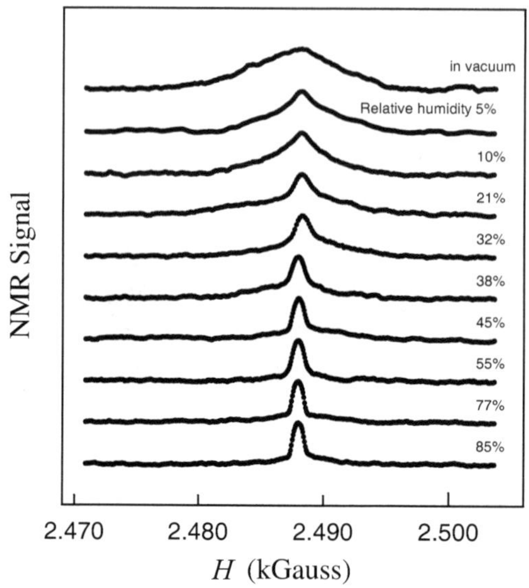

Figure 4.6 ^{1}H-NMR absorption line at various humidity in DNA-H. Reprinted with permision from Ref. [163], Copyright 2005, Elsevier.

of H_2O). In order to examine the humidity dependence of proton motion, we have carried out ^{1}H-NMR measurement.

Figure 4.6 shows the NMR absorption lines under various humidity conditions [163]. It is evident that the NMR absorption lines become narrow with the increase in humidity. Since the narrowing of NMR linewidth implies motional narrowing by proton migration, this result indicates that the proton transfer is realized under humidified conditions. Considering these results, the electrical conductivity in Fig. 4.5 is caused by the "proton conductivity."

Figure 4.7 shows the fine structures of ^{1}H-NMR absorption lines under vacuum and 32% RH. The NMR absorption line consists of two Gaussian lines under vacuum. We can see that the areas of these Gaussian lines are clearly different from each other. It is known that the area of ^{1}H-NMR line is proportional to the number of hydrogens. In Fig. 4.7, the area ratio of line (b) to line (a) is about 1:13. This ratio is in agreement with that of the number of hydrogens in the hydrogen bonds of bases to all others. Further, we can see that the

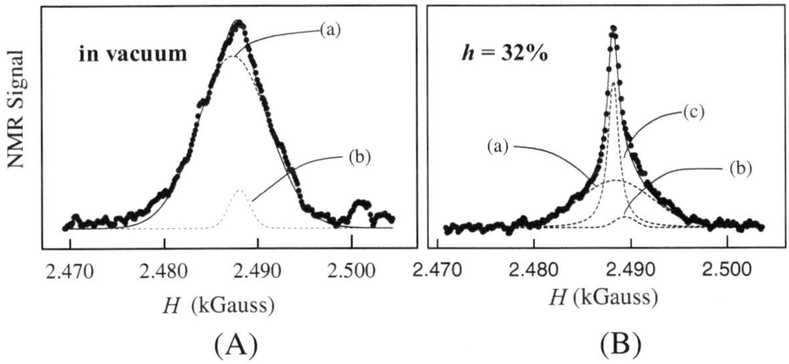

Figure 4.7 Fine structure of NMR absorption line in vacuum (A) and 32% RH (B). Reprinted with permision from Ref. [163], Copyright 2005, Elsevier.

linewidth of line (b) is 2.4 gauss, from which we can obtain the proton–proton distance by assuming a rigid lattice of the protons. In contrast, it is difficult to estimate the proton–proton distance from line (a) because the line broadening of (a) is caused by protons of multiple molecules. It is well known that the linewidth ΔH of NMR absorption line for powder specimens is given by the following equation:

$$\Delta H^2 = \frac{3}{5}\gamma^2\hbar^2 I(I+1)\frac{1}{r^6}, \tag{4.4}$$

where γ is the gyromagnetic ratio, $\hbar$ is the reduced Planck constant, $I(I+1)$ is the eigenvalue of I^2 (angular momentum operator), and r is the averaged proton–proton distance. The proton–proton distance calculated from the NMR linewidth of 2.4 gauss is 2.5 Å. This value is consistent with the distance of ≈2.3Å between the protons of the hydrogen bonds in the A-T or G-C base pair. In this way, under low-humidity conditions, we can see the existence of rigid protons in sugars and bases. On the other hand, with increasing humidity, a new line (c) appears, whose line shape is Lorentzian with a narrow linewidth. This result indicates that mobile protons appear with increasing humidity. Thus, we can obtain the ratio R_m of mobile protons to all protons, by comparing the areas of NMR lines.

Figure 4.8 shows the ratio R_m, which begins around 17% RH, then increases sharply and continues to 60% RH with relative humidity. This result indicates that the mobile protons appear

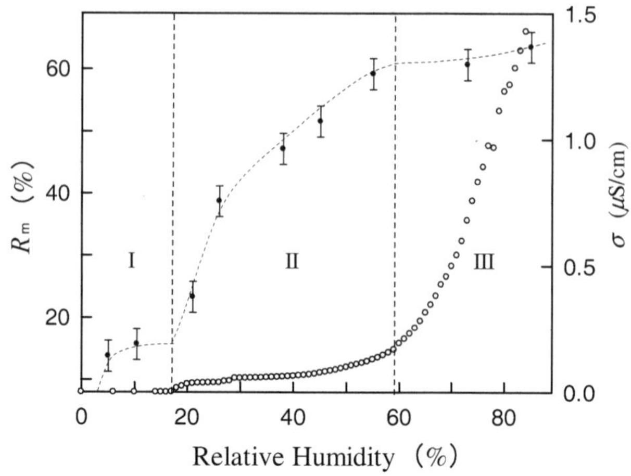

Figure 4.8 Relative humidity dependence of the ratio of mobile protons to all protons (closed circles) and the proton conductivity (open circles) in DNA-H. Reprinted with permision from Ref. [163], Copyright 2005, Elsevier.

around 17% RH, and thereafter the number of migrating protons increases with relative humidity. Since the electrical conductivity in Fig. 4.8 also begins to increase around 17% RH, the conductivity of DNA in humidified conditions above 17% RH is caused by proton conduction in DNA. It is known that DC measurement causes some errors in the results of conductivity due to the imperfect contacts of electrodes with specimens (ohmic, blocking, or Schottky). Therefore, to investigate the electrical conductivity of DNA in detail, we have carried out impedance measurement.

4.3.3 *AC Conductivity and Impedance Analysis*

Figure 4.9 shows the relative humidity dependence of the AC electrical conductivity at various frequencies [164]. The electrical conductivity increases not only with relative humidity but also with frequency.

Figure 4.10 shows the frequency dependence of AC electrical conductivity, which clearly increases with relative humidity, suggesting that the observed AC electrical conductivity also includes the contribution of proton conductivity. Moreover, it is noted

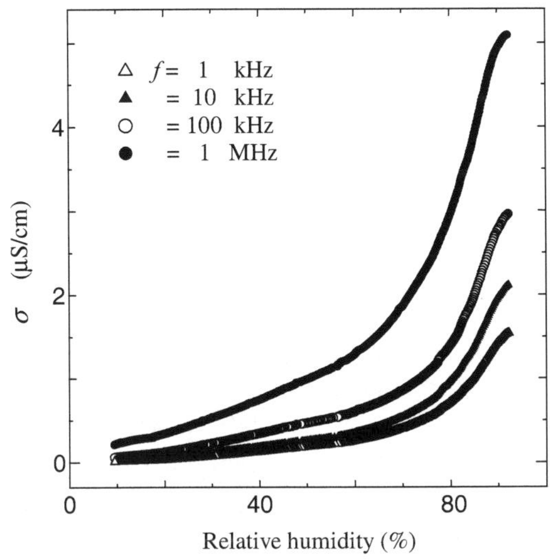

Figure 4.9 Relative humidity dependence of the AC electrical conductivity at various frequencies in a DNA film. Reprinted with permission from Ref. [164], Copyright 2005, Elsevier.

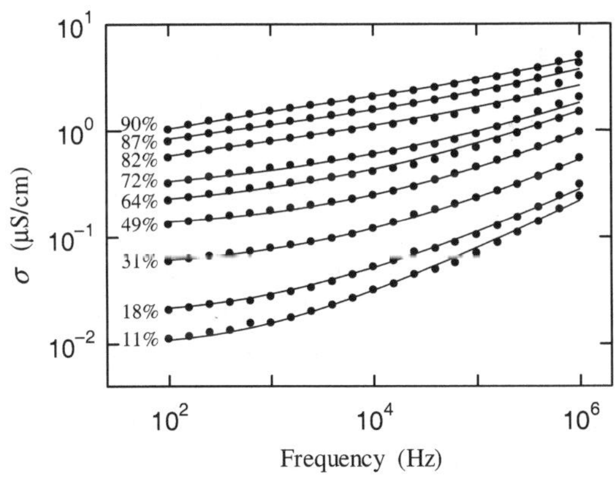

Figure 4.10 Frequency dependence of AC proton conductivity in a DNA film. The solid curves represent the fittings to Eq. (4.3). Reprinted with permission from Ref. [162], Copyright 2010, The Physical Society of Japan.

Table 4.2 DC conductivity and dielectric dispersion parameters. σ_{dis} is the second term in Eq. (4.3) and means the contribution of dielectric dispersion to AC conductivity

	σ_0 (μS/cm)	$\epsilon_s - \epsilon_\infty$	τ (sec)	β	σ_{dis} (μS/cm)
10%	7.5×10^{-3}	3.45×10^{1}	8.25×10^{-4}	0.50	7.44×10^{-4}
38%	8.6×10^{-2}	7.93×10^{2}	5.20×10^{-3}	0.62	3.34×10^{-3}
90%	0.45	2.40×10^{4}	3.00×10^{-3}	0.83	2.67×10^{-1}

that electrical conductivity above 49% RH shows a broad bump around 1 kHz, which means that AC proton conductivity cannot be described by the simple parallel equivalent circuit, in which the electrical conductivity obeys the equation $\sigma = \sigma_0 + \omega\epsilon_0\epsilon''$, where ϵ'' is the imaginary part of the dielectric constant and is frequency independent, as described in Section 4.2. If the frequency dependence of electrical conductivity contains step-like anomalies, as the broad bump in Fig. 4.10, the dielectric dispersion is included in AC electrical conductivity.

With Eq. (4.3) applicable to the proton conductivity including the dielectric dispersion, the DC conductivity and the dielectric dispersion parameters can be extracted, as summarized in Table 4.2. σ_{dis} is observed as a small peak for the existence of dielectric dispersion. The frequency dependence of electrical conductivity predicted by the extracted parameters with Eq. (4.3) is shown by the solid curves in Fig. 4.10, which are in agreement with the experimental results.

Figure 4.11 shows the humidity dependence of DC proton conductivity σ_0 and dielectric dispersion part σ_{dis}. σ_0 begins to increase around 17% RH up to ≈55% RH, and then drastically increases again above 70% RH. Similarly, the dielectric dispersion part σ_{dis} begins to gradually increase around 17% RH up to 60% RH and makes a rapid upturn above 80%. These results indicate that the appearance of proton conductivity is closely related to that of dielectric polarization.

In Fig. 4.12, we show the humidity dependence of dielectric relaxation time τ and β. The relaxation time τ markedly increases above 17% RH, forming a broad peak around 55% and then linearly decreases. β also increases with relative humidity up to 30% RH,

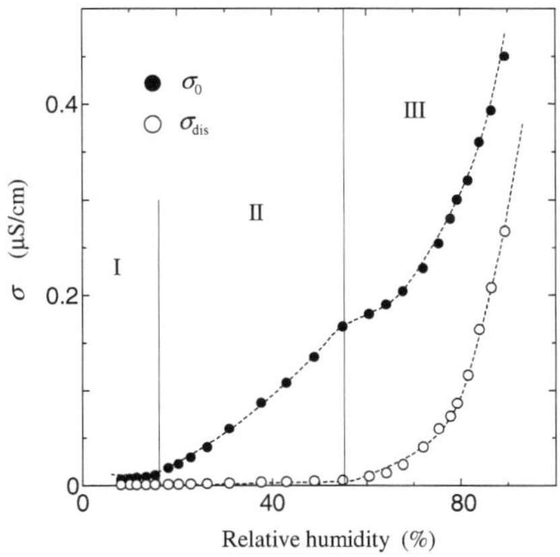

Figure 4.11 The DC conductivity σ_0 and the dielectric dispersion part σ_{dis} in a DNA film, extracted from Fig. 4.10 with Eq. (4.3).

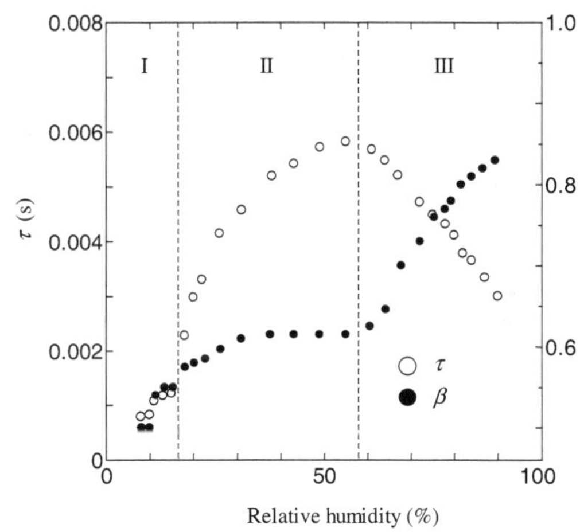

Figure 4.12 Relative humidity dependence of τ and β in DNA. Reprinted with permission from Ref. [162], Copyright 2010, The Physical Society of Japan.

levels off and increases again above 60%. The increase in τ suggests that the motion of molecules concerning dielectric polarization becomes slower. That is, the increase in τ above 17% RH implies the formation of long molecular structures with slow dynamics of the dielectric polarization. On the contrary, the decrease in τ

above 55% RH shows that the average size of effective molecular structures concerning dielectric polarization becomes smaller. This finding suggests that the number of water molecules interacting weakly with DNA is increasing rapidly above 55% RH. Since β indicates the degree of relaxation time distribution, for example, the relaxation time is unique for $\beta = 1$ without distribution and shows wider distribution for β approaching 0, the increase in β with the relative humidity in Fig. 4.12 shows that the orientational motion of dielectric polarization tends to become unique.

It is known that the dielectric relaxation times (τ) of 10^{-3} $\sim 10^{-4}$ s are observed in the dielectric dispersion of large molecules such as ethylene-co-vinyl acetate or polyamide [165]. Thus, in the present case, relatively long molecular clusters made of water molecules are related to dielectric relaxation. The water molecules are attracted to the bases and PO_4 ions of DNA by forming bonds with nitrogens and oxygens in them, which would generate oxonium ions or protons. Water molecules bind to DNA through hydrogen bonds and form "water bridges", as schematically described in Fig. 4.13. The water bridges are constructed between two PO_4 ions and between a PO_4 ion and a base. That is, the water bridges are classified into two kinds: (1) one between two backbones (PO_4–PO_4) and (2) the other between a base and a PO_4 ion. When the breaking and rearrangement of hydrogen bonds occur in these water bridges, the proton transfer is realized. These results show that the water bridge in DNA plays an important role in the appearance of proton

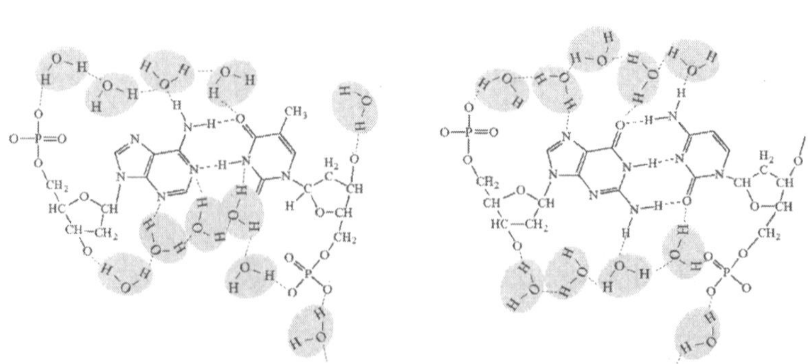

Figure 4.13 Schematic figure of water bridges in DNA.

conductivity in DNA. The process of water bridge formation will be shown in the experimental results of deuterated DNA.

4.3.4 *Phase Diagram of Proton Conduction*

It is noted that the humidity dependence of all the obtained parameters (σ, $\epsilon_s - \epsilon_\infty$, τ and β) changes discontinuously at 17% RH and markedly at 55% RH, and thus we can classify proton conductivity into three regions. Region I is below 17% RH, region II is between 17% and 55% RH, and region III is above 55% RH, as shown in Fig. 4.12. We summarize the characteristics of proton transfer in each region as follows.

In region I, both proton conductivity and dielectric polarization are very low, which mcans that the number of mobile protons is very small. It is also noted that β is approximately 0.5, which is the smallest in the three regions. Since $\beta = 1$ corresponds to single dielectric dispersion, the dielectric dispersion in region I is multi-dispersive. These results indicate that water bridges have just begun to be formed in this region. Thus, it is deduced that the transport path of protons is not formed enough in region I; for example, the water bridge is formed only between PO_4 and PO_4, which is the reason why proton conductivity becomes very low. To summarize, in region I water molecules and the formation of water bridges inside and outside the DNA strand are very small, and the proton transfer path is not completed yet, and therefore the number of mobile protons is very small and proton conductivity is low.

In region II, both proton conductivity and the number of mobile protons increase with relative humidity. The relaxation time τ increases up to 6×10^{-3} s, which indicates the presence of relatively long molecules as discussed in the previous section, that is, the development of water bridges. In addition, β increases with relative humidity, suggesting that the dielectric dispersion in region II approaches the single dispersion. Considering these results in region II, the water bridges around phosphate groups, such as the PO_4–PO_4 water bridge and the base–PO_4 water bridge, increase and are connected with each other. Thus, in region II proton conductivity increases with relative humidity because of the development of proton transport paths with the formation of water bridges such as the PO_4–PO_4 water bridge and the base–PO_4 water bridge.

As shown in Fig. 4.12, the relaxation time τ decreases above 55% RH in region III, whereas β increases with relative humidity, that is, the dielectric dispersion approaches the single dispersion of $\beta = 1$ with a faster dielectric relaxation than that at 55% RH. These results indicate that a new dielectric dispersion appears in region III, that is, new dielectric polarizations are formed. It is known that the spines of hydration exist under high humidity conditions as in region III [104]. Thus, it is deduced that the spines of hydration are formed and the proton transfer paths are fully connected in this region. As a result, in region III, protons can easily move by virtue of the formation of spines of hydration, and proton conductivity markedly increases in region III.

In this way, the proton conductivity in DNA strongly depends on the number and formation of water bridges, and the conducting paths of protons are dominated by the structure of water bridges.

4.4 Ionic Conductivity in Deuterated DNA

In the previous section, we clarified that the proton conductivity in DNA appears under humidified conditions and is caused by the water bridges between PO_4 molecules or base–PO_4 molecules. In the present section, we will show the experimental results of deuterated DNA (D-DNA). Under the humid condition of heavy water (deuterium oxide: D_2O), the hydrogen atoms of DNA are gradually replaced by the deuterium atoms D in the surrounding heavy water. Therefore, by the D_2O humidity dependence of the NMR absorption line, we can obtain information about the replacement of H by D and deduce the formation process of water bridges. Furthermore, by comparing these results with conductivity, we discuss the humidity dependence of proton-transfer paths.

The deuterated DNA specimens were prepared from DNA-H films by controlling D_2O humidity in the same manner as that in Section 4.3.1. The deuterium NMR (^{2}D-NMR) absorption lines were observed at 46.12 MHz using the Chemagnetics CMX Infinity 300 spectrometer.

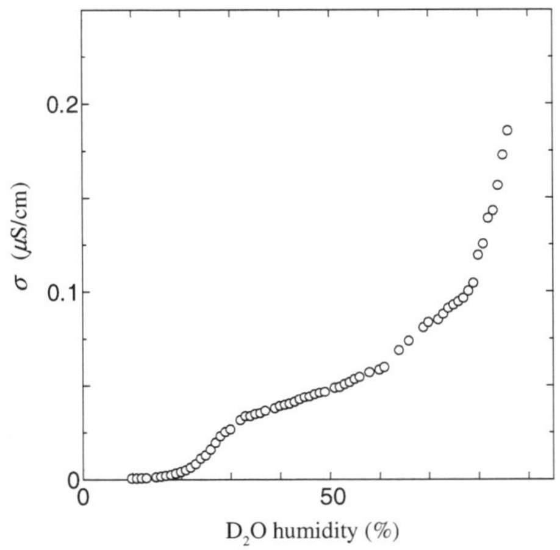

Figure 4.14 Humidity dependence of electrical conductivity in deuterated DNA.

Figure 4.14 shows the relative humidity dependence of electrical conductivity, which begins to increase around 20% RH, changes the slope at 32% RH, and increases monotonically up to 80% RH, above which it increases steeply. This result indicates that D-DNA also shows ionic conduction in D_2O humidified conditions as same as the proton conduction in DNA-H films in H_2O humidified conditions. Moreover, it is noted that the electrical conductivity in D-DNA is extremely small, although the humidity dependence of the electrical conductivity of D-DNA is similar to that of DNA-H. This result is reasonable if the ionic conduction of D-DNA is originated by the deuteron motion in the D_2O water bridges, because the mass of a deuteron is twice that of a proton and therefore the transportation of deuterons is less effective than that of protons.

In order to investigate the conduction mechanism of D-DNA, we show the ^{2}D-NMR absorption lines under various D_2O humidity conditions in Fig. 4.15. It is evident that the ^{2}D-NMR absorption line becomes narrow with relative humidity, which indicates that the mobile deuterons appear by D_2O humidification. Thus, it is deduced that electrical conductivity in Fig. 4.14 is caused by deuteron conduction.

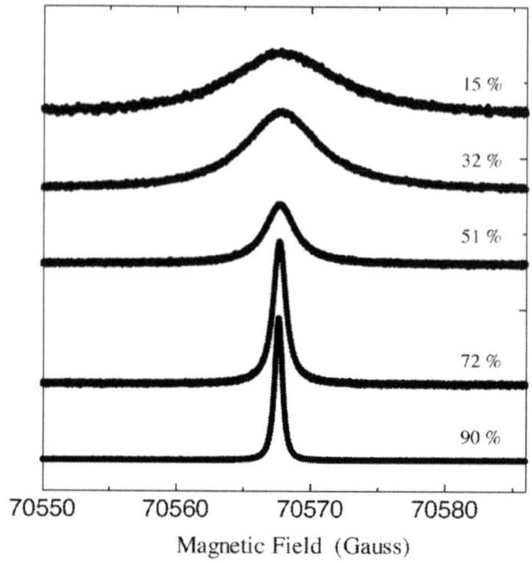

Figure 4.15 ^{2}H–NMR absorption lines in various D_2O humidity conditions.

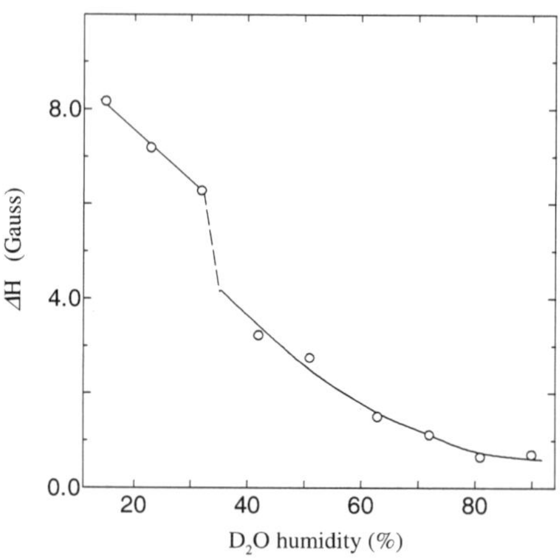

Figure 4.16 D_2O humidity dependence of ^{2}D-NMR line width ΔH.

Figure 4.16 shows the humidity dependence of ^{2}D-NMR linewidth ΔH, which is a peak-to-peak linewidth of a derivative curve of the NMR absorption line. We can see that the linewidth ΔH decreases with relative humidity, which clearly shows the motional narrowing of the ^{2}D-NMR absorption line by the migration

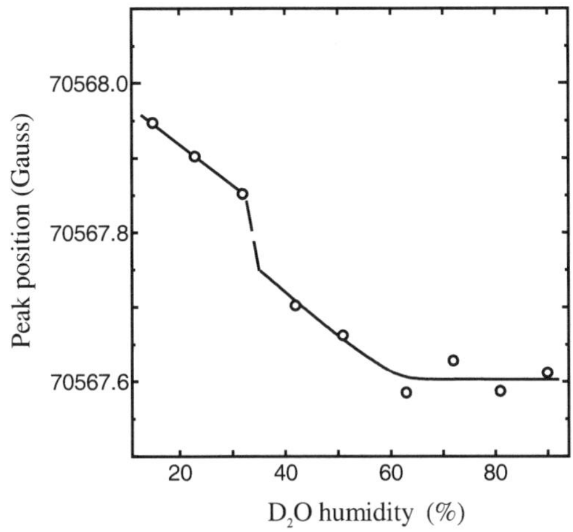

Figure 4.17 D_2O humidity dependence of the peak position of the ^{2}D-NMR line.

of deuterons. Moreover, it is noted that the anomalous decreases are observed around 32% RH. These results indicate that around 32% RH, the motion of the mobile ions becomes active.

Figure 4.17 shows the humidity dependence of the peak position of the ^{2}D-NMR absorption line, which gives information on the environmental change of deuterons, for example, chemical shifts, or the appearance of new dynamics (or new sites) of deuterons. The peak position moves to lower frequencies with the D_2O humidity and becomes constant above 80% RH. In addition, it is evident that the steep decrease is also observed in the ^{2}D-NMR peak position around 32% RH, similar to the linewidth in Fig. 4.16. In this way, the humidity dependence of the peak position shows the marked behaviors around 32% and 60% RH. It is noteworthy that the conductivity in Fig. 4.14 also shows marked qualitative changes in these relative humidity conditions, where the anomalous behaviors of the ^{2}D-NMR linewidth and the peak positions are observed.

Next, in order to investigate the dynamics of deuterons in detail, we analyzed a fine structure of the ^{2}D-NMR absorption line under several D_2O relative humidity conditions in Fig. 4.18. The observed ^{2}D-NMR absorption lines can be reasonably reproduced by Lorentzian and Gaussian lineshapes, as demonstrated in Fig. 4.18.

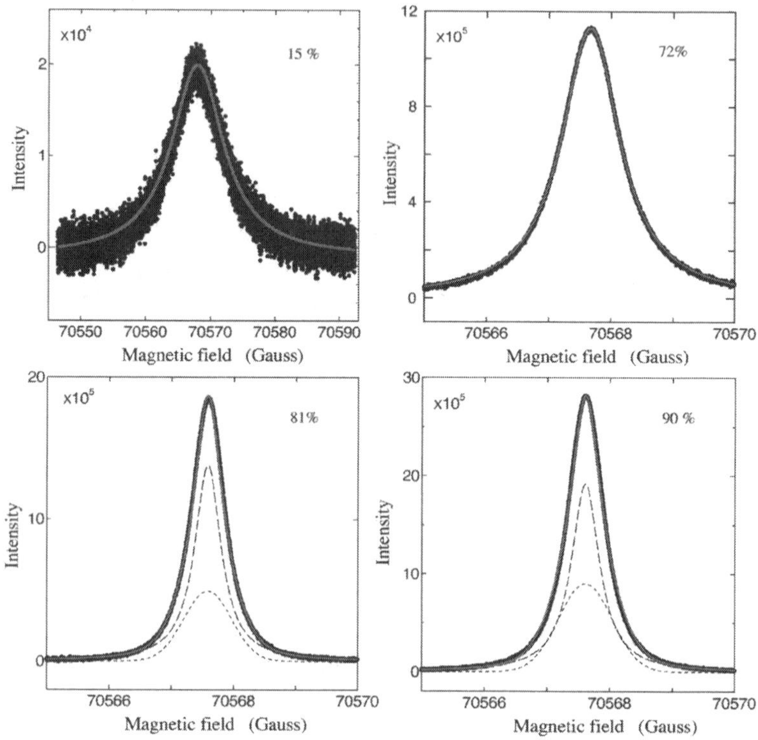

Figure 4.18 Fine structure in the ^{2}D-NMR lines at various D_2O humidity conditions.

The NMR absorption line is expressed by a single Lorentzian curve below 72% RH. On the other hand, above 81% RH, the NMR absorption line cannot be described by a single Lorentzian curve but by two lines, as shown in Fig. 4.18. One is a Lorentzian curve (dashed line) corresponding to the mobile deuterons, and the other is a Gaussian curve (dotted line) corresponding to the rigid sites of deuterons. The sum of these can reproduce well all the NMR lines observed above 81% RH, which indicates that not only the mobile deuterons but also the rigid sites of the deuterons exist above 81% RH in D-DNA. In summary, mobile deuterons are observed above 15% RH and their migration becomes faster with increasing relative humidity. However, the rigid deuteron sites appear above 81% RH in D-DNA.

In order to investigate the new Gaussian line that appears above 81% RH, we can estimate the deuteron–deuteron distance from the NMR linewidth. As seen in Fig. 4.18, the linewidth of the new Gaussian curve is approximately 0.6 gauss. From Eq. (4.4) and the NMR linewidth of 0.6 gauss, we can obtain the average deuteron–deuteron distance to be 2.4 Å, which corresponds to the average proton–proton distance of the DNA bases. Since the NMR line with Gaussian curve suggests the existence of rigid deuteron sites, it is speculated that the protons of the bases are interchanged with deuterons above 81% RH, at which the water bridge formation between PO_4 or base and D_2O was completed.

In summary, the conductivity and ^{2}D-NMR data in D-DNA clarify that D-DNA becomes a deuteron conductor under D_2O humidified conditions. From the markedly small conductivity of D-DNA and the conductivity increase with the D_2O relative humidity, it is speculated that the deuteron conduction is originated by the deuteron transport via the D_2O bridges. First, the D_2O bridges are formed only around PO_4 ions below 32% RH. Thereafter, the D_2O bridges are formed between PO_4 and OH in sugar above 32% RH, that is, the deuteration of OH in sugar occurs. Finally, accompanied by the deuteration of the bases, the D_2O bridges are formed between PO_4's and the bases.

4.5 DNA-Na and DNA-Li

A DNA-Na film is stable compared with a DNA-H film because the PO_4 ions are bonded to the Na ions in the DNA-Na film. However, the Na ions are easily dissociated from the PO_4 ions by hydration. Therefore, the stability of the DNA-Na film is changed and proton conductivity is influenced by humidification. In the present section, the ionic conductivity in DNA-Na is discussed with electrical conductivity and ^{1}H-NMR and ^{23}Na-NMR measurements, and the condition of proton conduction is investigated.

A DNA-Na film was prepared from DNA-Na powder with the casting method discussed in Section 4.3.1. Figure 4.19 shows the DNA-Na film, which is more stable than the DNA-H film. The ^{23}Na-NMR absorption lines were observed using the Chemagnetics

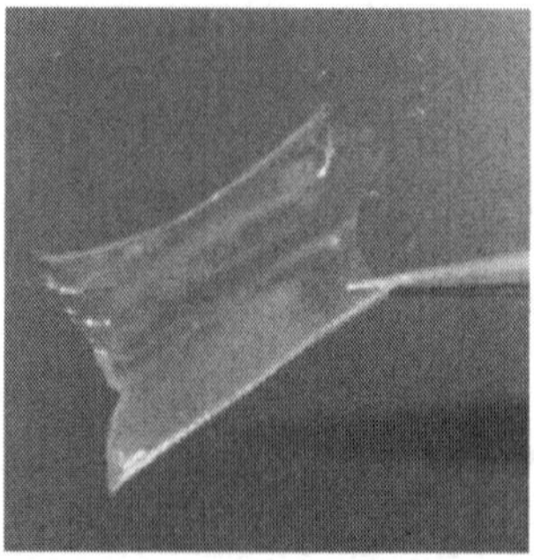

Figure 4.19 A DNA-Na film.

CMX Infinity 300 spectrometer at 79.48 MHz. The method of humidification is the same as that in Section 4.3.1.

Figure 4.20 shows the relative humidity dependence of electrical conductivity σ, which increases with relative humidity and shows step-like increase around 42% RH. Thereafter, σ increases further with relative humidity. In this way, DNA-Na also shows ionic conduction by humidification.

In order to investigate the proton transfer, we show the ^{1}H-NMR absorption lines of DNA-Na at various relative humidity

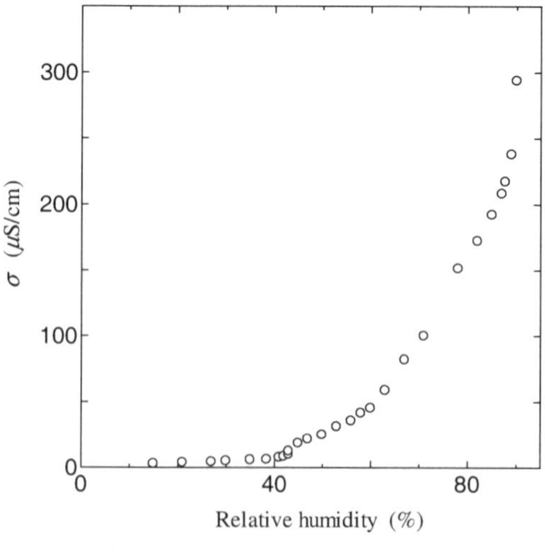

Figure 4.20 Humidity dependence of conductivity σ in DNA-Na.

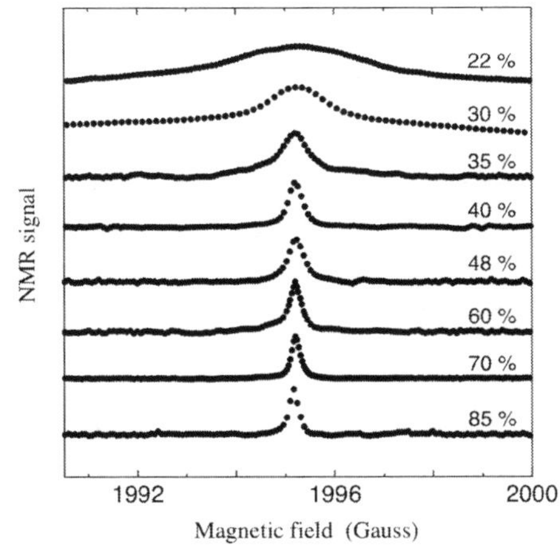

Figure 4.21 ^{1}H-NMR absorption lines of DNA-Na at various humidity conditions.

conditions as shown in Fig. 4.21. The ^{1}H-NMR absorption line becomes narrower with relative humidity, which indicates that the mobile protons appear in DNA-Na under humidified conditions. Considering the mechanism of proton conductivity in DNA-H, it is deduced that water bridges are formed in DNA-Na under humidified conditions between bases, base-PO_4 (dissociated from Na ion by hydration), and PO_4–PO_4. As a result, proton transfer is also realized via the water bridges in DNA-Na.

Next, in order to investigate the motion of Na ions, we analyze the ^{23}Na-NMR absorption lines at various humidity conditions in Fig. 4.22. It is evident that the ^{23}Na-NMR line at 35% RH is very similar to that at 20% RH, which are broadened not only by the dipole–dipole interaction but also by the quadrupole effect. If the Na ions migrate, the ^{23}Na-NMR line would show motional narrowing. Therefore, the broad ^{23}Na-NMR lines below 35% RH indicate that the Na ions are immobile. On the other hand, above 50% RH, the ^{23}Na-NMR absorption line actually shows narrowing, which is motional narrowing, and therefore the motion of Na ions begins to be active at relative humidity between 35% and 50% RH.

The ^{1}H-NMR absorption lines have been analyzed to obtain information on the mobile protons in DNA-Na. Figure 4.23 shows

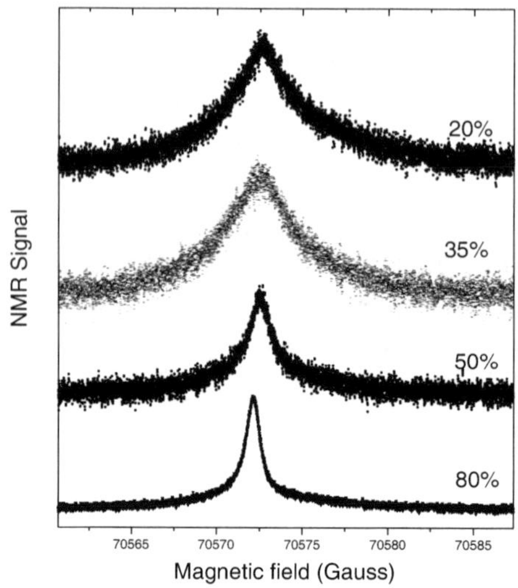

Figure 4.22 ^{23}Na-NMR line in DNA-Na at various humidity conditions.

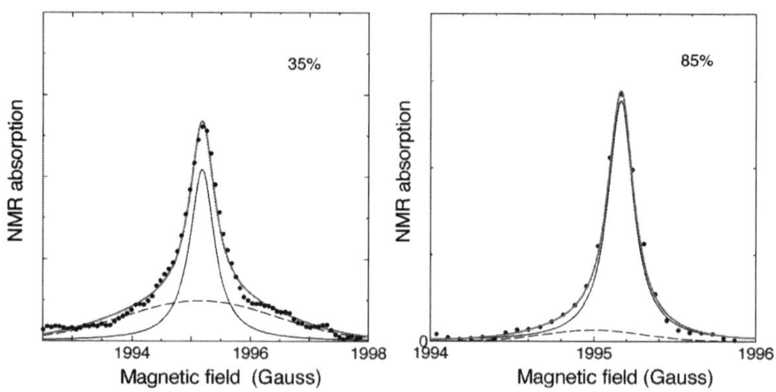

Figure 4.23 Fine structure of the ^{1}H-NMR absorption line in DNA-Na.

the fine structure analysis of the ^{1}H-NMR absorption lines at 35% and 85% RH. It is evident that the simulated NMR lines (bold solid line) are in agreement with the experimental ones (solid circle). The simulated NMR line consists of a sharp line for the mobile protons in water bridges (solid line) and a broad line for the other protons (dotted line). The ratio of the area of the sharp line to that of

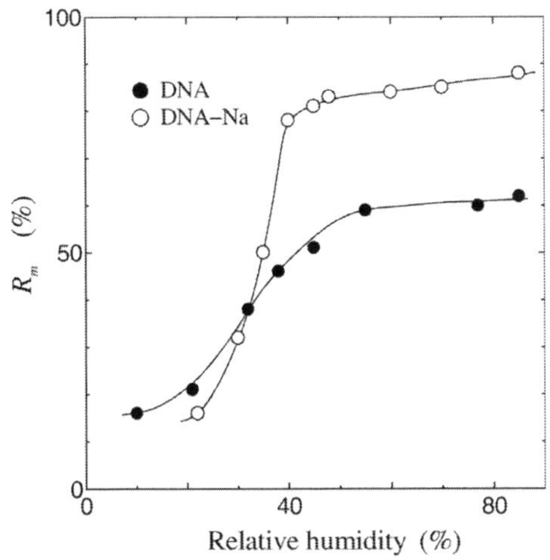

Figure 4.24 Ratio of the mobile protons R_m in DNA-Na.

the broad line changes with relative humidity, which indicates that the number of mobile protons in the water bridges increases with relative humidity.

Figure 4.24 shows the humidity dependence of the ratio of the mobile protons to all the protons of DNA-Na, together with that in DNA-H. The ratio of the mobile protons of DNA-Na is similar to that of DNA-H below 35% RH, but it is much larger than that of DNA-H above 35% RH, which suggests that the motion of the Na ions begins to be active above 35% RH. From these interesting results, it is deduced that the appearance of the mobile Na ions enhances the activation of proton transfer.

Figure 4.25 shows the humidity dependence of the linewidths of ^{23}Na-NMR and ^{1}H-NMR in DNA-Na. All the linewidths become narrow under highly humidified conditions, which indicates that proton conduction is realized not only in DNA-H but also in DNA-Na. Moreover, it is noted that the ^{23}Na-NMR linewidth begins to narrow around 40% RH. In contrast, the narrowing of the ^{1}H-NMR linewidth starts around 25% RH. In this way, the threshold humidity of the ^{1}H-NMR narrowing is lower than that of the ^{23}Na-NMR. This result indicates that mobile protons are generated around 25% RH, and after that the mobile Na ions appear in DNA-Na around 40%

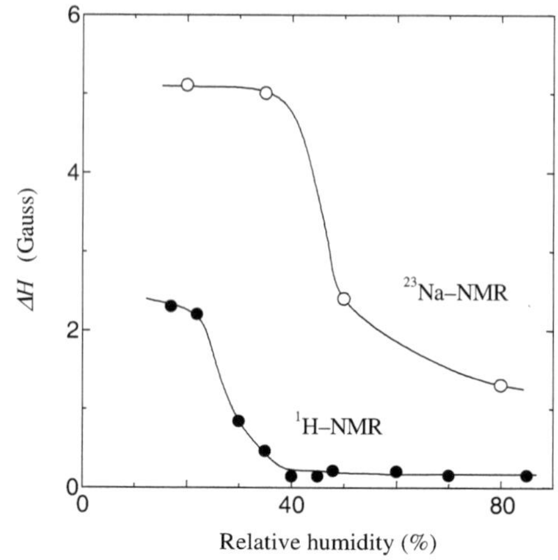

Figure 4.25 Humidity dependence of the linewidths of ^{23}Na-NMR and ^{1}H-NMR in DNA-Na.

RH. From the results of Figs. 4.24 and 4.25, it is deduced that the appearance of the mobile protons is needed for the appearance of the mobile Na ions and that the existence of the mobile Na ions facilitates further appearance of the mobile protons.

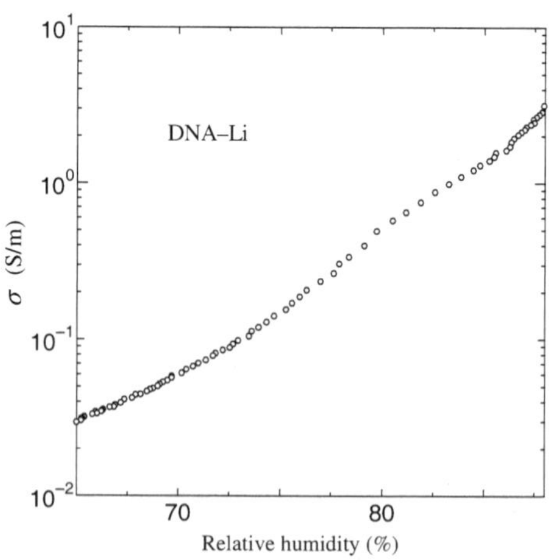

Figure 4.26 Humidity dependence of electrical conductivity in DNA-Li.

Finally, we show the humidity dependence of the AC electrical conductivity in DNA-Li in Fig. 4.26, which also increases with relative humidity. It seems that the mechanism of conductivity is the same as that of DNA-Na. Further results for the mechanism of conductivity in DNA-Li are presented in the future.

4.6 Conclusions

From impedance analysis and NMR study, it was found that DNA-H shows proton conduction, which is originated by the formation of water bridges in DNA-H films. Moreover, it was found that ^{2}D-DNA also shows deuteron conduction under D_2O humidified conditions. In DNA-Na, proton conduction appears even under relatively low-humidity conditions and a lot of protons move in DNA-Na, compared with DNA-H. From these results, the existence of Na bonded in DNA is important for the enhancement of mobile proton generation.

Acknowledgments

We would like to thank Mr. Go Kumasaka, Mr. Kazuya Inoue, Dr. Yukihiko Yoshida, and Mr. Hitoki Semizo for carrying out various measurements in DNA, ^{2}H-DNA, DNA-Na, and DNA-Li. Every result described in this chapter was accomplished with their help and support.

Chapter 5

DNA Molecules Studied by X-Ray Absorption Spectroscopy

Hiroki Wadati[a] **and Toshitaka Oka**[b]

[a] *Institute for Solid State Physics, University of Tokyo, Chiba, 277-8581 Japan*
[b] *Institute for Excellence in Higher Education, Tohoku University, Miyagi, 980-8576 Japan*
wadati@issp.u-tokyo.ac.jp

5.1 Introduction

There has been no general consensus on the electrical conductivity of deoxyribonucleic acid (DNA). Kasumov *et al.* [6] directly measured the electrical resistivity of double-stranded DNA with rhenium/carbon (Re/C) bilayer electrodes and found that DNA was metallic down to 50 mK and reported a proximity effect of the superconductivity of metallic rhenium. On the other hand, Porath *et al.* [4] have directly measured the electrical conductivity of synthetic DNA [poly(G)–poly(C)] with a length of 10 nm to find a semiconducting behavior with the temperature-dependent energy gap of 2–4 eV. Such controversial interpretations of the electronic states in DNAs can be considered to arise from the

DNA Engineering: Properties and Applications
Edited by Kenji Mizoguchi and Hirokazu Sakamoto

ISBN 978-981-4669-46-7 (Hardcover), 978-981-4669-47-4 (eBook)
www.panstanford.com

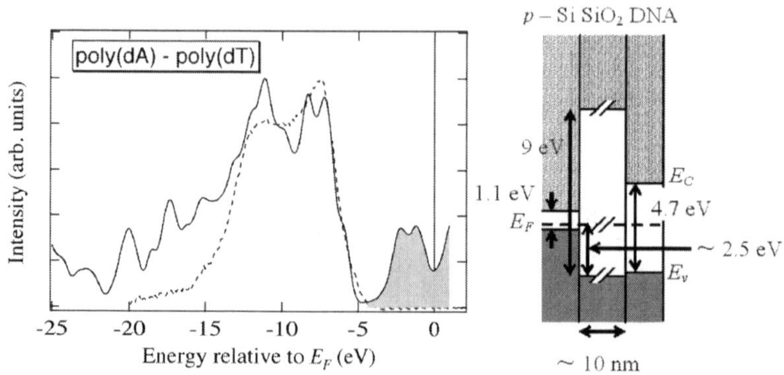

Figure 5.1 [Left] Comparison of the theoretical DOS (solid curve) and the photoemission spectrum (dashed curve) of poly(dA)–poly(dT). The hatched area corresponds to the unoccupied part of the calculated DOS. [Right] Schematic band diagram of poly(dA)–poly(dT) DNA on the Si/SiO_2 substrate. Reprinted with permission from Ref. [166], Copyright 2005, AIP Publishing LLC.

possible experimental errors in the process of attaching electrical contact to DNAs and/or flowing electric currents.

In these situations, we decided to apply the X-ray spectroscopic technique to DNAs to reveal the intrinsic electronic states by taking advantage of this technique without any need for electrical contact or current flow. For example, Wadati *et al.* [166] performed ultraviolet photoemission spectroscopy (UPS) study of artificially synthesized poly(dA)–poly(dT) DNA molecules on *p*-type Si substrates, as shown in Fig. 5.1. They showed from the UPS spectrum of DNA networks on the Si substrate that the Fermi level of the substrate is located in the middle of the band gap of DNA. These results show the completely insulating behavior of the DNA networks and the difficulty of carrier injection in the DNA networks because a bias of 2 eV must be applied for the networks on the *p*-type Si substrate.

On the other hand, soft X-ray has been used to understand the biological effect of irradiation on DNA [167, 168]. Soft X-ray photons interact with matter mainly through photoelectric effect. The inner-shell electrons of DNA constituent atoms (C, N, O, and P), which absorbed the soft X-ray energy, are excited/ionized and then

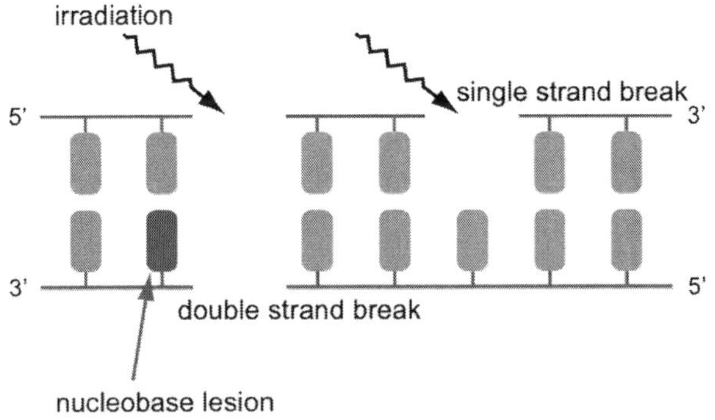

Figure 5.2 A nucleobase lesion, a single-strand break, and a double-strand break produced by the irradiation.

produce a nucleobase lesion, a single-strand break, a double-strand break, as shown in Fig. 5.2, etc. After the irradiation, a biological method such as electrophoresis is usually used to investigate DNA damage, but the X-ray spectroscopic technique is also a powerful tool to reveal the DNA damage.

Another topic is the doping of DNA with transition-metal cations [157]. Transition-metal cations are located in between the bases, making one-dimensional chains. One can expect one-dimensional charge transport, and the first step is determining the valence of the doped cations.

5.2 ESR under X-ray Absorption

Ionization of an electron from a molecule leaves an unpaired electron in the valence orbital. The unpaired electron in DNA is thought to play an important role in the reaction pathways to cause final damage. Electron spin resonance (ESR) is a powerful probe for studying the unpaired electrons of DNA radicals, and thus previous studies using a conventional ESR technique have given us useful knowledge about the stable DNA radicals that can exist for more than a few seconds after termination of the irradiation. In these studies, samples had to be kept at liquid nitrogen or helium temperatures during irradiation, transfer from an irradiator to an ESR spectrometer, and ESR measurement. Therefore, the very fast radical processes that are expected to be induced just after the photoelectric effect initiated from inner-shell photoionization in DNA are hardly observed because the unstable unpaired electron might decay during the irradiation, transfer, and measurement.

To overcome such a problem, Yokoya *et al.* have developed a unique ESR system at a synchrotron soft X-ray beamline BL23SU in SPring-8 (Japan) [169]. The ESR spectrometer was directly connected to the soft X-ray beamline, so that this system enables us to perform the ESR *in situ* measurement of an unstable short-lived unpaired electron of DNA and DNA-related molecules during soft X-ray irradiation in an ESR microwave cavity inside a high vacuum chamber [170].

Calf thymus DNA was dissolved in distilled water, cast on a clean substrate, and dried at room temperature. The DNA film for the ESR measurement was placed on an Au-plated Cu rod and in a high vacuum ESR chamber ($\approx 10^{-7}$ Pa). The DNA film for the transparent X-ray absorption fine structure (XAFS) measurement was sandwiched between sample holders, which has a hole larger than the beam size of the soft X-rays, and then placed in a high vacuum X-ray absorption chamber under $\approx 10^{-7}$ Pa. The intensities of the soft X-rays for the DNA sample, I, and the reference, I_0, without the DNA sample, were measured with a photodiode, and the soft X-ray absorbance A was obtained by using the Beer–Lambert law $A = \log(I/I_0)$ for each energy. Monochromatic soft X-ray photons with energies from 370 eV to 700 eV, including nitrogen and oxygen K-edge energies, were supplied by the soft X-ray beamline BL23SU. The monochrometer resolution was ≈ 50 meV at 500 eV, and the photon flux obtained by a photodiode at the sample position was of the order of 10^{11} per second per 100 mA ring current.

The short-lived unpaired electron of DNA obtained in this work shows a singlet ESR signal, as shown in Fig. 5.3, and it disappears immediately after the beam shutter is closed, whereas the long-lived DNA stable radicals have more than several lines in the ESR spectrum. The lifetime of such short-lived unpaired electron is estimated to be in the range of a few tens to several hundreds of μs [171]. A g-factor of 2.000, slightly smaller than that of a free electron (2.0023), was obtained for the short-lived unpaired electron by

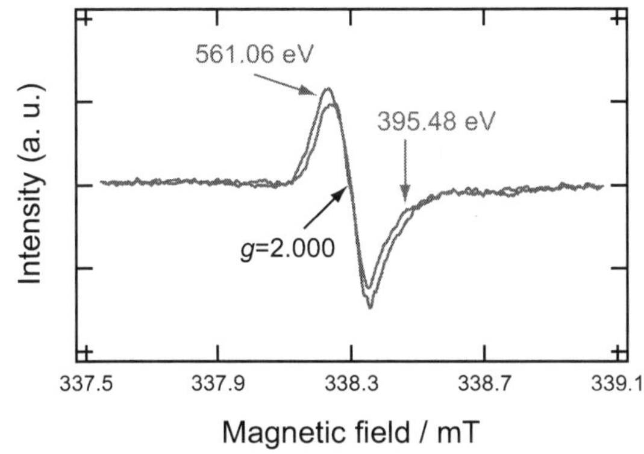

Figure 5.3 The ESR spectrum for calf thymus DNA during soft X-ray irradiation with an energy of 395.48 eV obtained at room temperature. The ESR spectrum for DNA around oxygen K-edge shows similar singlet signal. Reprinted with permission from Ref. [172], Copyright 2012 by the American Physical Society.

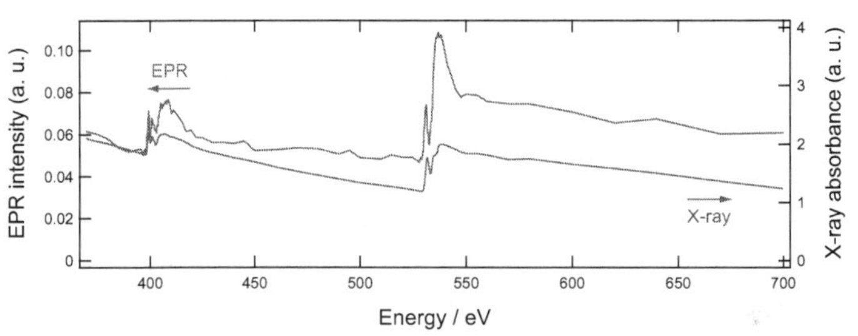

Figure 5.4 Dependence of the ESR intensity of the calf thymus DNA film on soft X-ray energy, from 370 eV to 700 eV, including the nitrogen and oxygen K-edges. The right-hand axis shows the XANES spectrum of the DNA film. The ESR intensities were significantly enhanced slightly above the K-shell thresholds of nitrogen and oxygen by photoabsorption. Reprinted with permission from Ref. [172], Copyright 2012 by the American Physical Society.

comparison with a standard ESR sample of Mn^{2+}/MgO (g-factors of 2.0337 and 1.981).

As shown in Fig. 5.4, the ESR intensity proportional to the spin concentration of the unpaired electrons for a DNA thin film shows a structure similar to the XAFS spectrum not only around the nitrogen K-edge but also around the oxygen K-edge. One can see

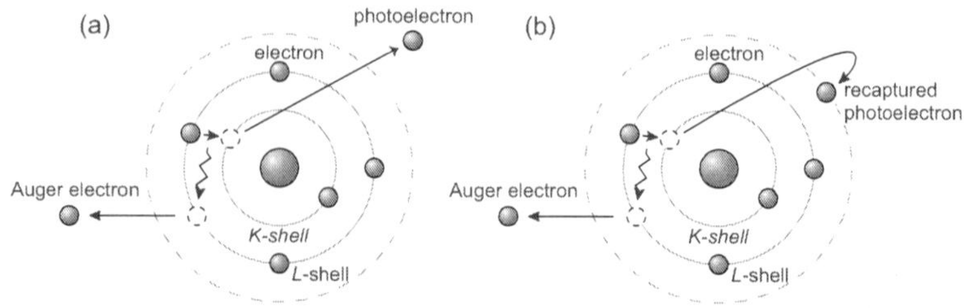

Figure 5.5 (a) Normal inner-shell ionization produces a doubly charged cation due to the emission of a photoelectron and an Auger electron. (b) Post-collision interaction produces a singly charged cation. The photoelectron is recaptured into an unoccupied orbital or a high-lying Rydberg orbital.

the significant enhancement of the ESR intensities slightly above the K-shell thresholds of nitrogen and oxygen by the photoabsorption [172].

On the creation of a K-shell hole of a light atom or molecule by soft X-ray absorption, a photoelectron is emitted from the inner shell and then Auger emission follows and the doubly charged parent cation is produced (Auger effect), as shown in Fig. 5.5(a). However, if the energy of the soft X-ray is slightly larger than the K-shell ionization potential, the charge state of the produced cation is somewhat altered due to the post-collision interaction (PCI). As is known, in PCI a photoelectron from an inner-shell orbital is significantly decelerated by the spontaneous Auger decay. The doubly charged parent cation core thus produced enhances Coulombic attraction to the receding photoelectron, so that the energy of a photoelectron ejected at a photon energy slightly above the inner-shell ionization threshold is often shifted to the lower side [173]. If the energy of the decelerated photoelectron is too small for it to escape from the potential of the doubly charged cation, the photoelectron is recaptured to yield a singly charged cation with an unpaired electron, as shown in Fig. 5.5(b). The recapture of slow photoelectrons by their parent ionic sites on the DNA into an unoccupied orbital or a high-lying Rydberg orbital in DNA results in the production of unpaired electrons, which may be detected by ESR.

In order to evaluate the PCI effect on the enhancement of the present ESR intensity, we calculated the recapture cross section after the Auger effect in DNA film by using a semiclassical approximation. The calculated result of the recapture probability for the production of a singly charged parent cation is almost unity at the very K-shell ionization threshold and steeply decreases to less than 10% at 7 eV above the threshold. The cross section for the recapture, which is the product of the recapture probability and the K-shell ionization cross section, shows a sharp increase and following gradual decrease with the soft X-ray energy above the ionizing threshold and reproduces the significant enhancement of the ESR intensity. This indicates that the enhancement of the ESR intensity at the K-shell thresholds can be attributed to the recapture of a photoelectron as a result of PCI effect.

The result suggests that the DNA itself prefers the formation of unpaired electrons through the excitation of enhanced electron recapturing by PCI effect. Unlike the well-known low-energy photoelectron attachment to an electronegative site in DNA followed by dissociation [174] or the recombination of slow electrons with ion holes, the observed ionization process induced by the soft X-ray absorption is a novel mechanism of DNA damage.

5.3 X-Ray Absorption of Metal Ion–Doped DNA

Here, we will introduce the XAFS study of transition-metal-doped DNA with hard X-rays, which only weakly interact with DNA, to reveal the valence states of doped transition-metal ions [157]. Soft X-ray spectroscopy at the transition-metal L-edge was reported in Refs. [175] and [176]. We used hard X-ray at K-edge, which is the transition mainly from $1s$ to $4p$ states. We decided to use the K-edge instead of the L-edge because hard X-rays have weaker interaction with materials than soft X-rays.

Figure 5.6(a) shows the Mn K-edge XAFS of the Mn-doped DNA with the reference data of Mn, Mn^{2+} (MnO), and Mn^{4+} (MnO_2). The spectrum of DNA is similar to that of MnO, meaning that the valence of the doped Mn is 2^+.

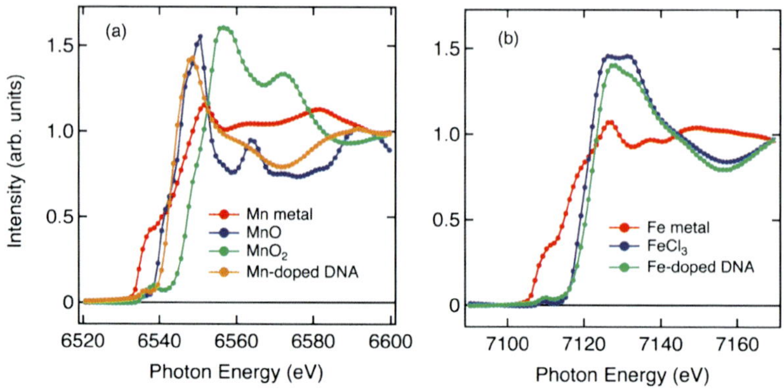

Figure 5.6 (a) Mn *K*-edge XAFS of the Mn-doped DNA with the reference data. (b) Fe *K*-edge XAFS of the Fe-doped DNA with the reference data.

Figure 5.6(b) shows the Fe *K*-edge XAFS of the Fe-doped DNA with the reference data of Fe and Fe^{3+} ($FeCl_3$). The spectrum of Fe-DNA is similar to that of $FeCl_3$, meaning that the valence of the doped Fe is 3^+. This result is also consistent with the ocher color of Fe-doped DNA. Pure Mn^{2+} and Fe^{3+} ions show insulating behaviors, so controlling the valence of such ions will be the key for realizing one-dimensional electronic transport in DNA. Doping other ions (Ti, V, Cr, Co, Ni, and Zn) will be interesting.

Chapter 6

The Hückel Theoretical Calculation for the Electronic Structure of DNA

Kazumoto Iguchi
Kazumoto Iguchi Research Laboratory (KIRL), 70-3 Shin-hari, Hari, Anan, Tokushima, 774-0003 Japan

6.1 Preliminaries

6.1.1 *Introduction*

Understanding the chemical reactions and electronic conduction of DNA in the academic point of view as well as in engineering and biochemical applications is essential. Although DNA has been studied by various methods in various areas for as long as 60 years since the discovery of the double-stranded structures of DNA [2], it is still far from being well understood. Our goal is to find how to calculate the electronic conduction of DNA [31, 177, 178]. For this at first we have to solve the problem of quantum mechanics of DNA as well as calculate the electronic states and electronic spectrum of DNA. Calculating the electronic states of DNA through quantum mechanics might sound simple in words, but practically

DNA Engineering: Properties and Applications
Edited by Kenji Mizoguchi and Hirokazu Sakamoto

ISBN 978-981-4669-46-7 (Hardcover), 978-981-4669-47-4 (eBook)
www.panstanford.com

it is extremely difficult because DNA has a complementary double-stranded helical structure made of complicated nucleotides, in which there are many atoms and electrons. Obviously, it is not a single-electron problem but a many-electron problem. In addition, since DNA consists of genetic information, it has a very irregular structure in itself. This means that we have to solve the Schrödinger equation for electrons moving on the complicated double strand of DNA.

Of the several methods for treating this problem, three main methods are as follows:

(i) The first is the un-empirical method of quantum chemistry. This approach calculates the electronic states of DNA by massive calculations using a super computer such that we treat the DNA as realistic as possible based on its raw atomic arrangements in nature [179–189].

(ii) The second is the reverse approach. We make complicated DNA as simple as possible up to the level of symbolic notation, and we calculate the electronic states of the model DNA based on its simple virtual DNA bases. In this case, the Hückel approximation in quantum chemistry [190–194] and the tight-binding models of solid state physics [195–197] are used. This has a long history. Although in most of such simplified DNA models, the linear chain model has been used [198–201], I have proposed the so-called decorated ladder model of DNA, which is a simple two-chain model of the double-stranded helical DNA structure [202–204].

(iii) The third approach is located in between the two. We approximate the global structure of DNA by the simple decorated ladder model of symbolic notation approach and use the rather realistic Hückel approximation for its internal structures of nucleotide bases [77].

Each method has its own advantages and disadvantages. The first approach of the method of first principle can only treat short segments of DNA, since it calculates all electrons in all the atoms in each segment of DNA. If we denote the total number of treated electrons in the base by N_i and the total number of bases of DNA by N_b, then the total number of all electrons, N_e, is given by

$N_e = N_b \times N_i$. Therefore, as long as the number N_e is finite, the number N_b decreases as the number N_i increases. Usually, considering the calculation prowess of recent super computers, the order of N_b might be taken as many as 50 up to 100.

In the second method, the number of electrons in the unit segment of the model DNA chain or the decorated ladder model of DNA is unity $\approx$ o(1). Therefore, contrary to the first approach, we can calculate the DNA model with a very long base arrangement as long as the order of $N_e \approx N_b \approx o(10^5)$. Therefore, this method is very useful for the electronic localization–delocalization problem in DNA [205–211].

As the third method is located in between, so are the merits and demerits. However, since it treats rather realistic DNA nucleotide bases, it can be thought that microscopically it reflects the situation of the system quite precisely. Especially, it would be a good approximation for the highest occupied molecular orbital (HOMO) and the lowest unoccupied molecular orbital (LUMO) [212, 213] of π-electrons in DNA molecules.

The present chapter reviews the third method. However, the goal of the present chapter is modest. We would like to apply the Hückel theory to electrons in one of the simplest DNA structures only. Even so, we can learn something important for studying the DNA electron transport problem [177, 178].

6.1.2 *Quantum Chemistry for Atoms in Biology*

Our life in nature is organized by the so-called biomolecules. Such biomolecules are made of organic materials such as carbon ($_6$C), nitrogen ($_7$N), oxygen ($_8$O), and phosphorus ($_{15}$P). Therefore, in the beginning, we have to understand the nature of such atoms. This is bio-quantum chemistry. I learned the main parts of the following arguments through the letters of Dr. Chikayoshi Nagata [214].

The problem here is that atoms that exist in vacuum are different from the same atoms in the biological environment. This is because coordination between the nearest atoms and the atoms under consideration causes a perturbation to the atoms under consideration as ligands, which then changes the electronic states of the isolated atoms to those of the atoms under the influence of ligands. This kind of perturbation provides the atoms with the geometry such as sp-hybrid (linear), sp^2-hybrid (triangular), and sp^3-hybrid (tetrahedral) orbitals. Hence, by such constraints of geometry, biomolecules should have particular geometry.

6.1.2.1 Carbon

Let us consider the electronic states of carbon atom (C) of atomic number 6 with 6 electrons. When the carbon atom is isolated, the electronic orbitals are given by

$$|1s\rangle,\ |2s\rangle,\ |2p_x\rangle,\ |2p_y\rangle,\ |2p_z\rangle, \tag{6.1}$$

respectively. The electron configuration in the ground state of carbon atom is now given by

$$\text{C:}\ (1s)^2(2s)^2(2p)^2. \tag{6.2}$$

Electrons in the most external shell are given by $(2s)^2(2p)^2$, which means $(2s)^{\uparrow\downarrow}(2p_x)^{\uparrow}(2p_y)^{\uparrow}(2p_z)$ such that two electrons occupy the $2s$-orbital, one electron occupies the $2p_x$-orbital, and one electron occupies the $2p_y$-orbital, respectively.

When the carbon atom is chemically linked with other atoms such as $-CH_2$ in biopolymers, the atomic orbitals are perturbed so as to form the sp^2-hybrid orbitals,

$$\left.\begin{aligned} |1, 0\rangle &= \tfrac{1}{\sqrt{3}}\,|2s\rangle + \tfrac{\sqrt{2}}{\sqrt{3}}\,|2p_x\rangle, \\ |-\tfrac{1}{2}, \tfrac{\sqrt{3}}{2}\rangle &= \tfrac{1}{\sqrt{3}}\,|2s\rangle - \tfrac{1}{\sqrt{6}}\,|2p_x\rangle + \tfrac{1}{\sqrt{2}}\,|2p_y\rangle, \\ |-\tfrac{1}{2}, -\tfrac{\sqrt{3}}{2}\rangle &= \tfrac{1}{\sqrt{3}}\,|2s\rangle - \tfrac{1}{\sqrt{6}}\,|2p_x\rangle - \tfrac{1}{\sqrt{2}}\,|2p_y\rangle. \end{aligned}\right\} \tag{6.3}$$

however. The sp^2-hybrid orbitals can form the σ-orbitals for the σ-bonds between the nearest atoms, while the left p_z-orbital forms a π-orbital. Here there are four electrons such that one electron occupies each orbital of the sp^2-hybrid orbitals and one electron occupies the π-orbital. The configuration is drawn in Fig. 6.1(A). This situation can be written symbolically as $-\dot{C}H_2$, where dot $(\cdot)$ means an electron in the π-orbital (i.e., π-electron).

Similarly, let us consider the carbon atom chemically linked with other atoms such as –C–. In this case, the atomic orbitals are perturbed to form the sp-hybrid orbitals:

$$\left.\begin{aligned} |1\rangle &= \tfrac{1}{\sqrt{2}}\,|2s\rangle + |2p_x\rangle, \\ |-1\rangle &= \tfrac{1}{\sqrt{2}}\,|2s\rangle - |2p_x\rangle, \end{aligned}\right\} \tag{6.4}$$

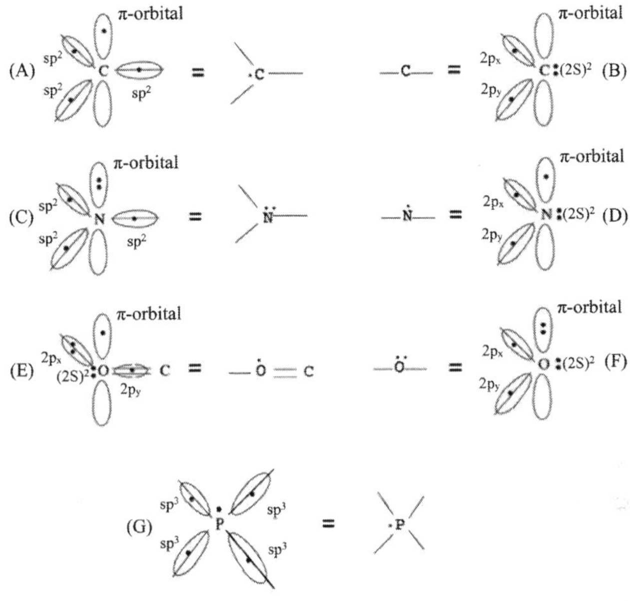

Figure 6.1 The electron configurations of C, N, O, and P, respectively. (A) The electronic state of a carbon with three bonds. (B) The electronic state of a carbon with two bonds. (C) The electronic state of a nitrogen with three bonds. (D) The electronic state of a nitrogen with two bonds. (E) The electronic state of an oxygen with a double bond. (F) The electronic state of an oxygen with two bonds. (G) The electronic state of a phosphorus with four bonds. Here dots (·) mean electrons, where electrons that occupy the π-orbitals are called π-electrons.

There are two electrons in the $2s$-orbital to make a *lone pair*, and one electron occupies the $2p_x$-orbital and one electron occupies the $2p_y$-orbital to form the σ-orbitals. Therefore, there is no electron in the π-orbital. This situation can be written symbolically such –C–, where there is no π-electron. The configuration is drawn in Fig. 6.1(B).

6.1.2.2 Nitrogen

Let us consider nitrogen atom (N) of atomic number 7 with 7 electrons. When the nitrogen atom is isolated, the electronic state in the ground state is given by

$$\text{N: } (1s)^2(2s)^2(2p)^3. \tag{6.5}$$

Electrons in the most external shell are given by $(2s)^2(2p)^3$, which means $(2s)^{\uparrow\downarrow}(2p_x)^{\uparrow}(2p_y)^{\uparrow}(2p_z)^{\uparrow}$, such that two electrons occupy the $2s$-orbital, one electron occupies the $2p_x$-orbital, one electron occupies the $2p_y$-orbital, and one electron occupies the $2p_z$ orbital.

When the nitrogen atom is chemically linked with other atoms such as $-NH_2$ in biopolymers, the atomic orbitals are perturbed to form the sp^2-hybrid orbitals, however. There are three electrons in the sp^2-hybrid orbitals. Each electron occupies each σ-orbital of the sp^2-hybrid orbital, and there are two electrons in the π-orbital. The configuration is drawn in Fig. 6.1(C). This situation can be schematically written as $-\ddot{N}H_2$, where the double dot means two π-electrons.

Let us next consider a nitrogen with two bonds such as –N–. The electronic configuration is given by Eq. (6.5) as before. Therefore, in this case, one electron occupies the $2p_x$-orbital, one electron occupies the $2p_y$-orbital, and two electrons occupy the single $2s$-orbital to make a lone pair, and one electron occupies the π-orbital. The configuration is drawn in Fig. 6.1(D). This situation can be symbolically written as $-\dot{N}-$, where the dot means a π-electron.

6.1.2.3 Oxygen

Let us consider an oxygen atom (O) of atomic number 8 with 8 electrons. When the oxygen atom is isolated, the electronic state in the ground state is given by

$$O: (1s)^2(2s)^2(2p)^4. \tag{6.6}$$

Electrons in the most external shell are given by $(2s)^2(2p)^4$, which means $(2s)^{\uparrow\downarrow}(2p_x)^{\uparrow\downarrow}(2p_y)^{\uparrow}(2p_z)^{\uparrow}$, such that two electrons occupy the $2s$-orbital, two electrons occupy the $2p_x$-orbital, one electron occupies the $2p_y$-orbital, and one electron occupies the $2p_z$ orbital.

When the oxygen atom is chemically linked with other atoms such as $-O{=}C$ in biopolymers, the atomic orbitals are perturbed, however. Two electrons occupy the $2p_x$-orbital, one electron occupies the $2p_y$-orbital, two electrons occupy the single $2s$-orbital to make a lone pair, and one electron occupies the π-orbital. The configuration is drawn in Fig. 6.1(E). This situation can be symbolically written as $-\dot{O}{=}C$, where the dot means a π-electron.

When the oxygen atom is chemically linked with other atoms such as -O-, one electron occupies the $2p_x$-orbital, one electron occupies the $2p_y$-orbital, two electrons occupy the single $2s$-orbital to make a lone pair, and two electrons occupy the π-orbital, respectively. The configuration is drawn in Fig. 6.1(F). This situation can be symbolically written as $-\ddot{\mathrm{O}}-$, where two dots $(\cdot\cdot)$ mean two π-electrons.

6.1.2.4 Phosphorus

Let us consider phosphorus atom (P) of atomic number 15 with 15 electrons. When the phosphorus atom is isolated, the electronic state in the ground state is given by

$$\mathrm{P}: (1s)^2(2s)^2(2p)^6(3s)^2(3p)^3. \tag{6.7}$$

Electrons in the most external shell are given by $(3s)^2(3p)^3$, which means $(3s)^{\uparrow\downarrow}(3p_x)^{\uparrow}(3p_y)^{\uparrow}(3p_z)^{\uparrow}$, such that two electrons occupy the $3s$-orbital, one electron occupies the $2p_x$-orbital, one electron occupies the $2p_y$-orbital, and one electron occupies the $2p_z$ orbital.

When the phosphorus atom is chemically linked with other atoms such as $-\mathrm{PO_4}$ in biopolymers, the atomic orbitals are perturbed to form the sp^3-hybrid orbitals

$$\left.\begin{aligned}
|1, 1, 1\rangle &= \tfrac{1}{2}(\,|3s\rangle + |3p_x\rangle + |3p_y\rangle + |3p_z\rangle),\\
|1, -1, -1\rangle &= \tfrac{1}{2}(\,|3s\rangle + |3p_x\rangle - |3p_y\rangle - |3p_z\rangle),\\
|-1, 1, -1\rangle &= \tfrac{1}{2}(\,|3s\rangle - |3p_x\rangle + |3p_y\rangle - |3p_z\rangle),\\
|-1, -1, 1\rangle &= \tfrac{1}{2}(\,|3s\rangle - |3p_x\rangle - |3p_y\rangle + |3p_z\rangle),
\end{aligned}\right\} \tag{6.8}$$

however. Here there are five electrons in the sp^3-hybrid orbital. Each electron occupies each σ-bond of the sp^3-hybrid orbital to form four bonds between the nearest atoms. Therefore, since there is no left p_z-orbital, there appears no π-orbital. And there is an extra electron sitting at the center of phosphorus. The configuration is drawn in Fig. 6.1(G). This situation can be written symbolically as $-\dot{\mathrm{P}}\mathrm{O_4}$, where the dot means an electron at phosphorus site.

This is the starting point for considering the electronic states of bases in DNA. From this, we can count the number of π-electrons and the number of π-orbitals in the bases.

6.1.3 π-Electronic Configurations in Organic Molecules in Biology

6.1.3.1 π-Electronic configuration in benzene

We first consider the simplest case of benzene. If a carbon atom is isolated, there are four bonds around the atom, where each bond consists of one valence electron. But once a carbon atom meets other carbons, such as in benzene, the coordination of the atom is changed to the sp^2-hybrid orbital; it is a three-coordination number plus one π-orbital perpendicular to the plane of benzene.

Thus, six sp^2-hybrid orbitals are in benzene, which are triangular. They form the main hexagonal geometry of benzene. These bonds are called σ-bonds through sp^2-hybrid orbitals. And six π-orbitals are still left. These form the π-bonds between carbon atoms through the π-orbitals, which are schematically represented by double bonds.

Since three of the four valence electrons in carbon are consumed to form the sp^2-hybrid bonds, only one valence electron is left in the carbon, which can exist for the π-electron. This is shown in Fig. 6.2 as dot $(\cdot)$ near the carbon C.

6.1.3.2 π-Electronic configurations in A, G, C, T base molecules

Let us consider the electronic configurations in nitrogenous base molecules such as *adenine* (A), *guanine* (G), *cytosine* (C), and

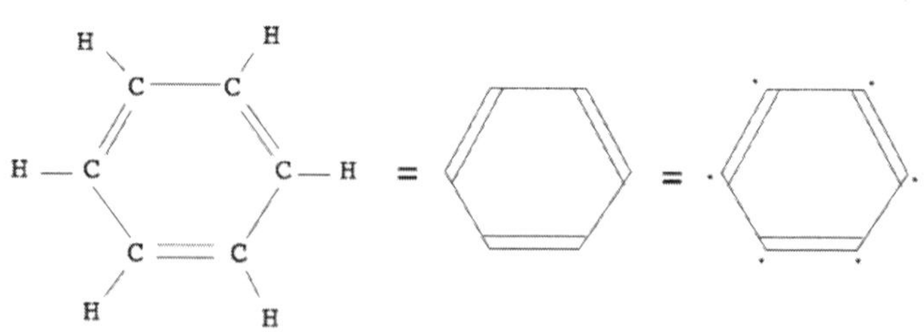

Figure 6.2 The π-electron configuration in benzene molecule. The leftmost figure is a schematic diagram of benzene molecule. C is not conventionally shown at the carbon atom sites as in the middle figure. As per the previous electron configuration for carbon atom, there is one π-electron at each carbon atom. Here dots $(\cdot)$ mean π-electrons.

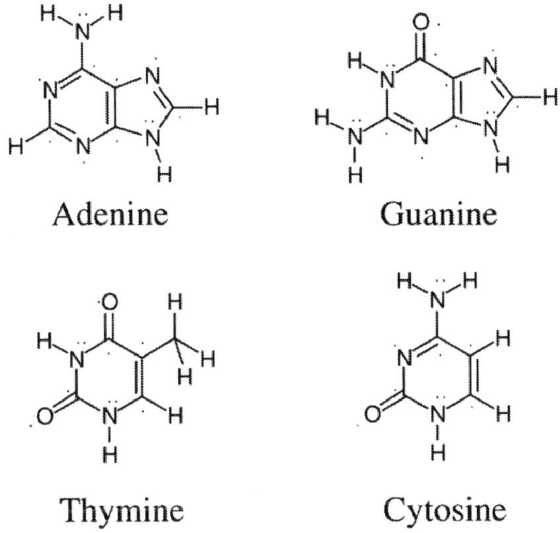

Figure 6.3 Electron configurations for the bases of A, G, C, and T, where there are 12 π-electrons and 10 π-orbitals in the A molecule; 14 π-electrons and 11 π-orbitals in the G molecule; 10 π-electrons and 8 π-orbitals in the C molecule; and 10 π-electrons and 8 π-orbitals in the T molecule. Here dots $(\cdot)$ mean π-electrons. Reprinted with permission from Ref. [77], Copyright 2004, World Scientific Publishing Company.

thymine (T). The nitrogenous base molecules in DNA are classified into two types: *purine* and *pyrimidine* [215]. A and G belong to purines, while C and T belong to pyrimidines. We may simply call the nitrogenous bases "the bases" if there is no confusion. The molecular structures of A, G, C, T base molecules using the bond model are schematically shown in Fig. 6.3.

In the A molecule, there are five nitrogen atoms, five carbon atoms, and four hydrogen atoms. In the G molecule, there are five nitrogen atoms, five carbon atoms, five hydrogen atoms, and one oxygen atom. In the C molecule, there are five hydrogen atoms, four carbon atoms, three nitrogen atoms, and one oxygen atom. In the T molecule, there are six hydrogen atoms, five carbon atoms, two nitrogen atoms, and two oxygen atoms. Let us denote the total numbers of N, C, and O atoms in the base molecules by $|N|$, $|C|$, and $|O|$, respectively.

Let us next consider the electron configurations in the π-orbitals of the nitrogenous bases of A, G, C, and T. Here we call such electrons in the π-orbitals *π-electrons* for the sake of simplicity. In this chapter, we are concerned with only the π-electrons and π-orbitals in the system.

Applying the basic knowledge of quantum chemistry to the biomolecules of nitrogenous bases, we find the following: There are 12 π-electrons and 10 π-orbitals in the A molecule; 14 π-electrons and 11 π-orbitals in the G molecule; 10 π-electrons and 8 π-orbitals in the C molecule; and 10 π-electrons and 8 π-orbitals in the T molecule. This is summarized in Fig. 6.3. Here we note that the carbon atom of CH_3 in the T molecule does not have any π-electron since it forms the sp^3-hybrid orbitals. This provides eight π-electrons for the T molecule. Thus, we can summarize the above as in Table 6.1.

6.1.3.3 Electronic configurations in molecules of sugar, phosphate, and triphosphate

Let us consider the electronic configurations in sugar, phosphate, triphosphate, sugar-phosphate, and sugar-triphosphate molecules [215], for later purposes [216–218]. Studying the electronic states of such molecules is an old problem. As early as in the 1960s, by using the Hückel theory [190–194], quantum chemists [216–218] studied

Table 6.1 The total numbers |N|, |C|, and |O| of N, C, and O atoms and of π-orbitals and π-electrons in the bases of A, G, C, and T, respectively. Here there are 12 π-electrons and 10 π-orbitals in the A molecule; 14 π-electrons and 11 π-orbitals in the G molecule; 10 π-electrons and 8 π-orbitals in the C molecule; and 10 π-electrons and 8 π-orbitals in the T molecule

Base	A	G	C	T
|N|	5	5	3	5
|C|	5	5	4	2
|O|	0	1	1	2
Total	10	11	8	9
π-orbitals	10	11	8	8
π-electrons	12	14	10	10

the electronic properties of complex molecules such as adenosine diphosphate (ADP) and adenosine triphosphate (ATP), in which sugar is a *ribose*.

Speaking naively, the Hückel theory can be applied only to systems with a plane geometry, such as benzene and nitrogenous bases such as A, G, C, and T [212, 219]. Also it is applicable to systems with a linear geometry, such as polyacetylene [220, 221], where the π-orbitals are the nearest neighbor orbitals to each other so that there exists an overlap between the adjacent π-orbitals. So systems with a tetrahedral geometry, such as phosphate ($-PO_4$), may not be a good system when we use the Hückel theory for π-orbitals, since the overlap between the π-orbitals is smaller than that in the planer molecule systems. And by the same reason, the system of a sugar is not a good system for applying the Hückel theory since the π-orbitals at the oxygen sites in the sugar are quite far from each other. Therefore, as the first approximation, chemists in the 1960s ignored the role of π-orbitals in the sugar molecule and treated only the π-orbitals in the phosphate sites [222, 223]. Later, this point was improved to include the role of the π-orbitals in the sugar molecule as well as in the phosphate sites [224].

As Fukui *et al.* [222, 223] considered, the orbital mixing (hybridization) between the sp^3-hybrid and d-orbital of phosphorus should be taken into account, since there are five valence electrons in the phosphorus so that four electrons occupy the sp^3-hybrid around the phosphorus atom and one electron occupies the orbital in the phosphorus atom. Therefore, one good approximation for treating the orbitals of phosphorus is that we put five orbitals for five electrons so that there are four orbitals forming the sp^3-hybrid and one orbital in the phosphorus site, where each orbital can be occupied by one electron (see Fig. 6.4).

Applying the above approximation for orbitals in phosphorus together with the knowledge of the previous subsections, we find the following: In sugar, there are four π-orbitals in oxygen sites where each has two electrons. In phosphate, there are five orbitals such that four orbitals are located in oxygen sites and one orbital is located in phosphorus site, where each of the three oxygen atoms with a single bond to phosphorus has two electrons, one oxygen atom with a double bond to the phosphorus has one electron, and

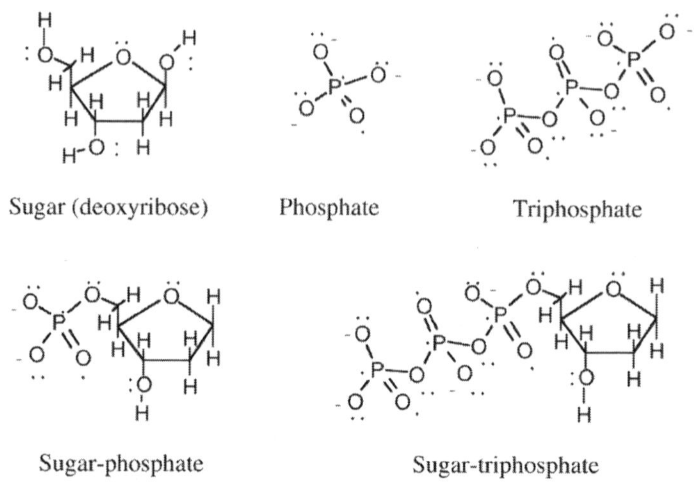

Figure 6.4 The electron configurations for sugar, phosphate, triphosphate, sugar-phosphate, and sugar-triphosphate molecules, respectively. Here dots (·) mean π-electrons and – means the negative ionicity of the site. The figures indicate the acid states of molecules. Reprinted with permission from Ref. [77], Copyright 2004, World Scientific Publishing Company.

the phosphorus atom consists of one electron. In triphosphate, there are 15 orbitals with 20 electrons. In sugar-phosphate, there are 7 orbitals with 12 electrons. And in sugar-triphosphate, there are 15 orbitals with 24 electrons. These are schematically shown in Fig. 6.3 and summarized in Table 6.2.

6.1.3.4 Electronic configurations in A, G, C, and T nucleosides

Let us consider the electronic configurations in the molecules of A, G, C, and T nucleosides. If the purines and pyrimidines are connected to a sugar, then all the molecules are called *nucleosides* [215]. For example, if an adenine A is combined to a sugar, then the molecule is called *A nucleoside*. The structures of such nucleoside molecules are schematically shown in Fig. 6.5.

Here an interesting aspect is that the geometry of the nucleosides is unique. That is, *the sugar is linked to the nucleoside almost perpendicular to each other.*

Table 6.2 The total numbers |C|, |O|, and |P| of C, O, and P atoms and the total numbers of π-orbitals and π-electrons in sugar, phosphate, triphosphate, sugar-phosphate, and sugar-triphosphate molecules, respectively. Here the following abbreviations have been used: S=Sugar, P=Phosphate, TP=Triphosphate, SP=Sugar-phosphate, STP=Sugar-triphosphate

Molecule	S	P	TP	SP	STP
|C|	5	0	0	5	5
|O|	4	4	10	6	12
|P|	0	1	3	1	3
Total	9	5	13	12	20
π-orbitals	4	5	13	7	15
π-electrons	8	8	20	12	24

Applying the previous knowledge of quantum chemistry, we find that the situation here is the same as that for the electron configurations in the nitrogenous base molecules, while the electron configuration in the sugar is given by the π-electrons at oxygen sites. Here each π-orbital in the oxygen atom consists of two π-electrons. This means that the π-orbitals are fully occupied by electrons so that the activity of oxygen can be ignored. This is also shown in Fig. 6.5. Thus, we can summarize the total numbers of orbitals and electrons in Table 6.3.

Table 6.3 The total numbers |N|, |C|, and |O| of N, C, and O atoms and of π-orbitals and π-electrons in A, G, C, and T nucleosides, respectively. Here there are 20, 22, 18, and 18 π-electrons for the 13, 14, 11, and 11 π-orbitals in the A, G, C, and T nucleosides, respectively. In this approximation, the π-orbitals in oxygen sites in the sugar have been taken into account

Nucleoside	A	G	C	T
|N|	5	5	3	5
|C|	10	10	9	7
|O|	3	4	4	5
Total	18	19	16	17
π-orbitals	13	14	11	11
π-electrons	20	22	18	18

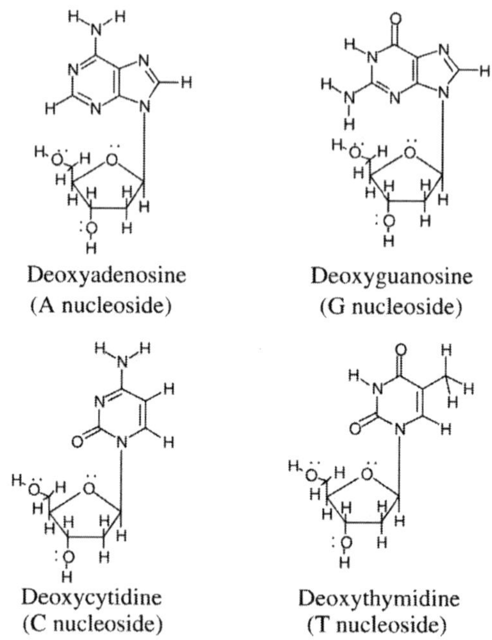

Figure 6.5 The nucleosides of A, G, C, and T. The A, G, C, and T molecules are linked with a sugar molecular group. These are called nucleosides. Here dot (·) means the π-electron occupied and – means the negative ionicity of the site. Reprinted with permission from Ref. [77], Copyright 2004, World Scientific Publishing Company.

6.1.3.5 Electronic configurations in dATP, dGTP, dCTP, and dTTP nucleotides

Let us consider the electronic configurations in dATP, dGTP, dCTP, and dTTP nucleotides. Nucleotide A is a phosphate ester of a nucleoside [215]. The most common site of esterification in naturally occurring nucleotides is the hydroxyl group attached to carbon-5 (C-5) of the sugar, where C-5 is located at the position where the phosphate is linked with sugar(i.e., β-D-2-deoxyribose) as shown in Fig. 6.4 [215]. This compound is called a *nucleoside 5-phosphate* or a *5′-nucleotide*. So when a nucleoside such as the A nucleoside is linked to a triphosphate, it is called a *deoxyadenosine 5′-triphosphate* or dATP. Here the d in dATP indicates that the sugar molecule is *deoxyribose*, distinguishing from ATP in which the sugar is *ribose*.

Let us consider the electronic configurations in the nucleotides of dATP, dGTP, dCTP, and dTTP. Applying the knowledge of quantum chemistry, we find the following: There is a double occupancy by the π-electrons in the oxygen site with two bonds such as -Ö-; a single occupancy by the π-electrons at the oxygen with a double bond such as -P=Ö; and a double occupancy by the π-electrons at the negatively ionized oxygen site with a single bond such as -Ö$^{-\delta}$ where δ means the charge on the oxygen. This situation is shown in Fig. 6.6, where "$-\delta$" is simply denoted by "$-$" such as -Ö$^{-}$.

6.2 Hückel Approximation

6.2.1 *Hückel Model*

In quantum chemistry, the famous Hückel model and its generalized model, called the extended Hückel model, are particularly significant [190–194]. We are concerned only with the π-orbitals of the system in this model. In this context, this theory is nothing but the so-called *tight-binding model* in solid state physics, where it concerns very localized orbitals at atomic sites such as the Wannier function [195, 196]. So it has been extensively applied to various polymer systems successfully, such as polyacetylene [220, 221]. Let us introduce the models [190–195, 212, 213].

Let us define the Hamiltonian H of the system. Denote by n the total number of electrons in the system. Denote the coordinates of the electrons by $\{\mathbf{r}_1, \ldots, \mathbf{r}_{n_e}\}$. Now, the Hamiltonian is given as the sum

$$H(\mathbf{r}_1, \ldots, \mathbf{r}_{n_e}) = \sum_{j=1}^{n_e} h(\mathbf{r}_j). \tag{6.9}$$

Here $h(\mathbf{r}_j)$ is the one-electron Hamiltonian for the j-th electron, defined by

$$h(\mathbf{r}_j) = -\frac{\hbar^2}{2m_e}\nabla_j^2 + U(\mathbf{r}_j), \tag{6.10}$$

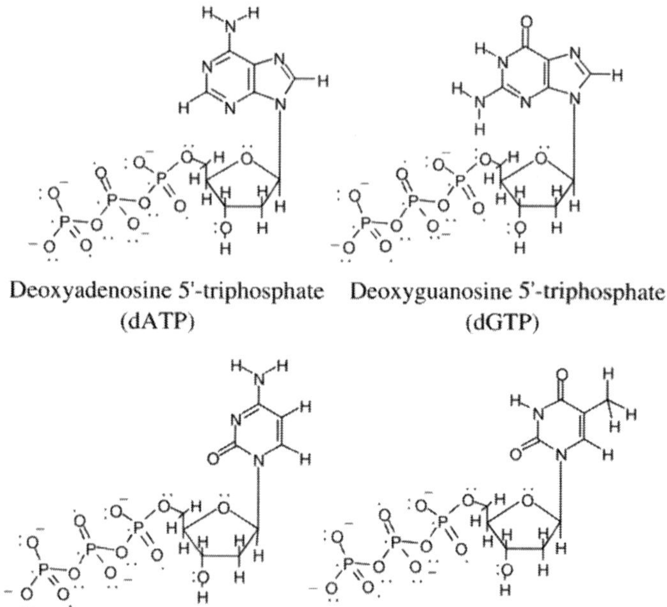

Deoxyadenosine 5'-triphosphate (dATP) | Deoxyguanosine 5'-triphosphate (dGTP)

Deoxycytidine 5'-triphosphate (dCTP) | Deoxythymidine 5'-triphosphate (dTTP)

Figure 6.6 The 5′-nucleotides of dATP, dGTP, dCTP, and dTTP. The dATP, dGTP, dCTP, and dTTP molecules are linked with a sugar and a triphosphate molecule, respectively. These are called the 5′-nucleotides. Here dot (·) means the π-electron occupied and – means the negative ionicity of the site. The figures indicate the acid states of the molecules. Reprinted with permission from Ref. [77], Copyright 2004, World Scientific Publishing Company.

where the first term means the kinetic energy of an electron of mass m_e and the second term is the averaged (i.e., the mean field) potential energy affected from other nuclei and electrons.

Let us consider the linear combination of atomic orbital–molecular orbital (LCAO-MO) method. Let φ be the molecular orbital and χ_r be the atomic orbital at the r-th site such that

$$\varphi(\mathbf{r}) = \sum_{r=1}^{n} c_r \chi_r(\mathbf{r}), \tag{6.11}$$

where n is the total number of atomic sites.

In the Hückel method, we consider the variational principle for c_r to find the minimal energy of

$$E = \frac{\int \varphi h \varphi d^3 r}{\int \varphi \varphi d^3 r}, \tag{6.12}$$

where the numerator and the denominator are given by

$$\int \varphi h \varphi d^3 r = \sum_r \sum_s c_r c_s h_{rs}, \tag{6.13}$$

$$\int \varphi \varphi d^3 r = \sum_r \sum_s c_r c_s S_{rs} = 1, \tag{6.14}$$

with the resonance integrals h_{rs} and the overlap integrals S_{rs}:

$$h_{rs} = \int \chi_r h \chi_s d^3 r,$$
$$S_{rs} = \int \chi_r \chi_s d^3 r. \tag{6.15}$$

By the variational problem $\frac{\partial E}{\partial c_l} = 0$, we obtain the eigenvalue equation

$$\sum_s (h_{rs} - E S_{rs}) c_s = 0. \tag{6.16}$$

In order for the set of c_s to have a nontrivial solution, Eq. (6.16) yields the secular equation:

$$\det |h_{rs} - E S_{rs}| = 0. \tag{6.17}$$

The nontrivial solution gives the molecular orbital of the j-th state,

$$\varphi_j = \sum_{r=1}^{n} c_r^j \chi_r, \tag{6.18}$$

for $j = 1, \ldots, n$. From this, the *charge density* q_r and the *bond order* p_{rs} are given by

$$q_r = \sum_{i=1}^{all} \nu_i (c_r^i)^2,$$
$$p_{rs} = \sum_{i=1}^{all} \nu_i c_r^i c_s^i, \tag{6.19}$$

respectively, where ν_i stands for the electron occupation number at the i-th state [225]; usually $\nu_i = 2$ for the occupied states, and $\nu_i = 0$ for the unoccupied states.

Then, in the so-called Hückel approximation [190–193], we adopt the orthogonality condition for the overlap integrals

$$S_{rr} = 1, \ S_{rs} = 0 \ (r \neq s), \tag{6.20}$$

and the resonance parameters $h_{r,s}$ are taken as the empirical parameters. On the other hand, in the extended Hückel approximation [194], we assume the special form of the resonance integrals [226]

$$h_{rs} = \frac{1}{2} K S_{rs} (h_{rr} + h_{ss}) \tag{6.21}$$

with $K = 1.75$. The overlap integrals $S_{rs} (r \neq s)$ are not necessary be diagonal, and otherwise, almost the same procedure is kept as the Hückel approximation.

6.2.2 *Hückel Parameters*

6.2.2.1 Introduction of parameters α and β

Let us now consider the resonance (i.e., hopping) parameters and the overlap integrals in the Hückel model [225–233]. These are simply called the *Hückel parameters*. Usually, the onsite $(r = s)$ resonance integrals are called the *Coulomb integrals* denoted by α_r, while the offsite $(r \neq s)$ resonance integrals are called the resonance integrals denoted by β_{rs} such that

$$\alpha_r = h_{rr}, \ \beta_{rs} = h_{rs} \ (r \neq s), \tag{6.22}$$

with having the conditions of Eqs. (6.20) and (6.21).

Here we would like to emphasize the following: In the sense of the Hückel theory, the parameters are taken *empirically*. This means that the parameters are *adjustable* and feasible to give consistent results with the experimental results or the *ab initio* calculation results. Therefore, the exact values of the parameters are neither so important nor should be taken so seriously in this framework. Because once one can obtain more precise values for the Hückel parameters, one can provide more plausible results from the Hückel theory.

Although many efforts of the *ab initio* calculations for DNA systems have been done [179–189], unfortunately there seem to be very few first principle calculations for such parameters in the DNA systems to fill this gap at this moment [232, 233]. Nevertheless, we must assign some values for the Hückel parameters to calculate the electronic properties of DNA in the framework of Hückel theory, for our purpose here. So we look back to the original method about time when the Hückel theory was conceived.

To use the standard Hückel theory, let us adopt some simple formulas for the Hückel parameters, which are defined as follows: Let X and Y be two different atoms. Denote the Coulomb integral at the X atom by α_X and the resonance integral between the X and Y atoms by β_{XY}:

$$\alpha_X = \alpha + a_X \beta, \tag{6.23}$$

$$\beta_{XY} = l_{XY} \beta. \tag{6.24}$$

Here a_X and l_{XY} are empirical parameters adopting from experimental data. And the parameters α and β are important; these can be thought of as the fundamental parameters in our problem of biopolymers. Conventionally, we take α as the Coulomb integral for the $2p_x$-orbital of carbon and β as the resonance integral between the $2p_x$-orbitals of carbon, such that

$$\alpha = \alpha_C \equiv 0, \quad \beta = \beta_{CC} \equiv 1. \tag{6.25}$$

This means that the energy level of a carbon atom is taken as the zero level, and the energy is measured in the unit of the resonance integral between carbon atoms. We note that the empirical values obtained from experiments are usually given by [216–218]

$$\alpha \approx -6.30 \sim -6.61 \text{ eV}, \quad \beta \approx -2.93 \sim -2.95 \text{ eV}. \tag{6.26}$$

There are some examples for the parameters of α_X and β_{XY}. They are summarized in Tables 6.4 and 6.5, respectively.

6.2.2.2 Convenient formulas for Hückel parameters

The above values of the Hückel parameters α and β are essentially those for the empty site without electrons. When π-electrons occupy

Table 6.4 The empirical parameters of the Coulomb integral parameters α_X of atom X in eV. For example, we adopt the values from Ref. [216–218]. Here $\alpha_C = \alpha = -6.61$ eV (which will be regarded as the zero energy level such that $\alpha_C = \alpha = 0$) and $\alpha_P = \alpha - \beta = -3.66$ eV has been assumed with $\beta = -2.95$ eV

Atom, X	C	O	N	P
α_X	−6.61	−8.73	−10.94	−3.66

Table 6.5 The empirical parameters of the resonance integrals, β_{XY}, between X and Y atoms in the unit of β. For example, we adopt the values from Ref. [216–218]

Atoms, XY	CC	OC	CN	OP
β_{XY}	1	2	1	≈ 1

an atomic site, there are a few good formulas for calculating the numerical values of α and β.

Denote by α_X the Coulomb integral of the atomic site X without any electron. Denote by $\alpha_{\dot{X}}$ the Coulomb integral of the atomic site X with one electron occupied. And denote by $\alpha_{\ddot{X}}$ the Coulomb integral of the atomic site X with two electrons occupied.

Sandorfy's formula
We now have

$$\alpha_{\dot{X}} = \frac{\chi_X}{\chi_C}\alpha = \alpha + \frac{\chi_X - \chi_C}{\chi_C} \cdot 4.1\beta, \tag{6.27}$$

where χ_X (χ_C) is *Pauling's electronegativity* of X (carbon) atom [231]. Since this method was proposed by Sandorfy [227], this is called Sandorfy's formula in quantum chemistry [217, 218].

Streitwieser's formula
To use Sandorfy's formula for $\alpha_{\ddot{X}}$, Streitwieser [228] introduced the following formula:

$$\alpha_{\ddot{X}} \cong \alpha_{\dot{X}} + \beta. \tag{6.28}$$

This is called Streitwieser's formula in quantum chemistry [217, 218].

Mulliken's formula
On the other hand, there are a couple of methods to evaluate l_{XY} in Eq. (6.24). One method was proposed by Mulliken [229, 230] such that

$$l_{CX} = \frac{S_{CX}}{S_{CC}}, \tag{6.29}$$

where S_{CX} stands for the overlap integral between carbon and X atoms and S_{CC} is the overlap integral between carbon atoms with assuming $S_{CC} = 0.25$. Another method uses the formula:

$$l_{CX} = \left(\frac{1.397}{R_{CX}}\right)^4, \tag{6.30}$$

where R_{CX} is the inter-distance between the C and X atoms [217, 218].

These formulas are just empirical, which can be derived by the *ab initio* method using a super computer. However, unless one can use such a super computer, one can obtain some appropriate values for the Hückel parameters below. The advantage of the above formulas is that before we use highly expensive tools, we can estimate the probable values.

6.2.2.3 Hückel parameters for biomolecules

For applying the Hückel model to DNA biomolecules, we have to set up the numerical values of Hückel parameters. Since biomolecules consist of C, N, O, and P atoms, let us find the plausible values of Hückel parameters for them.

The values of Pauling's electronegativity [231] for carbon (C), nitrogen (N), oxygen (O), and phosphorus (P) are as follows:

$$\chi_C \approx 2.55, \quad \chi_N \approx 3.0, \chi_O \approx 3.5, \quad \chi_P \approx 2.1. \tag{6.31}$$

Remembering our definition of the Hückel parameters for carbon [see Eq. (6.25)] and using Sandorfy's formula [227] [see Eq. (6.27)], we find

$$\alpha_{\dot{C}} = \alpha_C = \alpha,$$

$$\alpha_{\dot{N}} = \alpha + \frac{\chi_N - \chi_C}{\chi_C} \cdot 4.1\beta = \alpha + 0.723\beta,$$

$$\alpha_{\dot{\rm O}} = \alpha + \frac{\chi_{\rm O} - \chi_{\rm C}}{\chi_{\rm C}} \cdot 4.1\beta = \alpha + 1.53\beta,$$

$$\alpha_{\dot{\rm P}} = \alpha + \frac{\chi_{\rm P} - \chi_{\rm C}}{\chi_{\rm C}} \cdot 4.1\beta = \alpha - 0.724\beta, \tag{6.32}$$

Next using Streitwieser's formula [228] [see Eq. (6.28)], we obtain

$$\begin{aligned} \alpha_{\ddot{\rm O}} &= \alpha_{\dot{\rm O}} + \beta = \alpha + 2.53\beta, \\ \alpha_{\ddot{\rm N}} &= \alpha_{\dot{\rm N}} + \beta = \alpha + 1.72\beta. \end{aligned} \tag{6.33}$$

And using Mulliken's formula [229] [see Eq. (6.29)], we estimate the approximated values as

$$\begin{aligned} &\beta_{\dot{\rm C}-\dot{\rm C}} = \beta, \quad \beta_{\dot{\rm C}=\dot{\rm C}} = 1.1\beta, \\ &\beta_{\dot{\rm C}-\dot{\rm N}} = \beta_{\dot{\rm C}-\ddot{\rm N}} = 0.8\beta, \quad \beta_{\dot{\rm C}=\dot{\rm N}} = 1.1\beta, \\ &\beta_{\dot{\rm C}=\dot{\rm O}} = 1.7\beta. \end{aligned} \tag{6.34}$$

For example, the value of $\beta_{\dot{\rm C}-\dot{\rm C}}$ is obtained as follows: Substituting the distance between C–C bond $R_{\rm C-C} = 1.379$ Å into Eq. (6.30), we obtain $l_{\rm C-C} = 1$. Hence, $\beta_{\dot{\rm C}-\dot{\rm C}} = \beta$. Similarly, for $\beta_{\dot{\rm C}=\dot{\rm N}}$, we take $R_{\dot{\rm C}=\dot{\rm N}} = 1.37$ Å and substitute it into Eq. (6.30). We obtain $\beta_{\dot{\rm C}=\dot{\rm N}} = 1.1\beta$. For $\beta_{\dot{\rm C}=\dot{\rm C}}$, we have taken approximately the same value as $\beta_{\dot{\rm C}=\dot{\rm N}}$.

We will use the above parameters throughout this chapter. Here the adoption of these phenomenological values is just a convention for our purpose. If one can get accurate values from the *ab initio* calculations, we can always replace them by the new set of values [232, 233].

6.2.3 *Electronic States of Benzene C_6H_6*

Before considering the electronic states of biomolecules, let us consider the simplest case of benzene (Fig. 6.2) [213]. From this simple example, we can learn what will happen and how we can solve the problem [216–218].

6.2.3.1 Hückel matrix for benzene C_6H_6

Let us find the Hückel matrix for benzene, C_6H_6. For benzene, the resonance parameters are given by only $\alpha_{\dot{\rm C}} = \alpha_{\rm C} = \alpha = 0$ and $\beta_{\dot{\rm C}-\dot{\rm C}} = \beta_{\dot{\rm C}=\dot{\rm C}} = \beta = 1$ for simplicity. Numbering the sites of π-orbitals from 1 to 6, we obtain the following Hückel matrix:

$$\mathbf{H}_{\mathrm{C_6H_6}} = \begin{bmatrix} \alpha\ \beta\ 0\ 0\ 0\ \beta \\ \beta\ \alpha\ \beta\ 0\ 0\ 0 \\ 0\ \beta\ \alpha\ \beta\ 0\ 0 \\ 0\ 0\ \beta\ \alpha\ \beta\ 0 \\ 0\ 0\ 0\ \beta\ \alpha\ \beta \\ \beta\ 0\ 0\ 0\ \beta\ \alpha \end{bmatrix}. \tag{6.35}$$

6.2.3.2 Eigenequation for benzene C_6H_6

Defining the wavefunction $\varphi = \sum_{r=1}^{6} c_r \chi_r$ [see Eq. (6.11)], and from Eq. (6.16) together with the Hückel matrix for benzene C_6H_6, we obtain the eigenequation

$$(\mathbf{H}_{\mathrm{C_6H_6}} - E\,\mathbf{1}_6)\mathbf{c} =$$

$$\begin{bmatrix} \alpha - E & \beta & 0 & 0 & 0 & \beta \\ \beta & \alpha - E & \beta & 0 & 0 & 0 \\ 0 & \beta & \alpha - E & \beta & 0 & 0 \\ 0 & 0 & \beta & \alpha - E & \beta & 0 \\ 0 & 0 & 0 & \beta & \alpha - E & \beta \\ \beta & 0 & 0 & 0 & \beta & \alpha - E \end{bmatrix} \begin{bmatrix} c_1 \\ c_2 \\ c_3 \\ c_4 \\ c_5 \\ c_6 \end{bmatrix} = 0, \tag{6.36}$$

where c_r $(r = 1, \cdots, 6)$ in the vector $\mathbf{c}$ is the wavefunction for the r-th π-orbital at the r-th carbon site, and E is the energy of the six π-electrons.

6.2.3.3 Eigenvalues of Hückel matrix for benzene C_6H_6

Diagonalizing the eigenequation

$$\det|\mathbf{H}_{\mathrm{C_6H_6}} - E\,\mathbf{1}_6| =$$

$$\begin{vmatrix} \alpha - E & \beta & 0 & 0 & 0 & \beta \\ \beta & \alpha - E & \beta & 0 & 0 & 0 \\ 0 & \beta & \alpha - E & \beta & 0 & 0 \\ 0 & 0 & \beta & \alpha - E & \beta & 0 \\ 0 & 0 & 0 & \beta & \alpha - E & \beta \\ \beta & 0 & 0 & 0 & \beta & \alpha - E \end{vmatrix} = 0, \tag{6.37}$$

we find the six eigenvalues E_j and the corresponding six eigenfunctions φ_j. The result is very simple since there is symmetry in the molecule of benzene C_6H_6.

Solvi[ng] ...6) and (6.37), we obtain the following equation for c_s^j:

$$c_r^j = \sqrt{\frac{1}{n}} e^{i \frac{2\pi j}{n} r}, \tag{6.38}$$

... $= 1, \cdots, n$ and we set $n = 6$ for benzene C_6H_6. This ...ves the wave function at site r with the j-th state of energy ... E_j:

$$E_j = \alpha + 2\beta \cos\left(\frac{2\pi j}{n}\right), \tag{6.39}$$

with the corresponding j-th state being given by

$$\varphi_j = \sqrt{\frac{1}{n}} \sum_{r=1}^{n} e^{i \frac{2\pi j}{n} r} \chi_r. \tag{6.40}$$

6.3 Electronic States of Nucleotides

6.3.1 *Electronic States of Single Bases of A, G, C, and T*

In this section, we will study the electronic states of the single nitrogenous bases of A, G, C, and T in terms of the Hückel theory [190–194].

(i) First, we specify the molecule so that we can represent the matrix of the Hückel model (we will call the matrix the Hückel matrix) in terms of the Hückel parameters α_X and β_{XY} for the molecule.

(ii) Second, thanks to MATHEMATICA, we use it to solve directly the secular equation represented by the Hückel matrix to obtain its eigenstates and eigenvalues.

(iii) Third, we count the π-electrons and put them in the energy levels to know the HOMO and LUMO states of the system.

...1.1 Hückel matrices for A, G, C, and T

...g the above Hückel parameters, let us define the Hückel ...ices. Let us put the numbers on the sites for the π-orbitals in ...tructures of the nitrogenous bases A, G, C, and T. The structures ...hown in Fig. 6.7.

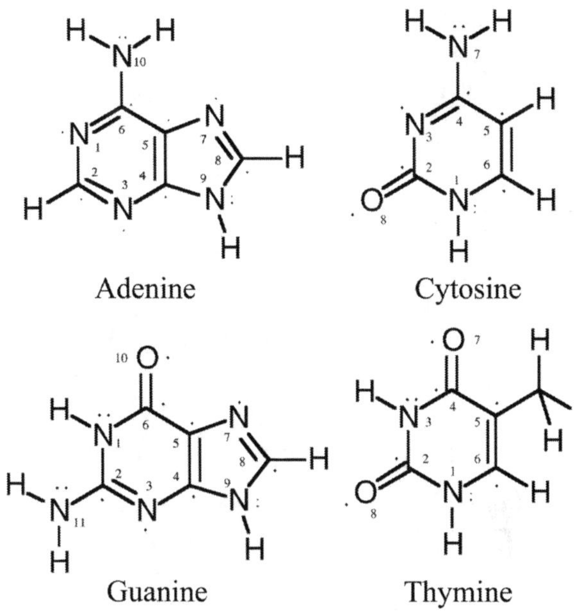

Figure 6.7 The electronic structures of A, G, C, and T. Here dots $(\cdot)$ mean π-electrons. The energies are measured in the unit of $|\beta|$. Reprinted with permission from Ref. [77], Copyright 2004, World Scientific Publishing Company.

Since we have already performed this method in Ref. [77], we only show the simplest case of thymine T. However, we do the same thing for any kind of them.

Considering the topology of the hopping of electrons on the π-orbitals, we now find the following Hückel matrices $\mathbf{H}_{\mathrm{T}}$ for T:

$$\mathbf{H}_{\mathrm{T}} = \begin{bmatrix} \alpha_{\ddot{\mathrm{N}}} & \beta_{\dot{\mathrm{C}}-\ddot{\mathrm{N}}} & 0 & 0 & 0 & \beta_{\dot{\mathrm{C}}-\ddot{\mathrm{N}}} & 0 & 0 \\ \beta_{\dot{\mathrm{C}}-\ddot{\mathrm{N}}} & \alpha_{\dot{\mathrm{C}}} & \beta_{\dot{\mathrm{C}}-\ddot{\mathrm{N}}} & 0 & 0 & 0 & 0 & \beta_{\dot{\mathrm{C}}=\dot{\mathrm{O}}} \\ 0 & \beta_{\dot{\mathrm{C}}-\ddot{\mathrm{N}}} & \alpha_{\ddot{\mathrm{N}}} & \beta_{\dot{\mathrm{C}}-\ddot{\mathrm{N}}} & 0 & 0 & 0 & 0 \\ 0 & 0 & \beta_{\dot{\mathrm{C}}-\ddot{\mathrm{N}}} & \alpha_{\dot{\mathrm{C}}} & \beta_{\dot{\mathrm{C}}-\dot{\mathrm{C}}} & 0 & \beta_{\dot{\mathrm{C}}=\ddot{\mathrm{O}}} & 0 \\ 0 & 0 & 0 & \beta_{\dot{\mathrm{C}}-\dot{\mathrm{C}}} & \alpha_{\dot{\mathrm{C}}} & \beta_{\dot{\mathrm{C}}=\dot{\mathrm{C}}} & 0 & 0 \\ \beta_{\dot{\mathrm{C}}-\ddot{\mathrm{N}}} & 0 & 0 & 0 & \beta_{\dot{\mathrm{C}}=\dot{\mathrm{C}}} & \alpha_{\dot{\mathrm{C}}} & 0 & 0 \\ 0 & 0 & 0 & \beta_{\dot{\mathrm{C}}=\ddot{\mathrm{O}}} & 0 & 0 & \alpha_{\dot{\mathrm{O}}} & 0 \\ 0 & \beta_{\dot{\mathrm{C}}=\dot{\mathrm{O}}} & 0 & 0 & 0 & 0 & 0 & \alpha_{\dot{\mathrm{O}}} \end{bmatrix}. \qquad (6.41)$$

Here we have used the notations for the Hückel parameters α's and β's. Using the values in the previous section, we can assign all the values of the Hückel parameters to the matrix elements. Taking the values of $\alpha = 0$ and $\beta = 1$, we have

$$\begin{aligned}
\alpha_{\dot{\mathrm{C}}} &= 0, \quad \alpha_{\ddot{\mathrm{C}}} = 1,\\
\alpha_{\dot{\mathrm{N}}} &= 0.72, \quad \alpha_{\ddot{\mathrm{N}}} = 1.72,\\
\alpha_{\dot{\mathrm{O}}} &= 1.53, \quad \alpha_{\ddot{\mathrm{O}}} = 2.53,\\
\beta_{\dot{\mathrm{C}}-\dot{\mathrm{C}}} &= 1, \quad \beta_{\dot{\mathrm{C}}=\dot{\mathrm{C}}} = 1.1,\\
\beta_{\dot{\mathrm{N}}-\dot{\mathrm{C}}} &= \beta_{\ddot{\mathrm{N}}-\dot{\mathrm{C}}} = 0.8, \quad \beta_{\dot{\mathrm{N}}=\dot{\mathrm{C}}} = 1.1,\\
\beta_{\dot{\mathrm{C}}=\dot{\mathrm{O}}} &= 1.7.
\end{aligned} \tag{6.42}$$

6.3.1.2 Eigenvalues and eigenvectors of secular equations

Denote by m_X the total number of sites in the nitrogenous molecule $X = \{\mathrm{A, G, C, T}\}$ such that $m_{\mathrm{A}} = 10$, $m_{\mathrm{G}} = 11$, $m_{\mathrm{C}} = m_{\mathrm{T}} = 8$. The above Hückel matrix yields the corresponding secular equation:

$$\det |\mathbf{H}_X - E\mathbf{1}_X| = 0, \tag{6.43}$$

where $\mathbf{1}_X$ stands for the unit matrices whose dimension is m_X.

To solve this, we substitute the Hückel parameters of Eq. (6.36) into Eq. (6.35) and use MATHEMATICA for the diagonalization of the matrix to find the eigenvectors and eigenvalues.

From this procedure, we obtain the following result: The eigenvalues E_j $(j = 1, \ldots, m_X)$ for thymine (T) are given by

$$E_j = \{3.167, 2.781, 2.045, 1.657, 0.919, -0.794, -1.351, -1.924\}. \tag{6.44}$$

Here the eigenvalues are measured in the unit of β such that the positive (negative) numbers listed in Eq. (6.38) mean negative (positive) energies since the value of β is negative, such as $\beta = -2.95$ eV. For convenience, we summarize the above results in Fig. 6.8. The corresponding eigenvectors for T are listed in Table 6.6. We can perform the same operations for the nucleotides other than T.

6.3.1.3 Energy gaps between HOMO and LUMO of A, G, C, and T

Pauli's exclusion principle tells us that each state with an energy level is occupied by a pair of electrons with spin up and down. So

Table 6.6 Eigenvectors (Ev) φ_j and eigenvalues (Eval) E_j of the Hückel matrix for thymine T. The upper numbers from 1 through 8 represent the atomic sites of the molecule, respectively. The φ_j stand for the molecular orbitals constructed from the π-orbitals χ_r at the sites such that $\varphi_j = \sum_n c_r^j \chi_r$. The c_r^j are listed in the middle as numbers, where for the sake of convenience, the values are shown within two digits only. The corresponding eigenvalues E_j are also shown in the unit of β

Ev				Atomic sites					Eval
φ_j	1	2	3	4	5	6	7	8	E_j
φ_1	0.31	0.43	0.44	0.37	0.16	0.14	0.38	0.45	3.167
φ_2	0.26	0.35	−0.07	−0.44	−0.15	0.02	−0.60	0.47	2.781
φ_3	0.75	−0.11	−0.27	0	0.22	0.41	−0.01	−0.36	2.045
φ_4	0.06	−0.03	0.80	−0.03	0	0.03	−0.41	−0.44	1.657
φ_5	−0.40	−0.06	−0.05	0.11	0.68	0.47	−0.32	0.18	0.919
φ_6	−0.20	−0.06	0.17	−0.47	−0.35	0.68	0.35	0.04	−0.794
φ_7	0.19	−0.79	0.15	0.22	−0.19	0.04	−0.13	0.47	−1.351
φ_8	0.13	−0.23	0.19	−0.62	0.53	−0.35	0.31	0.11	−1.924

π-electrons occupy the energy levels in the spectrum from the bottom at low temperature. Since the lower energy levels with one half of the total number of π-electrons can be occupied by the π-electrons, an energy separation appears between the occupied and unoccupied states, which is called the *energy gap*. Since the maximum energy level within the occupied negative energy levels is called HOMO and the minimum energy level within the unoccupied positive energy levels is called LUMO, the energy gap lies between the LUMO and HOMO energy levels.

Let us consider the total numbers of π-electrons and energy levels in each system of A, G, C, and T. From Table 6.1, we find the following:

- 12 π-electrons and 10 energy levels for A
- 14 π-electrons and 11 energy levels for G
- 10 π-electrons and 8 energy levels for C
- 10 π-electrons and 8 energy levels for T

Therefore, the π-electrons occupy the lower six levels for A, the lower seven levels for G, the lower five levels for C, and the lower five levels for T.

From our numerical results, we now find a remarkable fact that *the number of negative energies coincides with just half the number of π-electrons such that all the π-electrons can occupy the negative energy levels below the positive energies.*

If our calculations are correctly carried out [77], the energy levels of HOMOs and LUMOs are given by the following:

$$E_H = \{0.888, 0.699, 0.877, 0.919\},$$
$$E_L = \{-0.789, -0.855, -0.659, -0.794\}, \tag{6.45}$$

for A, G, C, and T, respectively, where energy is measured in the unit of β. From this, we find

$$E_H(\mathrm{G}) > E_H(\mathrm{C}) > E_H(\mathrm{A}) > E_H(\mathrm{T}),$$

$$E_L(\mathrm{G}) > E_L(\mathrm{T}) > E_L(\mathrm{A}) > E_L(\mathrm{C}). \tag{6.46}$$

Defining the energy gap between LUMO and HOMO, ΔE, by

$$\Delta E = E_L - E_H, \tag{6.47}$$

we can obtain the following results for A, G, C, and T, respectively:

$$\Delta E_\mathrm{A} = 1.677,\ \Delta E_\mathrm{G} = 1.555,\ \Delta E_\mathrm{C} = 1.535,\ \Delta E_\mathrm{T} = 1.713, \tag{6.48}$$

which provides

$$\Delta E_\mathrm{T} > \Delta E_\mathrm{A} > \Delta E_\mathrm{G} > \Delta E_\mathrm{C}, \tag{6.49}$$

where the energy gaps are measured in the unit of $|\beta|$. The above results are summarized in Fig. 6.8, where π-electrons at the levels are represented by dot $(\cdot)$.

6.3.1.4 Ground state energies of A, G, C, and T

Let us obtain the ground state energy of the systems by calculating the total energy of the π-electrons:

$$E_\mathrm{tot} = 2 \sum_{j=\mathrm{occ.states}} E_j. \tag{6.50}$$

The results are summarized as follows:

$$E_\mathrm{tot}(\mathrm{A}) = -20.74,\quad E_\mathrm{tot}(\mathrm{G}) = -26.47,$$

$$E_\mathrm{tot}(\mathrm{C}) = -19.05,\quad E_\mathrm{tot}(\mathrm{T}) = -21.14, \tag{6.51}$$

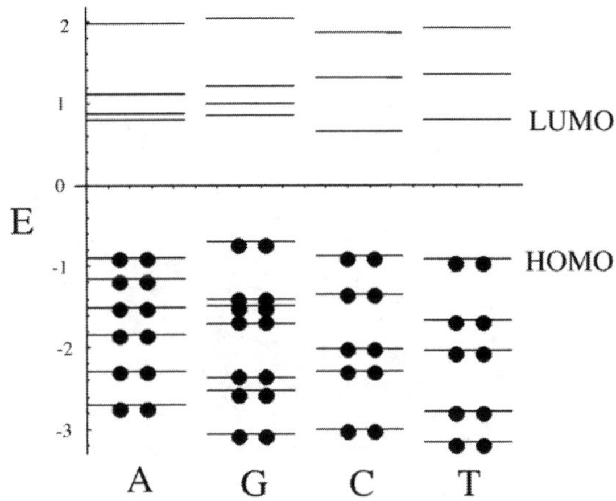

Figure 6.8 The electronic structures of A, G, C, and T. Here dots (●) mean π-electrons. The energies are measured in the unit of $|\beta|$. Reprinted with permission from Ref. [77], Copyright 2004, World Scientific Publishing Company.

which gives

$$E_{\text{tot}}(\text{C}) > E_{\text{tot}}(\text{A}) > E_{\text{tot}}(\text{T}) > E_{\text{tot}}(\text{G}), \tag{6.52}$$

where the energies are measured in the unit of $|\beta|$. This shows that the lower the ground state energy, the more stable the system; *the most stable molecule is G, while the most unstable molecule is C.*

Finally, we note that Eqs. (6.40), (6.45), and (6.46) qualitatively agree with the results in the early calculations of Ladik [199, 200] and in the recent *ab initio* calculations [179–189].

Therefore, even though our approach is phenomenological, using many empirical values for the Hückel parameters, the results obtained are quite encouraging. This situation gives us a starting point as the zeroth approximation for further study.

6.3.2 *Electronic States of a Single Sugar-Phosphate Group*

Now we will consider the electronic properties of a single sugar-phosphate group. Let us assume that there are six π-orbital sites,

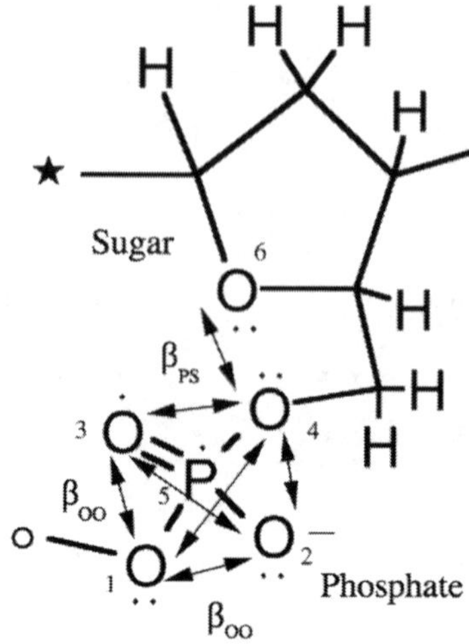

Figure 6.9 The π-orbitals of a sugar-phosphate group. We assume that there are six π-orbital sites in this group. Here dots ($\cdot$) mean π-electrons, arrows mean the electron hopping between the oxygen atoms, the black star ($\star$) means the position of the base, and the white circle ($\circ$) indicates the chain directions. Reprinted with permission from Ref. [77], Copyright 2004, World Scientific Publishing Company.

at five oxygen sites, and at one phosphorus site in this system, respectively. This is drawn in Fig. 6.9.

Using the Hückel model and according to the geometry of the sugar-phosphate group, we define the Hückel matrix as follows:

$$\mathbf{H}_{\mathrm{SP}} = \begin{bmatrix} \alpha_{\ddot{\mathrm{O}}} & \beta_{\ddot{\mathrm{O}}-\ddot{\mathrm{O}}} & \beta_{\dot{\mathrm{O}}-\ddot{\mathrm{O}}} & \beta_{\ddot{\mathrm{O}}-\ddot{\mathrm{O}}} & \beta_{\dot{\mathrm{P}}-\ddot{\mathrm{O}}} & 0 \\ \beta_{\ddot{\mathrm{O}}-\ddot{\mathrm{O}}} & \alpha_{\ddot{\mathrm{O}}} & \beta_{\dot{\mathrm{O}}-\ddot{\mathrm{O}}} & \beta_{\ddot{\mathrm{O}}-\ddot{\mathrm{O}}} & \beta_{\dot{\mathrm{P}}-\ddot{\mathrm{O}}} & 0 \\ \beta_{\dot{\mathrm{O}}-\ddot{\mathrm{O}}} & \beta_{\dot{\mathrm{O}}-\ddot{\mathrm{O}}} & \alpha_{\dot{\mathrm{O}}} & \beta_{\dot{\mathrm{O}}-\ddot{\mathrm{O}}} & \beta_{\dot{\mathrm{P}}=\dot{\mathrm{O}}} & 0 \\ \beta_{\ddot{\mathrm{O}}-\ddot{\mathrm{O}}} & \beta_{\dot{\mathrm{O}}-\ddot{\mathrm{O}}} & \beta_{\dot{\mathrm{O}}-\ddot{\mathrm{O}}} & \alpha_{\ddot{\mathrm{O}}} & \beta_{\dot{\mathrm{P}}-\ddot{\mathrm{O}}} & \beta_{\mathrm{PS}} \\ \beta_{\dot{\mathrm{P}}-\ddot{\mathrm{O}}} & \beta_{\dot{\mathrm{P}}-\ddot{\mathrm{O}}} & \beta_{\dot{\mathrm{P}}=\dot{\mathrm{O}}} & \beta_{\dot{\mathrm{P}}-\ddot{\mathrm{O}}} & \alpha_{\dot{\mathrm{P}}} & 0 \\ 0 & 0 & 0 & \beta_{\mathrm{PS}} & 0 & \alpha_{\ddot{\mathrm{O}}} \end{bmatrix}, \tag{6.53}$$

where β_{PS} means electron hopping between the oxygen atoms of the phosphate and the sugar.

Using the method explained in the previous section and assuming that $\alpha = 0$ and $\beta = 1$, we assign the following Hückel parameters:

$$\alpha_{\dot{\mathrm{O}}} = 1.53, \quad \alpha_{\ddot{\mathrm{O}}} = \alpha_{\dot{\mathrm{O}}} + \beta = 2.53,$$

$$\alpha_{\dot{\mathrm{P}}} = -0.72,$$

$$\alpha_{\dot{\mathrm{N}}} = 0.72, \quad \alpha_{\ddot{\mathrm{N}}} = \alpha_{\dot{\mathrm{N}}} + \beta = 1.72, \tag{6.54}$$

$$\beta_{\dot{\mathrm{O}}-\ddot{\mathrm{O}}} = \beta_{\ddot{\mathrm{O}}-\ddot{\mathrm{O}}} = \beta_{\mathrm{PS}} = 0.5\beta,$$

$$\beta_{\dot{\mathrm{P}}-\dot{\mathrm{O}}} = \beta_{\dot{\mathrm{P}}-\ddot{\mathrm{O}}} = 0.8\beta, \quad \beta_{\dot{\mathrm{P}}=\dot{\mathrm{O}}} = 1.1\beta, \tag{6.55}$$

respectively.

Using the above Hückel matrix, we consider the electronic spectrum of the following three systems:

(1) The phosphate group where the electron hopping between the oxygen sites is forbidden such that $\beta_{\mathrm{OO}} = \beta_{\mathrm{PS}} = 0$ (we denote this by PO_4').
(2) The phosphate group where the electron hopping between the oxygen sites is allowed such that $\beta_{\mathrm{PS}} = 0$ (we denote this by PO_4).
(3) The sugar-phosphate group where the electron hopping between the oxygen sites is not forbidden (we denote this by SP).

From the direct matrix diagonalization using MATHEMATICA, the energy levels E_j and the ground state energy E_{tot} are obtained as follows:

$$E_j(\mathrm{PO}_4') = \{3.1460, 2.53, 2.53, 1.771, -1.577\},$$

$$E_j(\mathrm{PO}_4) = \{4.414, 2.03, 2.03, 1.303, -1.377\},$$

$$E_j(\mathrm{SP}) = \{4.446, 2.740, 2.03, 1.804, -1.378, 1.286\}, \tag{6.56}$$

$$E_{\mathrm{tot}}(\mathrm{PO}_4') = 19.95, \quad E_{\mathrm{tot}}(\mathrm{PO}_4) = 19.55, \quad E_{\mathrm{tot}}(\mathrm{SP}) = 22.04, \tag{6.57}$$

respectively, where the energies are measured in the unit of β. The above results are shown in Fig. 6.10.

The corresponding wavefunctions can be calculated accordingly [77]. Here we show only one case of SP in Table 6.7.

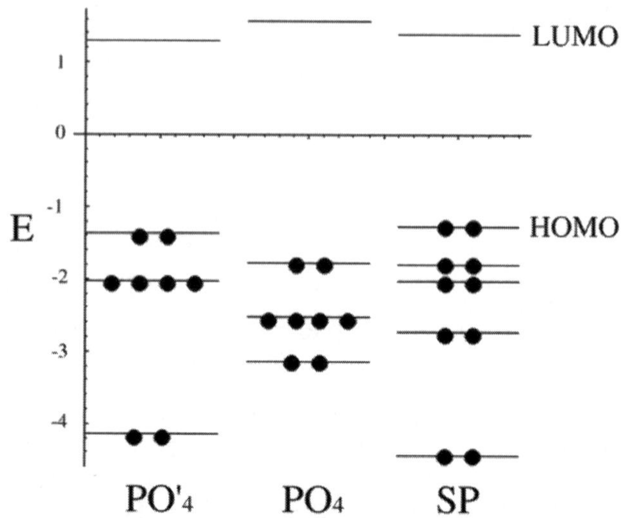

Figure 6.10 The spectrum of the π-orbitals of a sugar-phosphate group. PO_4 stands for the phosphate group where the electron hopping between the oxygen sites is taken into account, while PO'_4 stands for the phosphate group where the electron hopping between the oxygen sites is forbidden. SP stands for the sugar-phosphate group where the electron hopping between the oxygen sites is taken into account. Here dot (●) means the π-electron occupied and the level with four dots means the double degeneracy of the level. Reprinted with permission from Ref. [77], Copyright 2004, World Scientific Publishing Company.

Let us first consider the case of SP in Table 6.7. The wavefunction of the highest energy state is almost localized at the phosphorus site (number 5), while the wavefunctions of the other five states are distributed over the oxygen sites. Since the highest energy state is the only empty state, it is the LUMO state of this system. And the lower energies are filled by the π-electrons as in Fig. 6.10. The wavefunction of the HOMO state is localized at the number 3 oxygen site where there is a double bond to the phosphorus. This situation supports the picture in biology that there is a double bond between the oxygen and the phosphorus atoms. Of course, this is inexplicitly included in the values of Hückel parameters.

Table 6.7 Eigenvectors φ_j and eigenvalues E_j of the Hückel matrix for the sugar-phosphate. The upper numbers from 1 through 6 represent the atomic sites of the molecule, respectively. The φ_j stand for the molecular orbitals constructed from the π-orbitals χ_r at the sites such that $\varphi_j = \sum_n c_r^j \chi_r$. The c_r^j are listed in the middle as numbers, where for the sake of convenience, the values are shown only in two digits. The corresponding eigenvalues E_j are also shown in the unit of β

Eigenvectors			Atomic sites				Eigenvalues
φ_j	1	2	3	4	5	6	E_j
φ_1	0.49	0.49	0.37	0.52	0.31	0.13	4.446
φ_2	0.24	0.24	0.11	−0.36	0.07	−0.86	2.03
φ_3	−0.71	0.71	0	0	0	0	1.804
φ_4	0.36	0.36	−0.12	−0.70	−0.04	0.48	1.771
φ_5	−0.23	−0.23	0.87	−0.32	0.16	0.13	1.286
φ_6	−0.12	−0.12	−0.29	−0.12	0.93	0.02	−1.378

6.3.3 *Electronic States of a Single Nucleotide with a Single Sugar-Phosphate Group*

Let us consider the electronic properties of the molecules linked between the nitrogenous bases of A, G, C, T and the sugar-phosphate. As was discussed in Section 6.1.3, these molecules are called nucleotides. We use the geometry and numbering as in Fig. 6.11.

6.3.3.1 Hückel matrices for nucleotide molecules of A, G, C, and T

Let us first define the Hückel matrix for the system of the nucleotide molecules of A, G, C, and T. Use the same notation of the Hückel matrices $\mathbf{H}_X$ for the nitrogenous base of $X = \{A, G, C, T\}$ and $\mathbf{H}_{SP}$ for the sugar-phosphate group, as before.

Now we can write the Hückel matrix $\mathbf{H}_{nX}$ for the nucleotide of X as follows:

$$\mathbf{H}_{nX} = \begin{bmatrix} \mathbf{H}_X & \mathbf{V}_X \\ \mathbf{V}_X^t & \mathbf{H}_{SP} \end{bmatrix}, \tag{6.58}$$

where $\mathbf{V}_X$ and its transpose $\mathbf{V}_X^t$ are the matrices that describe the electron hopping between the sugar-phosphate and the nucleotide base groups. Since we take into account only the electron hopping

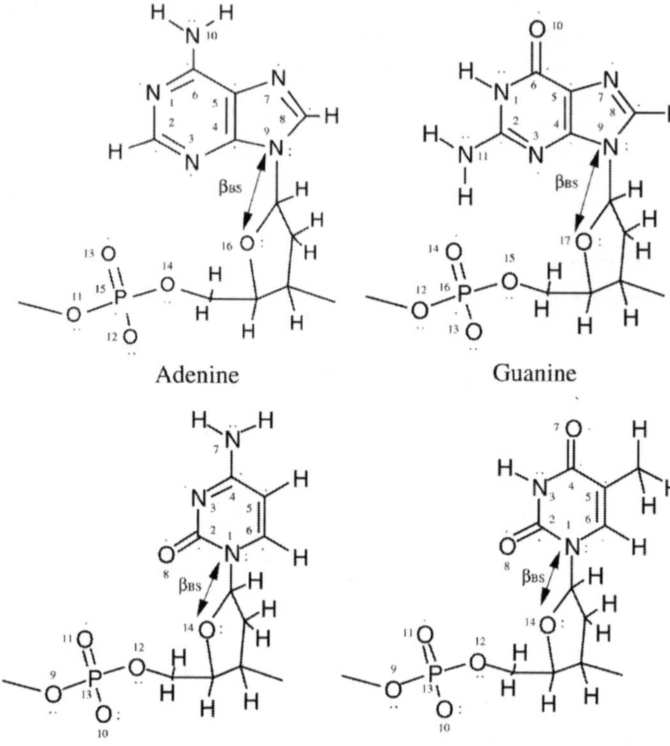

Figure 6.11 The π-orbitals of the nucleotides of A, G, C, and T. Here dot $(\cdot)$ means the π-electron occupied. Reprinted with permission from Ref. [77], Copyright 2004, World Scientific Publishing Company.

between the π-orbital of the nitrogen atom nearest to the carbon atom in the sugar and that of the oxygen atom in the sugar, we have only one nonzero component and all other components are zero. Therefore, we can define as

$$(\mathbf{V}_{\mathrm{A}})_{ij} = \beta_{\mathrm{BS}}\delta_{i9}\delta_{j16}, \quad (\mathbf{V}_{\mathrm{G}})_{ij} = \beta_{\mathrm{BS}}\delta_{i9}\delta_{j17},$$

$$(\mathbf{V}_{\mathrm{C}})_{ij} = \beta_{\mathrm{BS}}\delta_{i1}\delta_{j14}, \quad (\mathbf{V}_{\mathrm{T}})_{ij} = \beta_{\mathrm{BS}}\delta_{i1}\delta_{j14}. \tag{6.59}$$

Here, in our definition, $\mathbf{V}_{\mathrm{A}}$, $\mathbf{V}_{\mathrm{G}}$, $\mathbf{V}_{\mathrm{C}}$, and $\mathbf{V}_{\mathrm{T}}$ are 10×6, 11×6, 8×6, and 8×6 matrices, and $\mathbf{V}_{\mathrm{A}}^t$, $\mathbf{V}_{\mathrm{G}}^t$, $\mathbf{V}_{\mathrm{C}}^t$, and $\mathbf{V}_{\mathrm{T}}^t$ are their transpose 6×10, 6×11, 6×8, and 6×8 matrices, respectively. δ_{ij} stands for Kronecker's δ-function. And β_{BS} means the resonance integral between the base and the sugar. Thus, we find that $\mathbf{H}'_{n\mathrm{A}}$ is a 16×16 matrix, $\mathbf{H}'_{n\mathrm{T}}$ is a 14×14 matrix, $\mathbf{H}'_{n\mathrm{G}}$ is a 17×17 matrix, and $\mathbf{H}'_{n\mathrm{C}}$ is a 14×14 matrix.

6.3.3.2 Value of β_{BS}

To assign a value for β_{BS}, we need some considerations. The π-electron hopping β_{BS} between the oxygen sites in the base and in the sugar strongly depends on the geometry of the single strand of DNA. So if the distance between the oxygen sites of the base and the sugar is far apart, then we can think that the hopping is hard to occur. Hence, it may vanish as $\beta_{\mathrm{BS}} = 0$ in this case. On the other hand, if the distance is close enough to each other, we can think that hopping is possible to occur. Hence, we can assume a certain value for β_{BS}. Since the nearest neighbor hopping between the carbon atoms is β, we can take a value for it in between $0 \leq \beta_{\mathrm{BS}} \leq \beta$. Hence, we assume the intermediate value

$$\beta_{\mathrm{BS}} = 0.5\beta \tag{6.60}$$

as a reasonable assumption, where we choose $\beta = 1$. But this choice of the value is just a convention for the rest of this chapter.

6.3.3.3 Energy levels of nucleotide molecules of A, G, C, and T with a single sugar-phosphate group

The eigenvalues and eigenvectors of the secular equations for the Hückel matrices of H_{nX} with $X = \{\mathrm{A, G, C, T}\}$ have been calculated by direct diagonalization using MATHEMATICA. So we do not repeat it here. The results are as shown in Fig. 6.12 [77].

In Fig. 6.12, we show the spectrum of the π-orbitals of the nucleotides of A, G, C, and T, respectively. As discussed before, $\mathrm{PO_4}$ stands for the phosphate group where the electron hopping between the oxygen sites is taken into account, while PO'_4 stands for the phosphate group where the electron hopping between the oxygen sites is forbidden. SP stands for the sugar-phosphate group where the electron hopping between the oxygen sites is taken into account. nA, nG, nC, and nT stand for the nucleotides of A, G, C, and T, which are linked to the sugar-phosphate group of $\mathrm{PO_4}$, while A, G, C, and T stand for the nitrogenous bases of A, G, C, and T. Here dot $(\cdot)$

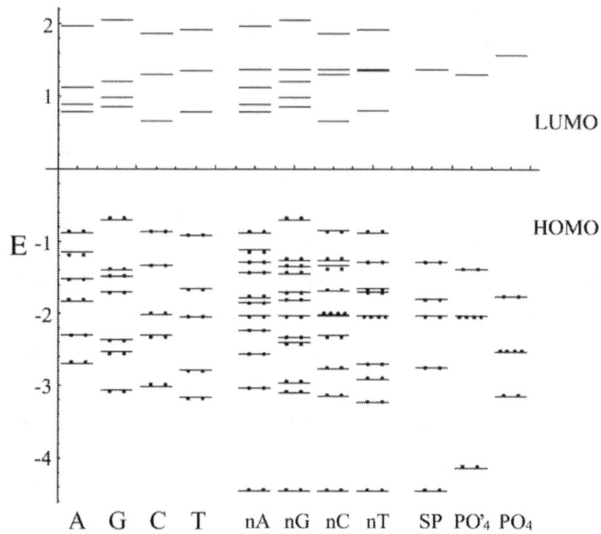

Figure 6.12 The spectrum of the π-orbitals of the nucleotides of A, G, C, and T. PO_4 stands for the phosphate group where the electron hopping between the oxygen sites is taken into account, while PO_4' stands for the phosphate group where the electron hopping between the oxygen sites is forbidden. SP stands for the sugar-phosphate group where the electron hopping between the oxygen sites is taken into account. nA, nG, nC, and nT stand for the nucleotides of A, G, C, and T, which are linked to the sugar-phosphate group of PO_4, while A, G, C, and T stand for the nitrogenous bases of A, G, C, and T. Here dot (·) means the π-electron occupied and the level with four dots means the double degeneracy of the level. Reprinted with permission from Ref. [77], Copyright 2004, World Scientific Publishing Company.

means the π-electron occupied and the level with four dots means the double degeneracy of the level.

This figure means the following: In our Hückel model here, there are six states in the sugar-phosphate group, while there are 10, 11, 8, and 8 states in the nitrogenous groups of A, G, C, and T, respectively. The spectra for the sugar-phosphate group are shown in the rightmost three columns, and the spectra for the nitrogenous groups of A, G, C, and T are shown in the leftmost four columns. And by the electron hopping between the sugar and the nitrogenous base (i.e., by the orbital mixing between the sugar and the base groups), the levels are affected to give the new spectra that are shown in the center four columns.

From the above result, we find the following:

(1) The energy gaps are essentially dominated by those of the base groups of A, G, C, and T, such that the energy levels of HOMO and LUMO in the system are almost equivalent to those of the HOMO and LUMO in the base groups. That is, *the effect of the sugar-phosphate group on the energy gap is weak*.
(2) The energy levels of the second HOMOs in the G, C, T bases are attributed to those of the HOMO in the sugar-phosphate group, while the energy level of the third HOMO in the A base is attributed to that of the HOMO in the sugar-phosphate group. Thus, *right below the energy level of the HOMO in the base group is the energy level of the HOMO in the sugar-phosphate group*.

On the other hand, as was discussed before [77], we find the following: (1) For the G, C, T nucleotides, the HOMO and LUMO of the π-orbitals are localized in the base group of the nucleotide, while the second HOMO is localized in the sugar-phosphate group of the nucleotide. (2) For the A nucleotide, the HOMO, the second HOMO, and the LUMO of the π-orbitals are localized in the base group of the nucleotide, while the third HOMO is localized in the sugar-phosphate group of the nucleotide.

Thus, we can conclude that *the energy gap of the nucleotide comes from that of the base group such that the energy levels of the HOMO and LUMO due to the base exist above the energy level of the second HOMO due to the sugar*.

6.4 Electronic States of DNA

6.4.1 *Decorated Ladder Models of a Single or Double Strand of DNA*

6.4.1.1 Geometry of a single or double strand of DNA

DNA is constructed as a double helix of two single strands of DNA chain. The single strand is constructed by a complicated repetition of the bases of A, G, C, and T and the sugar-phosphate alternatively. This is shown in Fig. 6.13.

Each chain in the helix goes in the opposite direction to one another as the negative–positive relationship. The two strands of DNA chains are linked by hydrogen bonds through the nucleotide

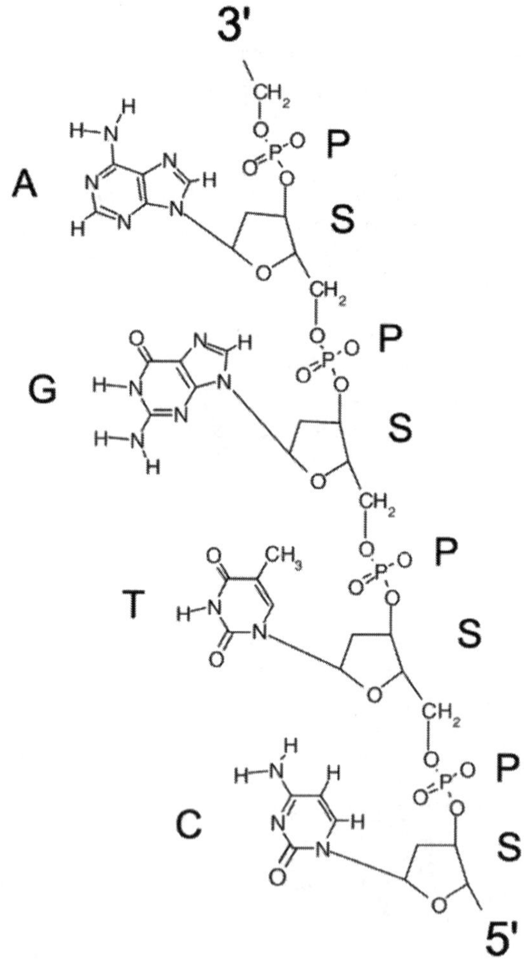

Figure 6.13 Example of a single strand of DNA. The chain structure is constructed by the nucleotides of A, G, T, and C. P and S denote the phosphate group and the sugar group, respectively.

bases A, G, C, and T at the center of the helix such that A-T and G-C bonds are naturally formed. This situation is shown in Fig. 6.14.

6.4.1.2 Simple modeling of a single or double strand of DNA

The modeling for the single or double strand of DNA is very important, since otherwise we cannot calculate the electronic properties of DNA at all. Some years ago, I have introduced a simple

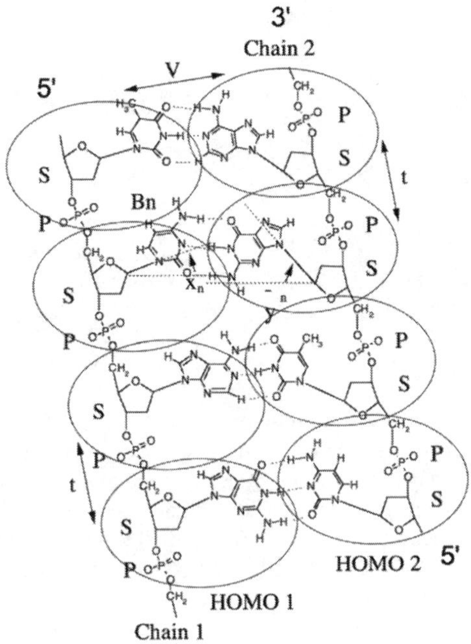

Figure 6.14 Example of a double strand of DNA. Each chain consists of the nucleotides of A, G, T, and C. The double strand is formed by hydrogen bonds between A and T and between G and C at the center of the double strand. P and S denote the phosphate group and the sugar group, respectively. B_n represents the n-th base group of the system. HOMO1 and HOMO2 schematically stand for the highest occupied states of the base groups in chains 1 and 2. V is the electron hopping (resonance integral) between the hydrogen bonded bases in the paired chains, and t is the electron (resonance integral) hopping between the base groups in the same chain. Reprinted with permission from Ref. [204], Copyright 2003, World Scientific Publishing Company.

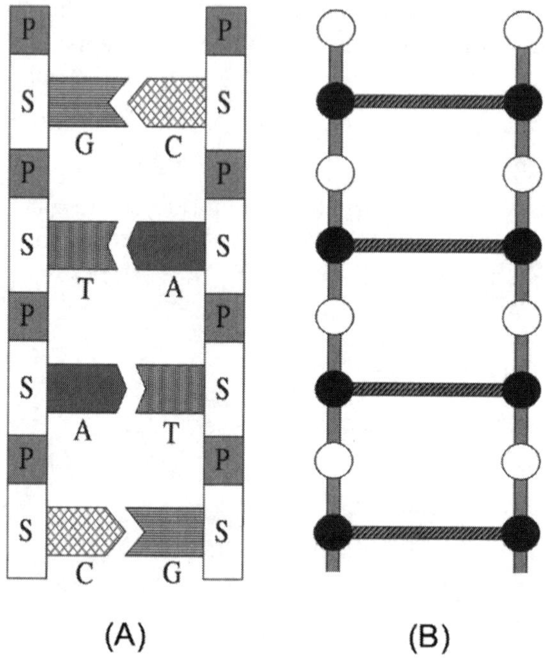

Figure 6.15 The decorated ladder model of DNA—the simplest model of a double strand of DNA is shown. (A) The schematic diagram of the DNA double chain. Here the repetition of the sugar (S) and phosphate (P) groups is emphasized. Hydrogen bonds are realized between nucleotide A and nucleotide T or between nucleotide G and nucleotide C. (B) The decorated ladder model. It is the simplest modeling for the double strand of DNA. Here the white circle stands for the phosphate site and the black circle stands for the sugar site. The bonds between the chains stand for the hydrogen bonds. This is called the decorated ladder model of DNA. Reprinted with permission from Ref. [202], Copyright 1997, World Scientific Publishing Company.

model for the DNA double chain [202–204, 207] as well as that for the DNA single chain [77]. This direction has been well developed by many people already [206, 208]. The model for the double strand of DNA is shown in Fig. 6.15. We also show some model for the single strand of DNA in Fig. 6.16.

Here, we would like to note the justification of ladder geometry. Generally speaking, unless the structure of a double helix of DNA

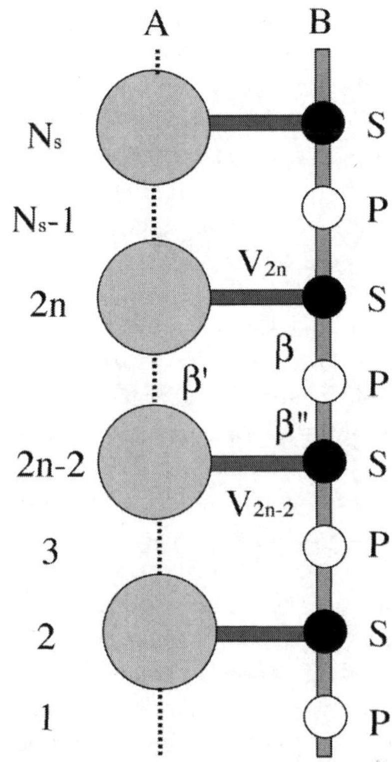

Figure 6.16 The decorated ladder model for a single strand of DNA. Here the large circles stand for the base groups and the small black (white) circles stand for the sugar (phosphate) groups. In this model, β and β'' mean the π-electron hopping (i.e., resonance) integrals between the sugar-phosphate groups, β' is the π-electron hopping integral between the base groups, and V_n is the n-th π-electron hopping integral between the base and the sugar-phosphate, respectively. Reprinted with permission from Ref. [77], Copyright 2004, World Scientific Publishing Company.

is constructed in the system, the system of a single strand of DNA may be very flexible such that it cannot make ladder geometry. Therefore, the ladder geometry of the single strand does not seem plausible for real situations in nature. However, I think it is not true.

For example, when we conduct an experiment, the single strand of DNA can be adsorbed on a solid surface such that the ladder geometry is realized by being lined up on the surface. Or even for the double strand of DNA, to understand the difference between the electronic properties of single and double strands of DNA, one needs to know the electronic property of ladder geometry as a *model* of the separated single chain of the DNA. Also in the literature of quantum chemists, the important role of ladder geometry has not been explicitly mentioned or fully studied previously [179–189]. Thus, we believe that the ladder geometry of DNA is quite important in DNA research.

The geometry of the single strand of DNA is schematically shown in Fig. 6.13. From the figure, we can recognize that the nucleotide base groups, represented as big molecules, are separated from each other, but the sugar-phosphate groups, represented as small molecules, are linked with each other. This geometry allows electrons to inevitably go through the backbone chain with an alternative repetition of sugar-phosphate groups in the system.

On the other hand, we can also allow electrons to hop between the nucleotide base groups in the system. Here since the nucleotide base groups are lined up or twisted in ladder geometry, the overlap of the π-orbitals between the nearest nucleotide groups can be realized. Otherwise, this electron hopping must be difficult to occur in the system of the single strand of DNA.

Thus, we can understand that the *topology of electron hopping* in the system is most important here such that *the essential nature of the geometry of the single strand of DNA can be regarded as a ladder*.

So we can mimic this ladder geometry by the decorated ladder models shown in Fig. 6.16, for the sake of simplicity. In the decorated ladder model, we treat the sugar-phosphate group as two sites and the nucleotide group as another site. We put just one π-orbital in each site and assume that electron hopping is allowed between the nearest neighbor sites. Iguchi [202–204] first introduced these models as the model for considering the electronic property of a single strand of DNA and later as the model for considering the electronic property of a double strand of DNA [77].

6.4.1.3 Tight-binding model for ladder systems

As was mentioned earlier, the decorated ladder model can be applied to the single and double strands of DNA, adopting suitable parameters. Therefore, we first introduce the mathematical part of the Hückel model for applying the decorated ladder models.

Let N_s be the total number of sites in each chain (in either Fig. 6.15 or Fig. 6.16). Since there is only one π-orbital per site, there are $2N_s$ π-orbitals in our model of the DNA single chain.

Denote by χ_n (χ'_n) the π-orbital at site n in the single chain A (B). By superposition of the π-orbitals, the wave function φ is given by

$$\varphi = \sum_{n=1}^{N_s} \{a_n \chi_n + b_n \chi'_n\}, \tag{6.61}$$

where a_n and b_n are the amplitudes of the wave function at site n in chains A and B, respectively. Adjusting with our model of the DNA single chain structure (Figs. 6.15 and 6.16), and applying to Eq. (6.12), we obtain

$$h_{n+1,n}a_{n+1} + h_{n,n-1}a_{n-1} + h_{n,n}a_n + v_n b_n = E a_n,$$

$$h'_{n+1,n}b_{n+1} + h'_{n,n-1}b_{n-1} + h'_{n,n}b_n + v'_n a_n = E b_n. \tag{6.62}$$

Here h_{ij}, h'_{ij} are the Hückel parameters of α_{ii}'s and β_{ij}'s for the system under investigation. And v_n and v'_n are the resonance integrals for the hydrogen bonds between the adjacent nucleotides. We would like to note that the conditions $h_{n+1,n} = h'_{n+1,n}$ and $v_n = v'_n$ are imposed without any problem in practical use. This reason has been explained in the literature [202, 203].

6.4.1.4 Transfer matrix method

Let us find the transfer matrix method for Eq. (6.62). Define the four-dimensional column vector $\Psi_n \equiv (a_n, a_{n-1}, b_n, b_{n-1})^t$. Together with trivial relations $a_n = a_n$ and $b_n = b_n$, Eq. (6.62) can be converted into the following form:

$$\Psi_{n+1} = M_n \Psi_n, \tag{6.63}$$

$$M_n = \begin{bmatrix} A_n & V_n \\ U_n & B_n \end{bmatrix}, \tag{6.64}$$

where M_n is the 4×4 transfer matrix with 2×2 matrices:

$$A_n \equiv \begin{bmatrix} \frac{E-\alpha_{n,n}}{\beta_{n+1,n}} & -\frac{\beta_{n,n-1}}{\beta_{n+1,n}} \\ 1 & 0 \end{bmatrix}, \quad B_n \equiv \begin{bmatrix} \frac{\varepsilon-\alpha'_{n,n}}{\beta'_{n+1,n}} & -\frac{\beta'_{n,n-1}}{\beta'_{n+1,n}} \\ 1 & 0 \end{bmatrix}, \tag{6.65}$$

$$U_n \equiv \begin{bmatrix} \frac{-v_n}{\beta_{n+1,n}} & 0 \\ 0 & 0 \end{bmatrix}, \quad V_n \equiv \begin{bmatrix} \frac{-v'_n}{\beta'_{n+1,n}} & 0 \\ 0 & 0 \end{bmatrix}. \tag{6.66}$$

According to the sequence of N_s segments, we have to take a matrix product M of the N_s transfer matrices M_n such that

$$M(N_s) \equiv M_{N_s} M_{N_s-1} \cdots M_1, \tag{6.67}$$

which is also a 4×4 matrix.

We now impose that the single strand of DNA is periodic. This means $a_{n+N_s} = a_n$ and $b_{n+N_s} = b_n$, (i.e., $\Psi_{n+N_s} = \Psi_n$). Let us use this in Eq. (6.64). We then have $\Psi_{n+N_s} = \Psi_n = M(N_s)\Psi_n$. This provides the condition:

$$\det[M(N_s) - I_4] = 0, \tag{6.68}$$

where I_4 is the 4×4 unit matrix. Equation (6.68) provides the wave vector k in the system such that $k = 2\pi j/N_s$ for $j = -N_s/2, \cdots, N_s/2$.

Next suppose that the system is arbitrarily large (i.e., $N_s \to \infty$) with the unit cell of N pairs of nucleotide base and sugar groups. Then we adapt the Bloch theorem to the system:

$$a_{n+N} = \rho a_n, \quad b_{n+N} = \rho b_n \tag{6.69}$$

with $\rho = e^{ikN}$. Applying Eqs. (6.68) to (6.64), we find a 4×4 determinant $D(\rho)$, which is a fourth-order polynomial of ρ:

$$D(\rho) \equiv \det[M(N) - \rho I_4] =$$
$$\rho^4 - s_1\rho^3 + s_2\rho^2 - s_3\rho + s_4 = 0, \tag{6.70}$$

where

$$M(N) \equiv M_N M_{N-1} \cdots M_1. \tag{6.71}$$

By the relation between roots and coefficients, the four roots ρ_1, ρ_2, ρ_3, and ρ_4 are given by

$$s_1 = \sum_{i=1}^{4} \rho_i \equiv tr\, M(N), \quad s_2 = \sum_{i<j=1}^{4} \rho_i \rho_j,$$
$$s_3 = \sum_{i<j<k=1}^{4} \rho_i \rho_j \rho_k, \quad s_4 = \det M(N) \equiv \rho_1\rho_2\rho_3\rho_4. \tag{6.72}$$

6.4.1.5 Symplectic property of transfer matrix

Let us solve the biquadratic equation $D(\rho) = 0$. As was proved previously [77, 202–204], if ρ is a solution to the biquadratic equation, so is ρ^{-1}. Hence, ρ^{-1} must be an eigenvalue of $D(\rho) = 0$ such that

$$D(\rho^{-1}) = \rho^{-4}(s_4\rho^4 - s_3\rho^3 + s_2\rho^2 - s_1\rho + 1) = 0. \tag{6.73}$$

This gives us the particular property of the matrix M, called the *symplectic structure* [234]:

$$M^{\dagger} J\, M = J, \tag{6.74}$$

$$J \equiv \begin{pmatrix} \mathbf{J} & \mathbf{0} \\ \mathbf{0} & \mathbf{J} \end{pmatrix}, \quad \mathbf{J} \equiv \begin{pmatrix} 0 & -1 \\ 1 & 0 \end{pmatrix}, \tag{6.75}$$

where $M^{\dagger}$ is the Hermitian conjugate of M and $\mathbf{0}$ is the 2×2 zero matrix. This yields

$$D(\rho) = \rho^4 D(\rho^{-1}), \tag{6.76}$$

from which we find

$$s_1 = s_3, \quad s_4 = 1. \tag{6.77}$$

Thus, M belongs to $SL(4, \mathbf{R})$.

Using this property and dividing $D(\rho)$ by ρ^2 reduces the biquadratic equation to the quadratic equation

$$x^2 - s_1 x + s_2 - 2 = 0,$$
$$x = \rho + \frac{1}{\rho} \tag{6.78}$$

Therefore, its two roots are given by

$$x_{\pm} = \frac{1}{2}\left(s_1 \pm \sqrt{D}\right),$$
$$D = s_1^2 - 4s_2 + 8. \tag{6.79}$$

As was shown in the literature [202, 203], by a simple manipulation using Eq. (6.72), we finally get

$$x_{\pm} = \frac{1}{2}[tr\,M \pm \sqrt{D}],$$
$$D \equiv 2tr(M^2) - (tr\,M)^2 + 8, \tag{6.80}$$

where tr is the trace of the matrix, which is the sum of diagonal elements, and $-$ $(+)$ channel means the bonding (antibonding) states between two parallel parts (i.e., the nucleotide base groups and the sugar-phosphate groups) in the single strand of DNA.

6.4.1.6 Scheme for obtaining energy bands and density of states

Now we can state a simple scheme to obtain the spectrum in the following: If an energy ε satisfies

$$x_{\pm} = 2\cos kN, \tag{6.81}$$

then the energy is allowed; otherwise, it is forbidden in channel $\pm$. This can be regarded as a generalized version of the Bloch condition for the single linear chain system with the 2×2 transfer matrix M where

$$tr\, M = 2\cos kN. \tag{6.82}$$

We can also calculate the density of states (DOS) $D_{\pm}(\varepsilon)$ using Eq. (6.80) together with Eq. (6.81) for each channel $\pm$, respectively:

$$\begin{aligned} dk_{\pm} &= \frac{1}{N} d\cos^{-1}[x_{\pm}(\varepsilon)/2] \\ &= -\frac{1}{N}\frac{\frac{\partial x_{\pm}}{\partial \varepsilon}}{\sqrt{4 - x_{\pm}^2}} d\varepsilon = D_{\pm}(\varepsilon) d\varepsilon. \end{aligned} \tag{6.83}$$

Therefore, the total DOS is given by the sum of $D_{+}(\varepsilon)$ and $D_{-}(\varepsilon)$:

$$D(\varepsilon) = D_{+}(\varepsilon) + D_{-}(\varepsilon), \tag{6.84}$$

where $D_{-}(\varepsilon)$ $[D_{+}(\varepsilon)]$ means the DOS contributed from the bonding (antibonding) channel $-$ $(+)$, respectively.

6.4.2 *Electronic Properties of a Single Strand of DNA*

6.4.2.1 π-Electronic energy spectrum of decorated ladder model for a single strand of DNA

Before considering the electronic spectrum of the single strand of DNA, let us first consider the decorated ladder model defined in Fig. 6.16. This can be regard as a prototype model for the more realistic DNA models.

In this model, the amplitudes of the wavefunction of the nucleotide bases b_n are defined on the even number sites only, while the amplitudes of the wavefunction of the sugar-phosphate group, a_n, are defined on every site. We take into account the π-electron hoppings as shown in Fig. 6.16. Here β and β'' are the

π-electron hopping (i.e., resonance) integrals between the sugar-phosphate groups, β' is the π-electron hopping integral between the base groups, and v_n is the n-th π-electron hopping integral between the base and the sugar-phosphate. In this set-up, we have the following equations:

$$\begin{aligned} E b_{2n} &= \alpha_{A,2n} b_{2n} + \beta'(b_{2n+2} + b_{2n-2}) + v_n a_{2n}, \\ E a_{2n} &= \alpha_{B,2n} a_{2n} + \beta'' a_{2n+1} + \beta a_{2n-1} + v_n b_{2n}, \\ E a_{2n+1} &= \alpha_{B,2n+1} a_{2n+1} + \beta a_{2n+2} + \beta'' a_{2n}. \end{aligned} \tag{6.85}$$

As the simplest case, when we assume that all the Hückel parameters in the base groups are identical to each other, the base group can be regarded as a unit cell for the whole strand of DNA system, with having a periodicity of $2a$, where a is the distance of the alternative repetition between the sugar (S) and the phosphate (P) groups. We now take the values as

$$\alpha_{A,2n} = \alpha_{\mathrm{A}}, \quad \alpha_{B,2n} = \alpha_{\mathrm{B}}, \quad \alpha_{B,2n+1} = \alpha'_{\mathrm{B}},$$

$$v_n = v. \tag{6.86}$$

Under this condition, we can adopt the Bloch theorem for both a_n and b_n:

$$a_{n+2} = e^{i2ka} a_n, \quad b_{n+2} = e^{i2ka} b_n. \tag{6.87}$$

Here the wavenumber k is defined by

$$-\frac{\pi}{2a} \le k \le \frac{\pi}{2a}. \tag{6.88}$$

Substituting these into Eq. (6.85), we obtain the following equations:

$$\begin{aligned} E b_{2n} &= (\alpha_{\mathrm{A}} + 2\beta' \cos 2ka) b_{2n} + v a_{2n}, \\ E a_{2n} &= \alpha_{\mathrm{B}} a_{2n} + (\beta'' + \beta e^{-i2ka}) a_{2n+1} + v b_{2n}, \\ E a_{2n+1} &= \alpha'_{\mathrm{B}} a_{2n+1} + (\beta e^{i2ka} + \beta'') a_{2n}. \end{aligned} \tag{6.89}$$

This can be converted into a matrix form:

$$\begin{bmatrix} h_{\mathrm{A}}(k) & -v & 0 \\ -v & h_{\mathrm{B}} & -\beta(k) \\ 0 & -\beta(k)^* & h_{\mathrm{B}} \end{bmatrix} \begin{bmatrix} b_{2n} \\ a_{2n} \\ a_{2n+1} \end{bmatrix} = 0. \tag{6.90}$$

where we have defined as

$$\begin{aligned} h_{\mathrm{A}}(k) &\equiv E - \alpha_{\mathrm{A}} - 2\beta' \cos 2ka, \\ h_{\mathrm{B}} &\equiv E - \alpha_{\mathrm{B}}, \\ \beta(k) &\equiv -(\beta'' + \beta e^{-i2ka}), \\ h'_{\mathrm{B}} &\equiv E - \alpha'_{\mathrm{B}}. \end{aligned} \tag{6.91}$$

From this, we get the eigenvalue equation

$$\begin{vmatrix} h_{\mathrm{A}}(k) & -v & 0 \\ -v & h_{\mathrm{B}} & -\beta(k) \\ 0 & -\beta(k)^* & h_{\mathrm{B}} \end{vmatrix} = 0. \tag{6.92}$$

Equation (6.92) provides, in general, three energy bands. We show such an example in Fig. 6.17.

6.4.2.2 Hückel matrices for a single strand of DNA with a single nucleotide base of A, G, C, and T

Let us apply the Hückel model to the single strand of DNA with the repetition of a single nucleotide base of A, G, C, and T. To do so, we show the electron hopping topology between the π-orbitals in Fig. 6.18. The main topology of the system in Fig. 6.18 is the same as the previous one in Fig. 6.16. Therefore, we can apply the Bloch theorem to this system as well:

$$c_{n+1} = e^{ika} c_n, \tag{6.93}$$

where n is the nth nucleotide base group.

In this situation, the only changes of the Hückel parameters in the Hückel matrix in Eq. (6.58) appear in those of the block-diagonal matrices, including the wave vector k. We now find the following Hückel matrices:

$$\mathbf{H}'_{nX} = \begin{bmatrix} \mathbf{H}'_X & \mathbf{V}_X \\ \mathbf{V}^t_X & \mathbf{H}'_{\mathrm{SP}} \end{bmatrix}. \tag{6.94}$$

Here $\mathbf{V}_X$ are the hopping matrices between the sugar-phosphate group and the nucleotide base of $X =$ A, G, T, C given by

$$\begin{aligned} (\mathbf{V}_{\mathrm{A}})_{ij} &= \beta_{\mathrm{BS}}\delta_{i9}\delta_{j16}, \quad (\mathbf{V}_{\mathrm{G}})_{ij} = \beta_{\mathrm{BS}}\delta_{i9}\delta_{j17}, \\ (\mathbf{V}_{\mathrm{T}})_{ij} &= \beta_{\mathrm{BS}}\delta_{i1}\delta_{j14}, \quad (\mathbf{V}_{\mathrm{C}})_{ij} = \beta_{\mathrm{BS}}\delta_{i1}\delta_{j14}, \end{aligned} \tag{6.95}$$

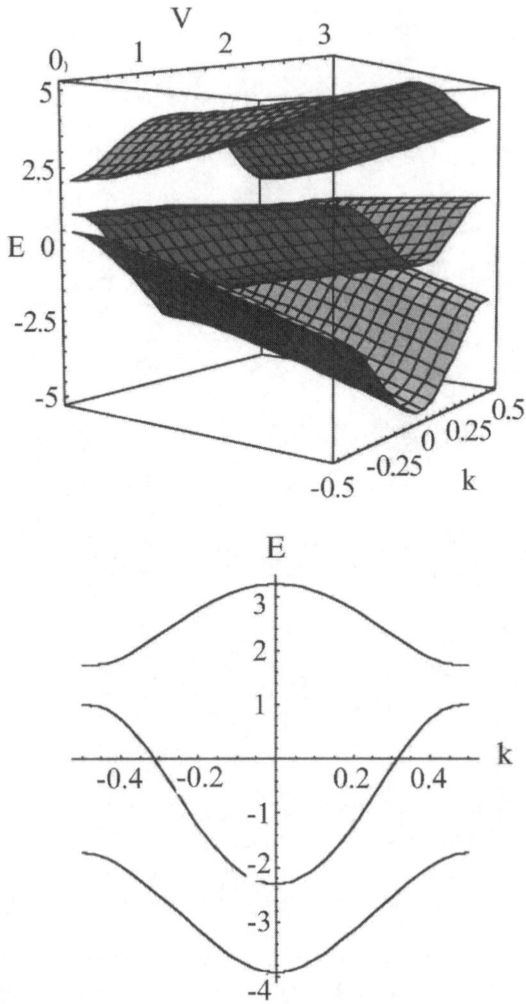

Figure 6.17 Energy bands of the decorated ladder model for a single strand of DNA. (A) The energy bands as a function of v. (B) Snapshot of the energy bands when $v = 1$. k is the wave vector in the unit of $\frac{\pi}{a}$ such that $-0.5 \leq k \leq 0.5$, v is the π-electron hopping integral between the base and the sugar-phosphate, and E is the energy in the unit of $\beta_{\mathrm{C}} = \beta = 1$. Here we have taken the values $\alpha_{\mathrm{A}} = -1$, $\alpha_{\mathrm{B}} = 1$, $\alpha'_{\mathrm{B}} = 1.5$, $\beta' = -1$, $\beta = -2$, and $\beta'' = -1.2$. Reprinted with permission from Ref. [77], Copyright 2004, World Scientific Publishing Company.

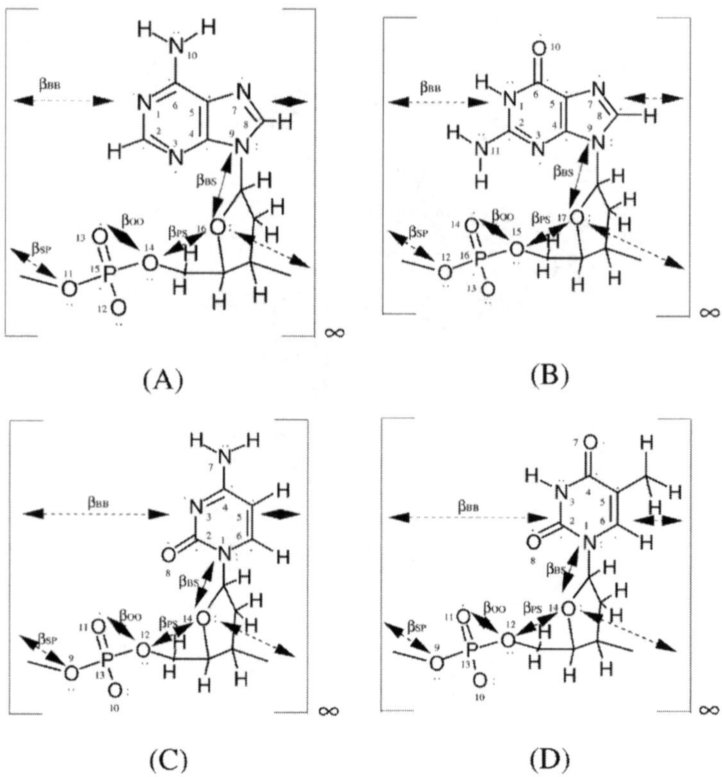

Figure 6.18 The single strands of DNA with the nucleotide bases of A, G, C, and T. Here $[\ ]_\infty$ means that the structure inside [ ] is repeated infinitely as the unit cell of the system. Dots $(\cdot)$ mean π-electrons, β_{BB} is the electron hopping between adjacent bases, β_{SP} (β_{PS}) is the electron hopping from the sugar site to the phosphate site (from the phosphate site to the sugar site), β_{SB} is the electron hopping between the sugar and the nucleotide base, and β_{OO} is the electron hopping between the oxygen sites in the phosphate. Reprinted with permission from Ref. [77], Copyright 2004, World Scientific Publishing Company.

where β_{BS} is the resonance integral between the base and the sugar and δ_{ij} represents Kronecker's δ-function, and $\mathbf{V}_X^t$ is defined as its transpose. We assume the value of $\beta_{BS} = 0.5\beta$ as taken in Eq. (6.60).

The Hückel matrices $\mathbf{H}'_X$ stand for $\mathbf{H}'_A$, $\mathbf{H}'_G$, $\mathbf{H}'_C$, and $\mathbf{H}'_T$ for the nucleotide bases of A, G, C, and T. These are explicitly given in Ref. [77]. Therefore, we do not show all of them but show only the thymine case as an example as before:

$$\mathbf{H}'_{\mathrm{T}} = \begin{bmatrix} \alpha'_{\ddot{\mathrm{N}}} & \beta_{\dot{\mathrm{C}}-\ddot{\mathrm{N}}} & 0 & 0 & 0 & \beta_{\dot{\mathrm{C}}-\ddot{\mathrm{N}}} & 0 & 0 \\ \beta_{\dot{\mathrm{C}}-\ddot{\mathrm{N}}} & \alpha'_{\dot{\mathrm{C}}} & \beta_{\dot{\mathrm{C}}-\ddot{\mathrm{N}}} & 0 & 0 & 0 & 0 & \beta_{\dot{\mathrm{C}}=\dot{\mathrm{O}}} \\ 0 & \beta_{\dot{\mathrm{C}}-\ddot{\mathrm{N}}} & \alpha'_{\ddot{\mathrm{N}}} & \beta_{\dot{\mathrm{C}}-\ddot{\mathrm{N}}} & 0 & 0 & 0 & 0 \\ 0 & 0 & \beta_{\dot{\mathrm{C}}-\ddot{\mathrm{N}}} & \alpha'_{\dot{\mathrm{C}}} & \beta_{\dot{\mathrm{C}}-\dot{\mathrm{C}}} & 0 & \beta_{\dot{\mathrm{C}}=\ddot{\mathrm{O}}} & 0 \\ 0 & 0 & 0 & \beta_{\dot{\mathrm{C}}-\dot{\mathrm{C}}} & \alpha'_{\dot{\mathrm{C}}} & \beta_{\dot{\mathrm{C}}=\dot{\mathrm{C}}} & 0 & 0 \\ \beta_{\dot{\mathrm{C}}-\ddot{\mathrm{N}}} & 0 & 0 & 0 & \beta_{\dot{\mathrm{C}}=\dot{\mathrm{C}}} & \alpha'_{\dot{\mathrm{C}}} & 0 & 0 \\ 0 & 0 & 0 & \beta_{\dot{\mathrm{C}}=\ddot{\mathrm{O}}} & 0 & 0 & \alpha'_{\dot{\mathrm{O}}} & 0 \\ 0 & \beta_{\dot{\mathrm{C}}=\dot{\mathrm{O}}} & 0 & 0 & 0 & 0 & 0 & \alpha'_{\dot{\mathrm{O}}} \end{bmatrix}. \tag{6.96}$$

Here the parameters β's are defined as before [see Eq. (6.34)]. And α''s are defined by

$$\alpha'_X \equiv \alpha_X + 2\beta_{\mathrm{BB}} \cos ka, \tag{6.97}$$

for X = A, G, T, C, where a is the interspacing between the nucleotide bases, β_{BB} is the π-electron hopping integral between the nucleotide bases shown in Fig. 6.18, and the wavenumber k is defined as $-\frac{\pi}{a} \leq k \leq \frac{\pi}{a}$. On the other hand, the Hückel matrix $\mathbf{H}'_{\mathrm{SP}}$ for the backbone chain is defined by

$$\mathbf{H}'_{\mathrm{SP}} = \begin{bmatrix} \alpha_{\ddot{\mathrm{O}}} & \beta_{\ddot{\mathrm{O}}-\ddot{\mathrm{O}}} & \beta_{\dot{\mathrm{O}}-\ddot{\mathrm{O}}} & \beta_{\ddot{\mathrm{O}}-\ddot{\mathrm{O}}} & \beta_{\dot{\mathrm{P}}-\ddot{\mathrm{O}}} & \beta_{\mathrm{SP}} e^{ika} \\ \beta_{\ddot{\mathrm{O}}-\ddot{\mathrm{O}}} & \alpha_{\ddot{\mathrm{O}}} & \beta_{\dot{\mathrm{O}}-\ddot{\mathrm{O}}} & \beta_{\ddot{\mathrm{O}}-\ddot{\mathrm{O}}} & \beta_{\dot{\mathrm{P}}-\ddot{\mathrm{O}}} & 0 \\ \beta_{\dot{\mathrm{O}}-\ddot{\mathrm{O}}} & \beta_{\dot{\mathrm{O}}-\ddot{\mathrm{O}}} & \alpha_{\dot{\mathrm{O}}} & \beta_{\dot{\mathrm{O}}-\ddot{\mathrm{O}}} & \beta_{\dot{\mathrm{P}}=\dot{\mathrm{O}}} & 0 \\ \beta_{\ddot{\mathrm{O}}-\ddot{\mathrm{O}}} & \beta_{\ddot{\mathrm{O}}-\ddot{\mathrm{O}}} & \beta_{\dot{\mathrm{O}}-\ddot{\mathrm{O}}} & \alpha_{\ddot{\mathrm{O}}} & \beta_{\dot{\mathrm{P}}-\ddot{\mathrm{O}}} & \beta_{\mathrm{PS}} \\ \beta_{\dot{\mathrm{P}}-\ddot{\mathrm{O}}} & \beta_{\dot{\mathrm{P}}-\ddot{\mathrm{O}}} & \beta_{\dot{\mathrm{P}}=\dot{\mathrm{O}}} & \beta_{\dot{\mathrm{P}}-\ddot{\mathrm{O}}} & \alpha_{\dot{\mathrm{P}}} & 0 \\ \beta_{\mathrm{SP}} e^{-ika} & 0 & 0 & \beta_{\mathrm{PS}} & 0 & \alpha_{\ddot{\mathrm{O}}} \end{bmatrix}, \tag{6.98}$$

where all parameters are kept as before [see Eqs. (6.54) and (6.55)] and β_{SP} (β_{PS}) represents the π-electron hopping from the sugar site to the phosphate site (from the phosphate site to the sugar site). These parameters are also shown in Fig. 6.18, respectively.

Here, we note the following: Compare Eq. (6.94) with Eq. (6.90). We can recognize that the k-dependence on the diagonal components comes from the onsite potentials of α's [Eq. (6.97)]. And that on the off-diagonal components comes from the off-diagonal terms for the π-electron hopping between the sugar and the phosphate such as $\beta_{\mathrm{SP}} e^{ika}$ in Eq. (6.97). Thus, the k-dependence on the components in the Hückel matrix of Eq. (6.94) comes from the ladder geometry of the single strand of DNA with an infinite repetition of the nucleotide base of A, G, T, and C. This is the reason why we studied the simple decorated ladder model in the previous section. Such models can be seen as a simplification of the Hückel model for the real single strand of DNA.

6.4.2.3 Energy bands of π-electronic states of a single strand of DNA

We show an example of the energy bands of the single strand of DNA for the simplest case of having all the same nucleotides of A, G, T, and C. This means that the chain structures are given artificial ones such as $\cdots$ AAA $\cdots$, $\cdots$ GGG $\cdots$, $\cdots$ TTT $\cdots$, and $\cdots$ CCC $\cdots$. If we can solve the eigenequation of the system by directly using MATHEMATICA, we can obtain the energy bands shown in Fig. 6.19.

Next, let us see the effect of the π-electron hopping integral β_{BB} between the nucleotide groups on the energy band structure. This is shown in Fig. 6.20 for taking some change of values.

The effect of the π-electron hopping integral β_{BS} between the nucleotide base and the sugar on the energy band structure is also shown in Fig. 6.21.

From the above results, we can find interesting tendencies of the energy band structure:

(1) As is shown in Fig. 6.19, if the strength of β_{BB} is increased, there appears *a transition from semiconductor to semimetal*. And if β_{BB} is very small, then the energy bands for the nucleotide base become *dispersionless*, while the energy bands for the backbone chain have dispersion. This means that *the states in the nucleotide base are very localized within the nucleotide base, while the states in the backbone chain are extended*. This is evident from (A)–(C).
(2) As is shown in Fig. 6.20, if β_{BS} is small, then the channel through the nucleotide base and the channel through the backbone chain are separated into two independent linear chain systems. Therefore, there are two independent classes of the energy bands: One is for the π-electron hopping through the sugar-phosphate backbone chain. Another is for the π-electron hopping through the nucleotide bases.

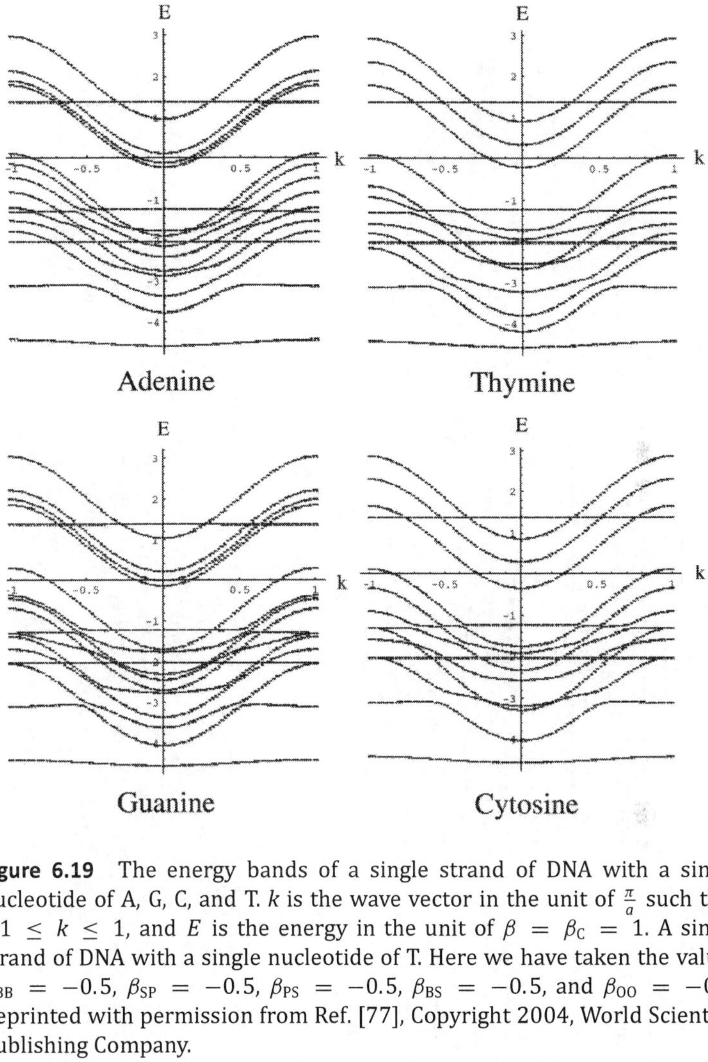

Figure 6.19 The energy bands of a single strand of DNA with a single nucleotide of A, G, C, and T. k is the wave vector in the unit of $\frac{\pi}{a}$ such that $-1 \leq k \leq 1$, and E is the energy in the unit of $\beta = \beta_C = 1$. A single strand of DNA with a single nucleotide of T. Here we have taken the values $\beta_{BB} = -0.5$, $\beta_{SP} = -0.5$, $\beta_{PS} = -0.5$, $\beta_{BS} = -0.5$, and $\beta_{OO} = -0.5$. Reprinted with permission from Ref. [77], Copyright 2004, World Scientific Publishing Company.

(3) As is shown in Fig. 6.21, if β_{SP} [or β_{PS}] is small, then the energy bands for the sugar-phosphate backbone chain become *dispersionless*. This means that *the states are localized within the sugar-phosphate group*.

(4) As is shown in Figs. 6.19-6.21, the level repulsion in the HOMO bands is much stronger than that in the LUMO bands.

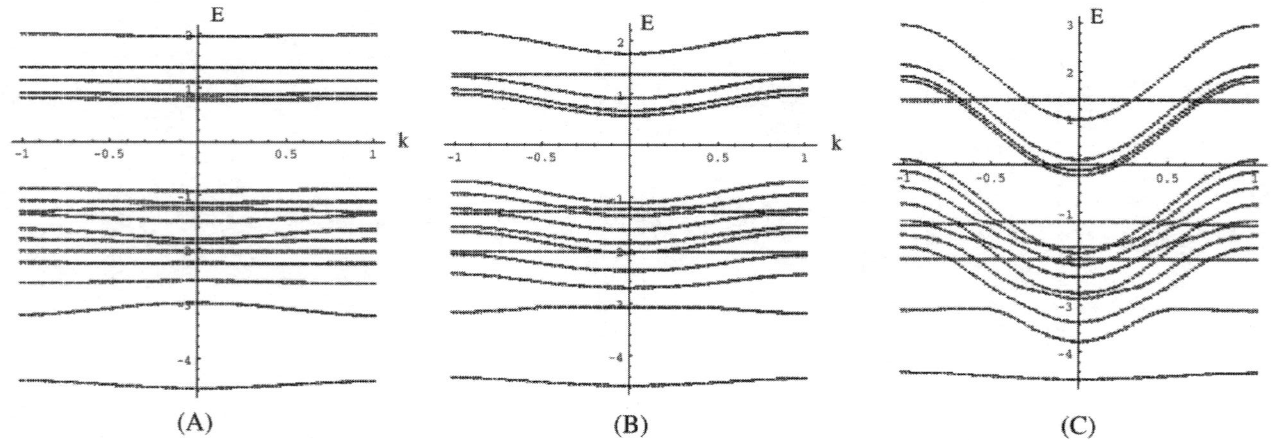

Figure 6.20 Effect of the π-electron hopping integral β_{BB} between the nucleotide groups on the energy band structure. This is shown for the single strand of DNA with a single nucleotide A. k is the wave vector in the unit of $\frac{\pi}{a}$ such that $-1 \leq k \leq 1$, and E is the energy in the unit of $\beta = \beta_C = 1$. Here we have taken the values (A) $\beta_{BB} = -0.01$, $\beta_{SP} = -0.5$, $\beta_{PS} = -0.5$, and $\beta_{BS} = -0.5$; (B) $\beta_{BB} = -0.1$, $\beta_{SP} = -0.5$, $\beta_{PS} = -0.5$, and $\beta_{BS} = -0.5$; (C) $\beta_{BB} = -0.5$, $\beta_{SP} = -0.5$, $\beta_{PS} = -0.5$, and $\beta_{BS} = -0.5$, where $\beta_{OO} = -0.5$. This shows that the larger the value of β_{BB}, the smaller the energy gap of the energy levels of the system, which means the system becomes more metallic. On the other hand, we see that the essential magnitude of the energy gap comes from the one for the nucleotide base. Therefore, the nucleotides can be seen as mother materials for giving the energy gap of the single strand of DNA. Reprinted with permission from Ref. [77], Copyright 2004, World Scientific Publishing Company.

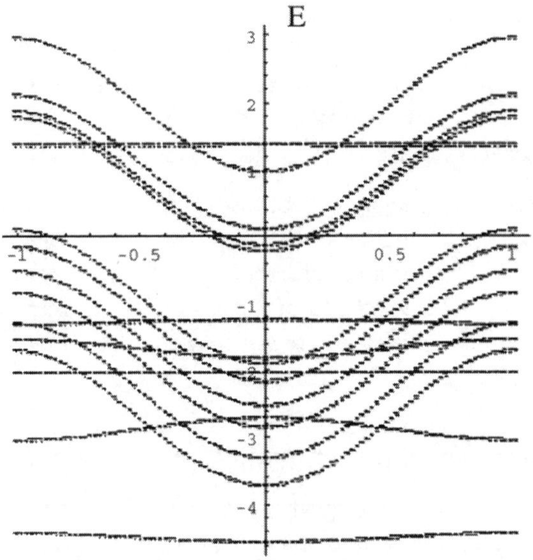

Figure 6.21 Effect of the π-electron hopping integral β_{BS} between the nucleotide base and the sugar on the energy band structure. This is shown for the single strand of DNA with a single nucleotide A. k is the wave vector in the unit of $\frac{\pi}{a}$ such that $-1 \leq k \leq 1$, and E is the energy in the unit of $\beta = \beta_{\text{C}} = 1$. Here we have taken the values $\beta_{\text{BB}} = -0.5$, $\beta_{\text{SP}} = -0.5$, $\beta_{\text{PS}} = -0.5$, and $\beta_{\text{BS}} = -0.1$, where $\beta_{\text{OO}} = -0.5$. This is compared with the case (C) in Fig. 6.20, where $\beta_{\text{BS}} = -0.5$. This shows that the smaller the value of β_{BS}, the more the energy levels of the nucleotide base and the sugars become independent from each other. Reprinted with permission from Ref. [77], Copyright 2004, World Scientific Publishing Company.

We can summarize the above results as follows:

(i) Suppose that $\beta_{\text{BS}} = 0$, maintaining $\beta_{\text{BB}} \neq 0$ and $\beta_{\text{PS}} \neq 0$. Then the system consists of two independent channels for the electronic transport: One channel is through the nucleotide base groups and the other is through the sugar-phosphate backbone chain.

(ii) Suppose that $\beta_{\text{BB}} = 0$, maintaining $\beta_{\text{BS}} \neq 0$ and $\beta_{\text{PS}} \neq 0$. Then the states in the nucleotide groups become localized within each base group, while the states in the backbone chain are extended through the entire chain.

(iii) Suppose that $\beta_{\mathrm{PS}} = 0$, maintaining $\beta_{\mathrm{BS}} \neq 0$ and $\beta_{\mathrm{BB}} \neq 0$. Then the states in the sugar-phosphate groups become localized within each sugar-phosphate group, while the states in the nucleotide groups are extended through the entire stacking of the nucleotide bases.

Thus, we can conclude that *the electronic spectrum of the system can be affected by both the π-electron hoppings through the nucleotide base groups and that through the sugar-phosphate backbone chain as a whole.* In other words, *the electronic states in the single-stranded DNA are very sensitive to the geometry change of the system.*

6.4.3 *Electronic Properties of a Double Strand of DNA*

6.4.3.1 π-Electronic energy spectrum of decorated ladder model for a double strand of DNA

Before considering the electronic spectrum of the double strand of DNA, we can apply the same decorated ladder model defined in Fig. 6.15 (as well as Fig. 6.16) to this system as well. This can be regarded as a prototype model for the more realistic models for the double strand of DNA.

For our propose here, the equations that we need have been presented in Section 6.4.1, by applying the Hückel method to this model. To use the equations, we need to assign the Hückel parameters for this model. We only consider the simplest case that all nucleotides are identical. In this case, we define the following:

$$\beta_{n+1,n} = \beta'_{n+1,n} = -\beta (= \beta'),$$

$$\alpha_{n,n} = \alpha'_{n,n} = 0,$$

$$v_n = v'_n = -v (= 0) \quad \text{at odd (even) } n \tag{6.99}$$

Applying this to Eq. (6.64), we have

$$M \equiv M_1 M_2 = \begin{bmatrix} \frac{E^2}{\beta\beta'} - \frac{\beta}{\beta'} & \frac{E}{\beta} & -\frac{vE}{\beta\beta'} & 0 \\ -\frac{E}{\beta} & -\frac{\beta'}{\beta} & \frac{v}{\beta} & 0 \\ -\frac{vE}{\beta\beta'} & 0 & \frac{E^2}{\beta\beta'} - \frac{\beta}{\beta'} & \frac{E}{\beta} \\ \frac{v}{\beta} & 0 & -\frac{E}{\beta} & -\frac{\beta'}{\beta} \end{bmatrix}, \tag{6.100}$$

Now we take $tr\,M$ and $tr\,M^2$ and obtain

$$\frac{1}{2}tr\,M = \frac{E^2}{\beta\beta'} - \left(\frac{\beta}{\beta'} + \frac{\beta'}{\beta}\right), \tag{6.101}$$

$$\frac{1}{2}tr\,M^2 = \frac{E^4}{\beta^2\beta'^2} - \left[2\left(\frac{1}{\beta^2} + \frac{1}{\beta'^2}\right) - \frac{v^2}{\beta^2\beta'^2}\right]E^2 + \frac{\beta^2}{\beta'^2} + \frac{\beta'^2}{\beta^2}, \tag{6.102}$$

Substituting these into Eq. (6.80), we obtain

$$x_\pm = \frac{E^2}{\beta\beta'} - \left(\frac{\beta}{\beta'} + \frac{\beta'}{\beta}\right) \pm \left|\frac{vE}{\beta\beta'}\right|, \tag{6.103}$$

where $x_\pm \equiv 2\cos 2k$. Solving Eq. (6.103) for the energy E, we obtain the energy bands

$$E_+^{(\pm)}(k) = \frac{1}{2}\left(v \pm \sqrt{v^2 + 4\left(\beta^2 + \beta'^2 + 2\beta\beta'\cos 2k\right)}\right),$$

$$E_+^{(\pm)}(k) = \frac{1}{2}\left(-v \pm \sqrt{v^2 + 4\left(\beta^2 + \beta'^2 + 2\beta\beta'\cos 2k\right)}\right), \tag{6.104}$$

where $E_\pm^{(+)}(k)[E_\pm^{(-)}(k)]$ stands for the upper (lower) energy bands in channels $\pm$, respectively. These are shown in Fig. 6.22.

We now observe that the spectrum is symmetric around $E = 0$ and $E_\pm^{(+)}(k) > 0 > E_\pm^{(-)}(k)$ with respect to k. The band gap at the center of the spectrum is given by

$$\Delta_0 = E_-^{(+)}(\pi/2) - E_+^{(-)}(\pi/2) = \sqrt{v^2 + 4(\beta - \beta')^2} - v, \tag{6.105}$$

and the overlapping of the two bands $E_+^{(+)}(k)$ and $E_-^{(+)}(k)$ is given by

$$\begin{aligned}\Delta_1 &= E_+^{(+)}(\pi/2) - E_-^{(-)}(0)\\ &= v + \frac{1}{2}\left[\sqrt{v^2 + 4(\beta - \beta')^2} - \sqrt{v^2 + 4(\beta + \beta')^2}\right],\end{aligned} \tag{6.106}$$

Let us consider the above energy bands at half filling, $E_{\mathrm{F}} = 0$. In this model, there is always a gap at $E = 0$ since $\Delta_0 > 0$ except the case of $\beta = \beta'$, contrary to the former case of the ladder structure. However, there are several distinct characters of the system classified in terms of Δ_0 and Δ_1.

(1) For $v = 0$, there are two doubly degenerate energy bands separated by the gap $\Delta_0 = 2|\beta - \beta'|$ with the band width $\Delta_1 = (\beta + \beta' - |\beta - \beta'|)/2$. So the spectrum is semiconductor-like around $E = 0$. Hence, one expects that the system is a *semiconductor* at half filling.

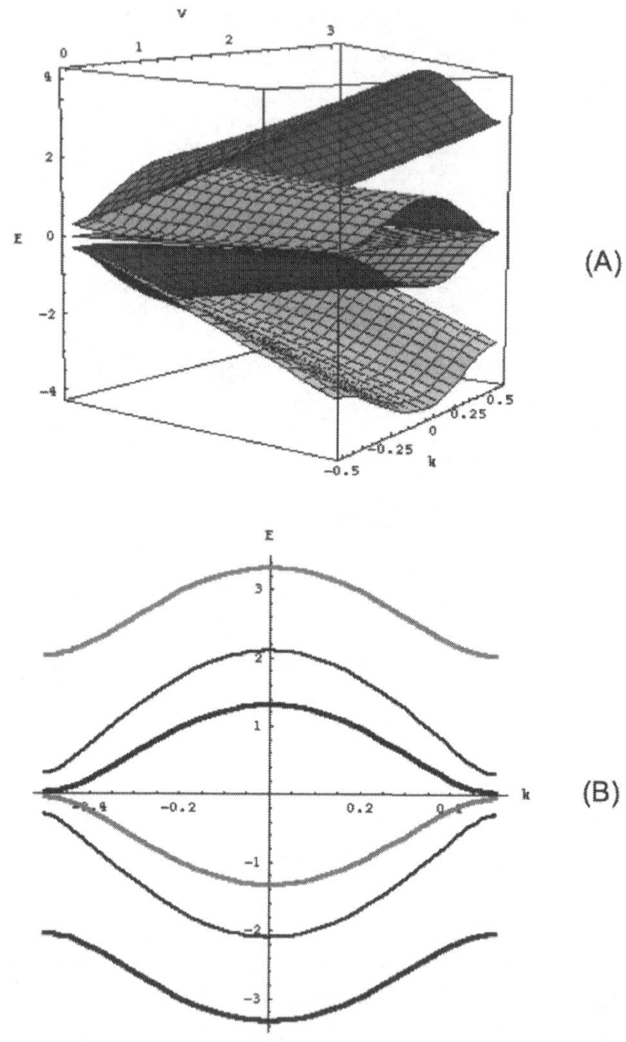

Figure 6.22 Energy bands of the decorated ladder model for a double strand of DNA. (A) The energy bands as a function of v ranging v from 0 to 3. (B) Snapshot of the energy bands when $v = 0$ (thin curves) and $v = 2$ (thick gray and black). Here v is the π-electron hopping integral between the base and the sugar-phosphate. The horizontal axis k is the wave vector in the unit of π/a such that $-0.5 \leq k \leq 0.5$, where a is the interspacing between the base groups. The vertical axis E is the energy in the unit of $\beta_C = \beta = 1$. The horizontal plane (i.e., $E_F = 0$) is the position of the Fermi energy. We have taken the values $\alpha = 0$, $\beta = 1.2$, and $\beta' = 0.9$.

(2) Suppose that

$$v_c \equiv \frac{2\beta\beta'}{\sqrt{\beta^2 + \beta'^2}}. \qquad (6.107)$$

For $0 < v \leq v_c$, the gap Δ_0 persists to exist and the overlapping Δ_1 is *semimetal-like* (i.e., $\Delta_1 < 0$) with the separation of v. Thus, the spectrum is again semiconductor-like at $E = 0$. Hence, one expects that the system is a *semiconductor* at half filling.

(3) For $v > v_c$, another gap Δ_1 appears so that $\Delta_1 > 0$ as well as the gap $\Delta_0 > 0$. However, the spectrum is once again semiconductor-like around $E = 0$. Hence, one expects that the system is a *semiconductor* at half filling.

In this way, if $\beta - \beta' \neq 0$, then the system is maintained to be a semiconductor for the whole range of v, in a strict sense that there is always a gap at $E = 0$ in this model. However, unless $v = 0$, 1 find from Eq. (6.105) that if $|(\beta - \beta')/v| \ll 1$, then

$$\Delta_0(v) \approx 2|\beta - \beta'|^2/v \approx 0 \ll \Delta_0(0) = 2|\beta - \beta'|. \qquad (6.108)$$

This tendency is enhanced as v is increased. Therefore, even if the center gap exists in the spectrum, it can be relatively small and the system can behave like a metal at finite temperature in the limit

$$\Delta_0 \ll k_B T < \Delta_0(0), \qquad (6.109)$$

where k_B is the Boltzmann constant and T is the temperature of the system. Thus, the system can show a kind of *semiconductor–metal transition* as v is increased. It is not an absolute insulator–metal transition in the strict sense, but it might be regarded as an effective one mediated at finite temperature.

6.4.3.2 Electronic conduction of decorated ladder model for a double strand of DNA

Let us now consider the electronic property of DNA. Evans and Gergely [47] as well as Meade and Kayyem [235] found peculiar properties of DNA a long time ago. The former found that the electric conduction of DNA follows the famous formula of semiconductor. The latter found that the electronic conduction in the double strand of DNA is enhanced by the formation of the double strand of DNA as large as 10^4, which is larger than that of the single strand of DNA.

Let us now consider the electronic property of the π-electrons in the decorated ladder model for the double strand of DNA [202, 203]. For this purpose, since we have the conduction formula for the current I with the band gap Δ

$$I = I_0 e^{-\frac{\Delta}{k_B T}}, \tag{6.110}$$

where I_0 is a constant, we just plague our expression of the center band gap into the energy gap Δ. This produces the current I_s when the double strand is broken to two separate chains such as

$$I_s = I_0 e^{-\frac{\Delta_0(0)}{k_B T}} = I_0 e^{-\frac{2|\beta-\beta'|}{k_B T}}. \tag{6.111}$$

This is the case when $\nu = 0$. Similarly, when the double strand is formed such that $\nu > 0$, we obtain the conduction formula for the current I_d as

$$I_d = I_0 e^{-\frac{\Delta_0(\nu)}{k_B T}} = I_0 e^{-\frac{2|\beta-\beta'|+\sqrt{\nu^2+4|\beta-\beta'|^2}-\nu}{k_B T}} \tag{6.112}$$

In this way, we finally have the ratio I_d/I_s as

$$\frac{I_d}{I_s} \propto e^{-\frac{\Delta_0(\nu)}{k_B T}} = I_0 e^{-\frac{\sqrt{\nu^2+4|\beta-\beta'|^2}-\nu}{k_B T}} \tag{6.113}$$

This was first obtained by the author [202, 203]. We show this result in Fig. 6.23. This result may explain the intriguing property of DNA conduction found by Meade and Kayyem [235].

6.4.3.3 Energy bands of π-electronic states of a double strand of DNA

When we consider the π-electronic states of the *double strand* of DNA going beyond considering the π-electronic states of the *single strand* of DNA, we may use the method introduced in Section 6.4.2. While a single strand of DNA is constructed such as

$$5' - \cdots \mathrm{AAA} \cdots - 3', \quad 5' - \cdots \mathrm{TTT} \cdots - 3',$$

$$5' - \cdots \mathrm{GGG} \cdots - 3, \quad 5' - \cdots \mathrm{CCC} \cdots - 3',$$

a double strand of DNA is constructed such as

$$\begin{matrix} 5' - \cdots \mathrm{AAA} \cdots - 3' & 5' - \cdots \mathrm{GGG} \cdots - 3' \\ 3' - \cdots \mathrm{TTT} \cdots - 5'' & 3' - \cdots \mathrm{CCC} \cdots - 5' \end{matrix} \tag{6.114}$$

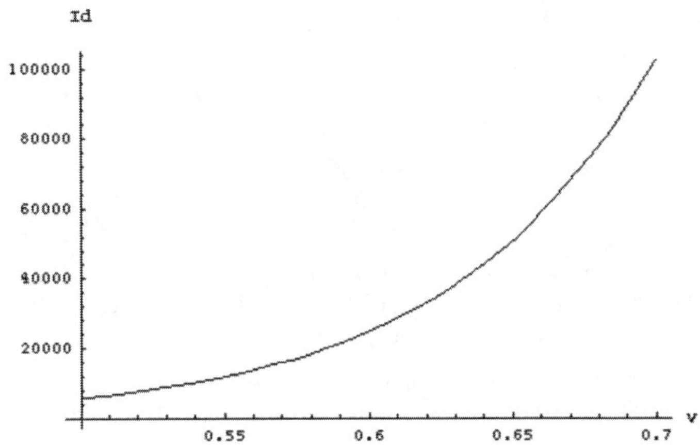

Figure 6.23 The relative current I_d/I_s. It is shown for $\Delta_0(0) = 2.4$ eV and $k_\mathrm{B}T = 0.026$ eV. For the range of $v \approx 0.45 - 0.64$ eV, it is of the magnitude of 10^4. The vertical axis means the relative current I_d in the unit of I_s, while the horizontal axis means v. Reprinted with permission from Ref. [202], Copyright 1997, World Scientific Publishing Company.

Therefore, if we want to apply the Hückel model to the double strand of DNA, we have to enlarge the Hückel matrices such as

$$\mathbf{H}_{\mathrm{AT}} = \begin{bmatrix} \mathbf{H}'_{n\mathrm{A}} & \mathbf{V}_{\mathrm{AT}} \\ \mathbf{V}_{\mathrm{TA}} & \mathbf{H}'_{n\mathrm{T}} \end{bmatrix} \quad \text{or} \quad \mathbf{H}_{\mathrm{GC}} = \begin{bmatrix} \mathbf{H}'_{n\mathrm{G}} & \mathbf{V}_{\mathrm{GC}} \\ \mathbf{V}_{\mathrm{CG}} & \mathbf{H}'_{n\mathrm{C}} \end{bmatrix}, \tag{6.115}$$

where $\mathbf{H}'_{nX}$ (for $X =$ A, G, T, C) are defined by Eq. (6.94). $\mathbf{V}_{\mathrm{AT}}$ and $\mathbf{V}_{\mathrm{GC}}$ are the π-electron hopping resonance parameters through hydrogen bonds between the nucleotide bases. Since $\mathbf{H}'_{n\mathrm{A}}$ is a 16×16 matrix and $\mathbf{H}'_{n\mathrm{T}}$ is a 14×14 matrix, $\mathbf{H}'_{\mathrm{AT}}$ is a 30×30 matrix. And since $\mathbf{H}'_{n\mathrm{G}}$ is a 17×17 matrix and $\mathbf{H}'_{n\mathrm{C}}$ is a 14×14 matrix, $\mathbf{H}'_{\mathrm{GC}}$ is a 31×31 matrix. Therefore, $\mathbf{V}_{\mathrm{AT}}$ becomes a 16×14 matrix and $\mathbf{V}_{\mathrm{GC}}$ a 17×14 matrix.

Using the numbering in Fig. 6.11 and considering the paring in hydrogen bonds, we see that only a few matrix elements can survive such that $h_{1,19}$, $h_{2,23}$, and $h_{10,24}$ are nonzero for $\mathbf{V}_{\mathrm{AT}}$ and $h_{1,20}$, $h_{10,24}$, and $h_{11,25}$ are nonzero for $\mathbf{V}_{\mathrm{GC}}$. So we can define as

$$\begin{aligned} (\mathbf{V}_{\mathrm{AT}})_{ij} &= v_{1,19},\ v_{2,23},\ v_{10,24}, \\ (\mathbf{V}_{\mathrm{GC}})_{ij} &= v_{1,20},\ v_{10,24},\ v_{11,25}. \end{aligned} \tag{6.116}$$

We can take them all the same as v.

We omit the calculation since it goes beyond the scope of this chapter. Even though we can carry out such calculations, as long as we take into account the energy levels around their HOMO and LUMO near the Fermi level of the π-electrons, we believe that a property similar to that of the simple decorated ladder model of the double strand of DNA will appear. This means that the energy gap of the double strand of DNA is dominated by the gap of the mother materials of nucleotides of DNA and that the main feature of electronic property of DNA turns out to be semiconductor. Very recently, this expectation has been supported by Tsuburaya, Sakamoto, and Mizoguchi [27, 157]. So this problem seems very interesting. But we would like to leave this for future study.

6.5 Further Problems on DNA

6.5.1 *Screw Symmetry of DNA*

Since the discovery of DNA, we all know that DNA is a double helix—the double helical structure of DNA in Fig. 6.24 [2]. However, to make a theory for DNA conduction, we always have to assume that DNA does not have any twist and that DNA is a ladder or a ribbon. Otherwise, it is difficult to calculate the electronic properties of DNA. So there appears a problem: what is the effect of the twists of DNA on the system?

Long ago right after the discovery of the DNA double helix structure [2], early theoretical chemists studied this helical twist problem, using the Hückel theory [197, 199, 200, 217, 218].

In their point of view, since such effect results from the twist of DNA on the overlap integrals between the adjacent base groups along the double helix, they modified the π-electron hopping to include the twist angle ϕ of DNA. This introduces the system to a kind of commensurability between the period of the twist of DNA with angle ϕ and that of the repetition of the base groups along the DNA double helix. They called it the *screw symmetry* [197]. For example, we define the g-fold screw symmetry of the DNA helix system as follows: Suppose that there are g base groups per turn, where one turn defines the unit cell with pitch g of base groups.

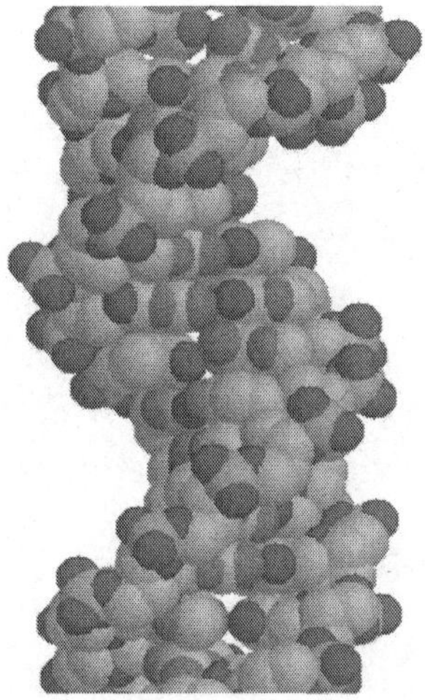

Figure 6.24 Double helix of DNA.

Suppose that there are P repetitions of base groups such that $N = gP$, where P is the number of complete turns of the helix. Then, the n-th base group along the axis of helix can alternatively be expressed as the q-th base group in the p-th turn (cell) by the commensurability relation [197]

$$n = gp + q, \tag{6.117}$$

where p, q are some integers satisfying the relation

$$-\frac{P}{2} < p \leq \frac{P}{2}, -\frac{g}{2} < q \leq \frac{g}{2}. \tag{6.118}$$

From this, we can simply define the twist angle

$$\phi = 2\pi \frac{P}{L} = \frac{2\pi}{g}. \tag{6.119}$$

Hence, the size of the unit cell of the system can be enlarged to impose the Bloch theorem:

$$c_{n+g} = e^{ikl} c_n, \tag{6.120}$$

where $l \equiv ga$ with a being the inter-distance between the base groups. This reduces the Brillouin zone as

$$-\frac{\pi}{ga} \leq k \leq \frac{\pi}{ga}. \tag{6.121}$$

And correspondingly, the energy bands $E_j(k)$ for the j-th molecule can be given by

$$E_j(k) = \alpha_j + \beta_j \cos\left(ka + \frac{2\pi s}{g}\right), \tag{6.122}$$

where $s = 1, \cdots, g - 1$ [199, 200, 217, 218]. This result is not affected by the change in the π-electron hopping parameters. Therefore, the energy bands degenerate at the zone boundary such that there are no band gaps between them [216].

To consider this effect, let us consider the π-electron hopping, $\beta_{B_i.B_j} \equiv \beta_{ij}$, between the i-th and j-th base groups. We can simply assume the approximation

$$\beta_{ij} = kS_{ij}, \tag{6.123}$$

where k is some constant and S_{ij} is the overlap integral of the π-orbitals between the i-th and j-th base groups. If we can use this assumption, then the twist angle ϕ between the base groups enters the expression through the overlap integral such as

$$S_{ij}(\phi) = S_{ij}^{\sigma\sigma} \cos^2\theta - S_{ij}^{\pi\pi} \sin^2\theta. \tag{6.124}$$

Here $S_{ij}^{\sigma\sigma}$ is the σ-bond-like overlap integral and $S_{ij}^{\pi\pi}$ is the π-bond-like overlap integral of the π-orbitals of the i-th and j-th base groups. And θ is related to the twist angle ϕ by

$$\cos\theta = \frac{\sqrt{a^2 - (r\phi)^2}}{a}, \tag{6.125}$$

where a is twice the interspacing between the sugar-phosphate group along the backbone of the helix and r is the distance from the center axis of the helix to the position of the base group.

This can be derived from geometrical consideration [236]. From Eq. (6.123), we find

$$\beta_{ij}(\phi) = \beta_{ij}^{\sigma\sigma} \cos^2\theta - \beta_{ij}^{\pi\pi} \sin^2\theta. \tag{6.126}$$

Here $\beta_{ij}^{\sigma\sigma}$ is the σ-bond-like resonance integral and $\beta_{ij}^{\pi\pi}$ is the π-bond-like resonance integral of the π-orbitals of the i-th and j-th base groups.

When the commensurability between the twist and the change of the Hückel parameters is realized, band gaps may appear at the zone boundary. Therefore, the match between the pitch of DNA and the overlapping of π-orbitals is crucial for having band gaps in the spectrum. However, at that time they assumed that the system is a linear object of DNA, like a line segment for the foundation of the theory. Therefore, in their theory even though they included the twist of the system, it was just like a torsion of the linear system. Hence, not much difference appeared in the results from those of the standard linear chain system.

6.5.2 *Effect of Half-Twist of DNA*

We will discuss some peculiar mathematical problems related to DNA, which are important for further study. We encounter a problem when we think of the half-turns of the ladder or ribbon models. This has not been treated in the early theoretical works discussed earlier. Consider the ladder structure of DNA in Fig. 6.15. For our previous calculations, we always assumed the periodic boundary condition. This can be described as follows: Suppose that both ends of DNA are open ends with numbering 1 and N such as in Fig. 6.15. If we impose the periodicity to make a ring geometry of DNA, then we have the periodicity of N for the amplitudes of wave function as

$$a_n = a_{n+N}, \quad b_n = b_{n+N}, \tag{6.127}$$

or equivalently

$$\begin{bmatrix} a_{n+N} \\ b_{n+N} \end{bmatrix} = \begin{bmatrix} 1 & 0 \\ 0 & 1 \end{bmatrix} \begin{bmatrix} a_n \\ b_n \end{bmatrix}, \tag{6.128}$$

where a_n are the amplitudes of wave function for helix A and b_n for helix B as in Eq. (6.62). This assures that when the unit cell of base group is simple enough, we hold the Bloch theorem for the system such as

$$a_{n+1} = e^{ika} a_n, \quad b_{n+1} = e^{ika} b_n, \tag{6.129}$$

where a stands for the spacing between the nucleotide base groups and $-\pi/a \leq k \leq \pi/a$ with $k_j = 2\pi j/N$ for $-N/2 \leq j \leq N/2$. This means that there are $2N$ states for two channels, where for each channel there are N states [202, 203].

Let us now introduce a half-turn to the DNA ladder or DNA ribbon to make a ring. Hence, we encounter the problem of topology of DNA [236] such that we have a *Möbius strip* of DNA ladder or ribbon [237, 238]. In this case, Eq. (6.127) is no longer correct. The boundary condition must be given by

$$a_n = b_{n+N}, \quad b_n = a_{n+N}, \tag{6.130}$$

or equivalently

$$\begin{bmatrix} a_{n+N} \\ b_{n+N} \end{bmatrix} = \begin{bmatrix} 0 & 1 \\ 1 & 0 \end{bmatrix} \begin{bmatrix} a_n \\ b_n \end{bmatrix}. \tag{6.131}$$

Obviously, it is identical to the following:

$$c_n = c_{n+2N}, \tag{6.132}$$

which means that there is no difference between a_n for chain A and b_n for chain B for the system of the Möbius strip. In other words, the double chain of length N in the ladder is enlarged to the single chain of length $2N$ so that the short range interactions $V_{n,n} \equiv v_n$ through bridges between the neighbor chains become the long distance interaction $V_{n,n+N}$ between site n and site $n+N$ in numbering.

Since we now have the Bloch theorem:

$$c_{n+1} = e^{ika} c_n, \tag{6.133}$$

we also have

$$c_{n+N} = e^{ikNa} c_n, \tag{6.134}$$

where $k = \frac{2\pi p}{2Na}$ with p being a nonzero integer between $-N$ and N such that $-\frac{\pi}{a} \leq k \leq \frac{\pi}{a}$. Hence, we have a complex factor in front of the long-range interaction $V_{n,n+N}$ as

$$V_{n,n+N} = e^{ikNa} V_{n,n} \equiv e^{ikNa} v_n. \tag{6.135}$$

This produces the onsite potential. The Möbius strip has only one surface (i.e., no surface direction). Therefore, we should be able to define only one direction for the hopping. After one going around the strip, the direction of the hopping becomes reverse. This breaks the *parity* of the system such that the energy bands break parity as well.

To see this point, we compare the energy bands of the simplest ladder model where the Hückel parameters are all the same [202, 203]. For the ladder without half-turn, we have

$$E_{\pm}(k) = 2\beta \cos ka \pm v, \tag{6.136}$$

where $k = \frac{2\pi p}{Na}$ for p integers of $-\frac{N}{2} \leq p \leq \frac{N}{2}$ such that $-\frac{\pi}{a} \leq k \leq \frac{\pi}{a}$. And for the ladder model with one half-turn, we have

$$E_{\pm}(k) = 2\beta \cos ka + ve^{ikNa} = 2\beta \cos ka + (-1)^p v. \qquad (6.137)$$

Here $k = \frac{2\pi p}{2Na}$ for p nonzero integers of $-N \leq p \leq N$ such that $-\frac{\pi}{a} \leq k \leq \frac{\pi}{a}$; since the chain consists of $2N$ sites in this case, p cannot be zero. Hence, parity can be broken. This situation may discriminate between the right-handed half-turn and the left-handed half-turn.

This indicates that there are two energy bands for two channels of bonding and antibonding states in the ladder model without half-turn. On the other hand, there is only one energy band for one channel of bonding in the ladder model with one half-turn. But in the latter, the energy level rapidly alternates with respect to $k = \frac{2\pi p}{2Nl}$, whether p is even or odd. If $p =$, even then the level lies in the upper band, and if $p =$ odd, then the level lies in the lower band. This again produces seemingly two bands for two channels. In other words, the wave vector k in the upper band shifts that in the lower band by the amount of $\Delta k = \frac{\pi}{Na}$. This peculiar aspect is not so well known, however.

The ladder model with many half-turns can be generalized in the same way. We will leave these problems for future study, for intelligent people like you.

6.5.3 *Aperiodicity of DNA*

The generalization of the system with aperiodic arrangement of DNA is also important. As mentioned earlier, the spectrum of DNA may consist of many band gaps when the commensurability of the pitch of twist of DNA and the period of arrangement of base groups is realized. This problem goes back to the era of Schrödinger—What is life? [239].

Let me recall my personal academic experience. In the autumn of 1984, near the end of my graduate school at the Osaka University, Shechtman [240] made the big discovery of quasicrystal. Earlier in 1983, Kohmoto, Kadanoff, and Tang [241] formulated the theory of Fibonacci lattice. I carefully read and studied these topics. I had to graduate unsuccessfully to receive PhD, however. Hopefully, I became a company man.

When I was in the company, I found a Japanese edition book—Gödel, Esher, Bach—written by D. Hofstadter [242], which later became a famous book in the world. He drew back to Schrödinger's work of aperiodic very long polymer like DNA and asked how to calculate the electronic properties of DNA. He then called the problem *Schrödinger's dream*. Since then, I became aware of the problem and I started calling it Schrödinger's dream as well. Considering both quasicrystal and DNA, I found that if we can formulate the theory of aperiodic lattices, we can apply the theory to DNA and protein. So when I quit the company, I decided to go to the USA and applied to many academic institutions, including the University of Utah. Thankfully, I became a graduate student of the University of Utah in the autumn of 1986, when Prof. Mahito Kohmoto was there, who had helped me to go there.

At the end of the first year, during the Thanksgiving Day holiday, I had an idea and applied the trace map method of Fibonacci lattice of Kohmoto, Kadanoff, and Tang [241] to optical multilayers. It worked out well. Next week, I discussed my idea with Prof. Bill Sutherland, although my spoken English was poor at that time. He discussed Kohmoto later. The following week, he came to me and told me that the idea was very nice. Next week, I received from him a reprint of a paper on optical Fibonacci multilayers—Localization in Optics: Quasiperiodic Media—which was published in 1987 [243].

After Kohmoto left Salt Lake, Utah, for Tokyo, Japan, in 1988, I studied quasiperiodic systems under Sutherland. In the spring of 1989, luckily we found the way to go beyond the Fibonacci lattice. We succeeded to generalize the theory of Fibonacci lattice to that of an arbitrary quasiperiodic lattice. Some of the essential parts of this work had been published by Sutherland [244] in the summer of 1989. After writing my thesis [245] for about a year, I received my PhD. Since my job search in the USA failed, I returned to Japan in the autumn of 1990. My work [246] was published in 1991.

The main results of the work were published in Ref. [247, 248]. I showed the spectrum of the one-dimensional quasiperiodic lattices (i.e., chains), which are made by only two species of atoms A and B,

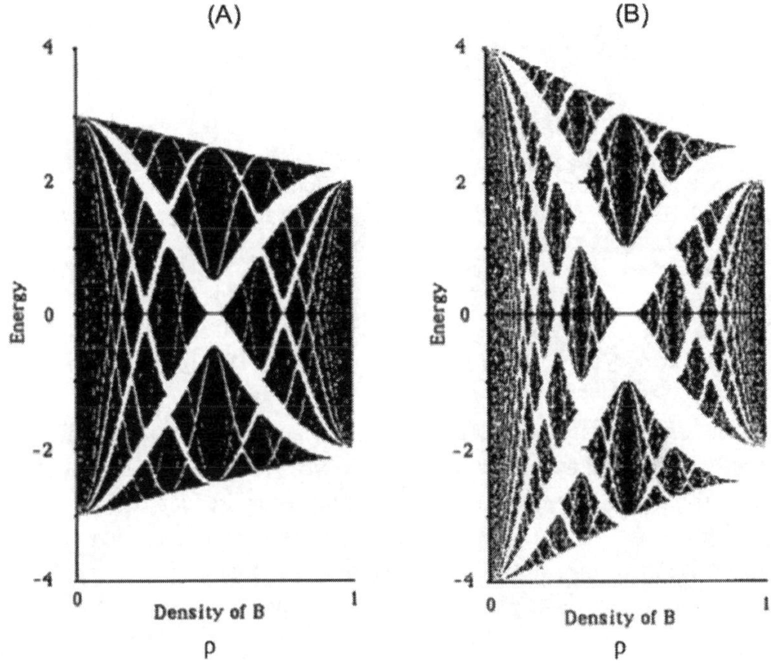

Figure 6.25 The energy spectrum of one-dimensional quasiperiodic lattices. The spectrum is shown for (A) $T_a/T_b = 1.5$ and (B) $T_a/T_b = 2$, where T_a (T_b) is the hopping integral from A atom to B atom (from B atom to A atom). The horizontal line indicates the density ρ of B atoms in the chain, which varies from 0 to 1. The vertical line indicates the energy E. Reprinted with permission from Ref. [247]. Copyright 1992, AIP Publishing LLC.

such that the ratio between A and B, λ, becomes fractional; hence, the density of B in the quasi periodic chain is given by $\rho = 1/(1+\lambda)$. In this case, the system shows the quasiperiodicity characterized by the fractional ρ. Although I skip the details since they are given long ago, I would like to show the results in Fig. 6.25.

Later in 1995, I applied the above method of one-dimensional quasiperiodic lattices to the electronic spectrum of ternary alloys of $AB_{1-x}C_x$ [249]. In this model, the local coordination in the system is assumed to be three types of bonds—AB ($\equiv$ AC), CA, and BA bonds—which are defined to be long, short, and shorter bonds, as described in Fig. 6.26. According to this, the hopping integrals can be given by

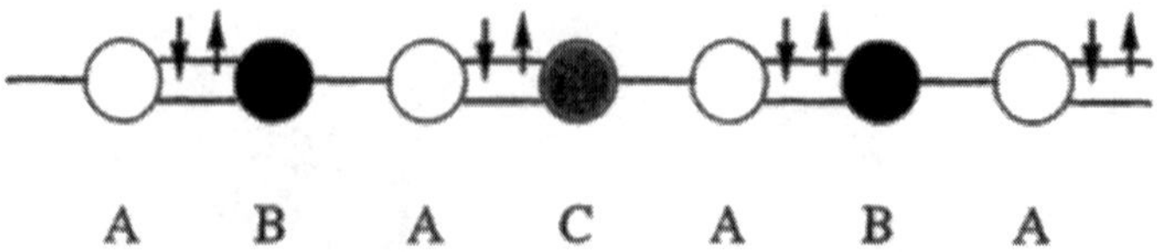

Figure 6.26 The one-dimensional quasiperiodic ternary alloys of $AB_{1-x}C_x$. Reprinted with permission from Ref. [249]. Copyright 1995, Physical Society of Japan.

T_a, T_b, and T_c, respectively. We assume the onsite potential $V_n = 0$. And to adjust with the three types of bonds, we assume that $0 < T_a < T_c < T_b$.

When x is fractional or irrational, we can assume that the system of one-dimensional quasiperiodic ternary alloys of $AB_{1-x}C_x$ is nothing but one-dimensional quasiperiodic lattices of two types of dimers, AB and AC, where the density of dimer AC is given by x. Hence, we can apply the theory of one-dimensional quasiperiodic lattices to this alloy type as well. In this model, the interspacing a_x is given by

$$a_x = (1 - x)a_{AB} + xa_{AC}, \tag{6.138}$$

where a_{AB} (a_{AC}) means interspacing for pure AB (AC) alloy.

We would like to show the results in Fig. 6.27. We see that when the difference of hopping integrals T_b and T_c is not so large, the center energy gap Δ_x for $AB_{1-x}C_x$ follows the fitting law

$$\Delta_x = (1 - x)\Delta_{AB} + x\Delta_{AC}, \tag{6.139}$$

where Δ_{AB} (Δ_{AC}) is the center energy gap for pure AB (AC) alloy.

The above aspects of the theory seem to have the same features as those of aperiodic DNA. Can one apply this theory to DNA? Roché [205] and Yamada [207] have already made such efforts. However, they applied the aperiodicity of DNA to either the one-dimensional chain models or the ladder models as simple as the linear chain. So we need to apply the theory of one-dimensional quasiperiodic alloys (or lattices) to the more realistic models of DNA double helix structure such as the Hückel theory of DNA with base pairing of nucleotides of DNA. If this can be done, then Schrödinger's dream comes true [250]. Any kind of generalization is welcome.

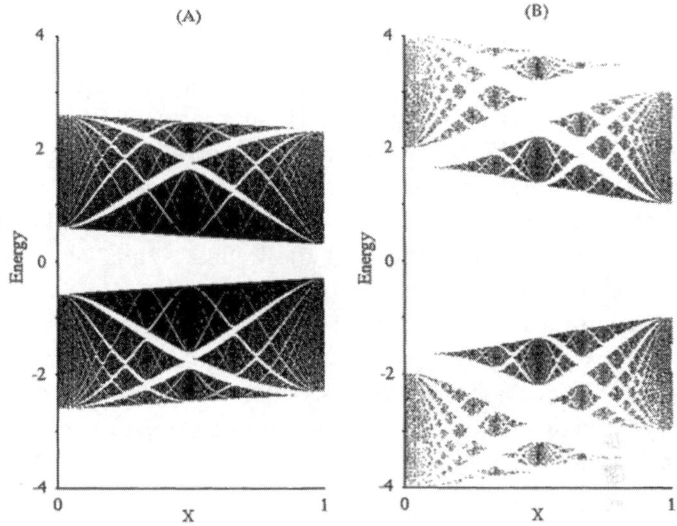

Figure 6.27 The energy spectrum of one-dimensional quasiperiodic alloys of $AB_{1-x}C_x$ with x ranging from 0 to 1. The hopping integrals are taken as (A) $(T_a, T_b, T_c) = (1, 1.4, 1.2)$ and (B) $(T_a, T_b, T_c) = (1, 3, 2)$. Reprinted with permission from Ref. [249]. Copyright 1995, Physical Society of Japan.

6.6 Conclusion

We have discussed the Hückel theoretical approach for π-electrons in DNA to know the electronic properties of DNA. We have reviewed it from our point of view.

In Section 6.1.2, we summarized the quantum chemistry for atoms frequently appearing in biology. In Section 6.1.3, we discussed the π-electronic configurations in organic molecules in biology. In Section 6.2.1, we introduced the useful and famous Hückel theory in short. In Section 6.2.2, we derived the values of Hückel parameters using the famous empirical formulas known in quantum chemistry for our purpose. In Section 6.2.3, we applied the Hückel theory to calculate the energy states of benzene for an example. In Section 6.3.1, we calculated the electronic states of single nucleotide bases of A, G, T, and C. In Section 6.3.2, we calculated the electronic states of a single sugar-phosphate group. In Section 6.3.3, we calculated the electronic states of a single nucleotide with a single sugar-phosphate group. In Section 6.4.1, we summarized the decorated ladder models of a single or double strand of DNA. In Section 6.4.2, we calculated the electronic properties of the single strand of DNA, using the decorated ladder models. In Section 6.4.3, we calculated the electronic properties of the double strand of DNA, using the decorated ladder models. From the energy gap obtained, we calculated the conductivity of the double strand of DNA. In Section 6.5, we discussed the effect of twists of DNA.

Although much of the results were obtained in the past, we find it valuable that we surveyed what we have done so far from the point of view of the decorated ladder model of DNA. If this piece of work helps your understanding of DNA, then we will be very happy. Good luck to you all.

Acknowledgements

I would like to thank Dr. Kenji Mizoguchi for giving me the chance to summarize my work. I also thank Dr. Hiroaki Yamada for sharing time to study DNA problems. And I would like to thank Dr. Eugeni B. Starikov for sharing with us information on DNA research and for communicating as a friend for a long time. Finally, I would like to express my special thanks to my wife, Kazuko Iguchi, for her financial support as well as encouragement. Unless her cordial and continuous support, I would not have been able to continue my research because I have been an independent researcher and playing the role of a house husband for as long as 18 years.

Part III

Metal Ion–Incorporated DNA

Chapter 7

Basic Properties of *M*-DNA

Kenji Mizoguchi and Hirokazu Sakamoto
Department of Physics, Tokyo Metropolitan University, Hachioji, Tokyo, 192-0397 Japan
mizoguchi@phys.se.tmu.ac.jp, sakamoto@phys.se.tmu.ac.jp

7.1 Introduction

As confirmed in Chapter 2, natural DNA is a semiconducting material, and as discussed in Chapter 4, the electrical conduction of DNA with macroscopic scale is mainly carried out by the proton conduction of water molecules contained in the double-helical DNA. One of the attempts to inject electronic charge carriers into DNA is the insertion of a divalent metal ion between the bases of a base pair of DNA, as first reported by Lee *et al.* [16], which is introduced in Section 1.2.2. After that, Rakitin *et al.* measured the I–V characteristics of Zn-DNA with nanotechnology and found the disappearance of the characteristic feature of semiconducting materials: the threshold voltage against the current flow, which has been observed in B-DNA [17]. Thus, they concluded that Zn-DNA was a promising candidate for conducting nanowires. Following their report, many researches on Zn-DNA and other *M*-DNAs have been

DNA Engineering: Properties and Applications
Edited by Kenji Mizoguchi and Hirokazu Sakamoto

ISBN 978-981-4669-46-7 (Hardcover), 978-981-4669-47-4 (eBook)
www.panstanford.com

carried out experimentally [19, 24, 25, 27, 28, 46, 93–95, 144, 157, 175, 176, 251–253] and theoretically [21–23]. Does Zn-DNA really have metallic nature? The valence of Zn ions in Zn-DNA is known to be 2+, that is, charge transfer from Zn to DNA does not occur. Nanotechnological measurement of electrical conduction [17] is affected by several factors, which are sometimes difficult to control. Thus, it is meaningful to study the electronic states of *M*-DNA by means of other techniques, especially non-contact techniques, such as optical absorption and magnetic property. On the other hand, does *M*-DNA have the same electronic structure even with the samples synthesized by different methods? These points on *M*-DNA will be described in Section 7.2. In this chapter, UV/Vis absorption and the magnetic property of *M*-DNA will be discussed.

7.2 UV/Vis Absorption of *M*-DNA

7.2.1 *Sample Preparation for UV/Vis Study*

Salmon sperm DNA and single-stranded (SS) 30mer-DNAs (poly(dG), poly(dA), poly(dC), and poly(dT) with 30 base pairs) were provided by Wako Pure Chemical Ind. Ltd. and Hokkaido System Science Co. Ltd., respectively. As introduced in Section 1.2.2, Lee *et al.* reported the cooperative synthesis of *M*-DNA with the excess divalent metal ions of Zn, Ni, or Co in a tris-HCl buffer solution at pH = 8 [16]. We started *M*-DNA synthesis following the method described in their report and successfully prepared Zn-DNA, Ni-DNA, and Co-DNA. We need to prepare Mn-DNA as a characteristic system with a magnetic probe of Mn, but pH = 8 is not suitable for Mn-DNA synthesis since Mn forms manganese oxide at this pH. Thus, we tried to use a mixture of a pure water solution of DNA and another water solution of metal dichlorides (MCl_2).

The relative molar concentration of the divalent metal ions is required to be 10 times higher than that of DNA, which is crucial for *M*-DNA synthesis. The divalent metal ions with much higher concentration than that of Na^+ mono-cations cause the ion exchange for the counter-cations of PO_4^- anions from Na^+ to divalent ions M^{2+}. After stirring for 30 min, the excess ethanol at −20°C is poured into the mixture solution of M-DNA and MCl_2. The M-DNA precipitate is collected by a disposable wooden chopstick and is washed out thoroughly to remove unreacted MCl_2 with pure ethanol, in which DNA is insoluble (the ethanol precipitation method). After that, the obtained M-DNA is dried in air and stored in a refrigerator. Optical absorption measurements are carried out in an aqueous solution of the stored M-DNA with concentrations around 0.05 mM/L. This technique is, however, not applicable for Fe-DNA because the precipitate is insoluble in water. Instead of the ethanol precipitation method, dialysis is used to remove excess $FeCl_2$ from an aqueous solution of Fe-DNA and $FeCl_2$ for optical measurements. A Shimadzu UV-1700 spectrometer is used for optical absorption measurements.

7.2.2 M-*DNA Solution*

7.2.2.1 Comparison with B-DNA

As discussed in Section 2.2, natural DNA has the semiconducting electronic state with the energy gap of $\approx$4 eV. DNA reacts with divalent metal ions in solutions and forms a complex M-DNA, as schematically shown in Fig. 7.1. If the insertion of metal ions changes the electronic states, the optical absorption spectra will show the

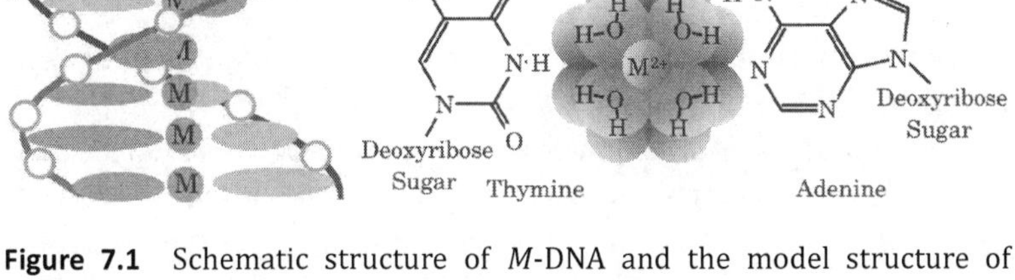

Figure 7.1 Schematic structure of M-DNA and the model structure of thymine-M^{2+}-adenine of M-DNA prepared with the ethanol precipitation method. Metal ions are hydrated by water molecules and form ionic coupling with two PO_4^- ions in DNA backbones. Reprinted with permission from Ref. [27], Copyright 2014 by the American Physical Society.

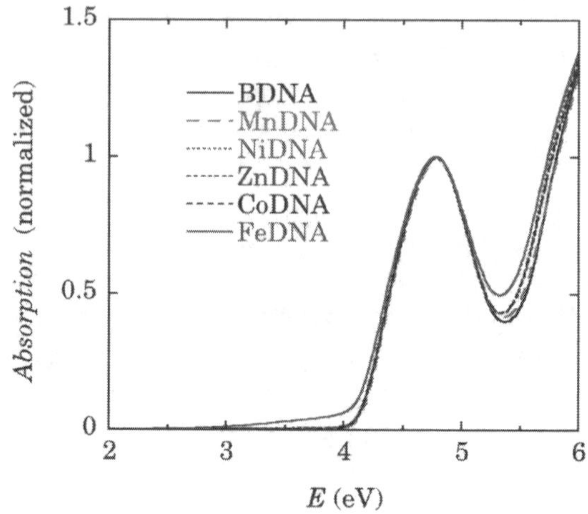

Figure 7.2 Absorption spectra in aqueous solutions of Mn-DNA, Ni-DNA, Zn-DNA, Co-DNA, and Fe-DNA, together with that of B-DNA. The peak intensity is normalized to compare the spectral shape. All the *M*-DNAs, except Fe-DNA, show almost the same spectra as B-DNA. Fe-DNA behaves differently as discussed in Section 7.2.3. Omerzu *et al.* reported redshift of the peak position up to 0.1 eV in Zn-DNA [46]. Such a large shift is not found in the *M*-DNAs, including Zn-DNA prepared with the ethanol precipitation method.

corresponding change. Figure 7.2 shows several *M*-DNAs prepared with the ethanol precipitation method except for Fe-DNA. All the spectra, except for Fe-DNA, agree fairly well with that of B-DNA. The absorption spectra in the energy range in Fig. 7.2 are known to originate in the electrons of the base pairs, as described in Section 2.2. Thus, these agreements with that of B-DNA suggest that the insertion of divalent metal ion into DNA with the ethanol precipitation method does not affect the electronic states of DNA base pairs.

Omerzu *et al.* found that the absorption spectrum of Zn-DNA prepared with the tris-HCl buffer in pH $= 8$ redshifted by $\approx$0.1 eV from that of B-DNA, which suggested a difference in the electronic states from each other. Lee *et al.* [16] and Rakitin *et al.* [17] suggested the other model structures of guanine-Zn^{+}-cytosine and adenine-Zn^{+}-thymine, as shown in Fig. 7.3. Omerzu *et al.*, however, confirmed the valence state of Zn ion in the Zn-DNA as 2+. Thus,

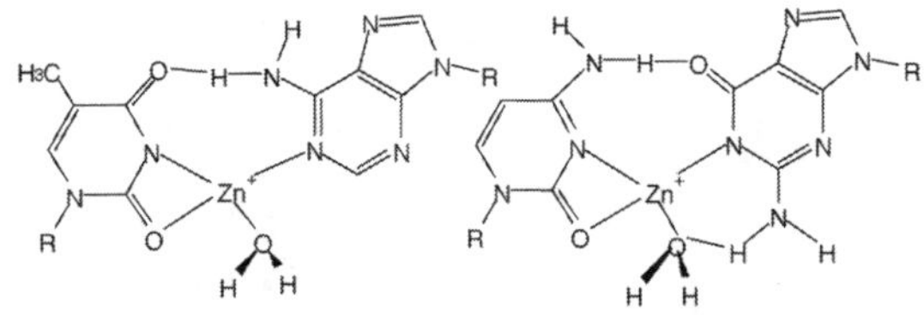

Figure 7.3 Model base pairing for Zn-DNA. The imino protons with coordination to the N3 position of thymine and the N1 position of guanine are replaced by the Zn^{2+} ion, giving rise to a charge-doped Zn^{+}. However, Omerzu *et al.* confirmed that the valence of Zn ions in the Zn-DNA prepared with tris-HCl buffer solution at pH $=$ 8 was 2+ [46]. Thus, this structure cannot be that of Zn-DNA. Reprinted with permission from Ref. [17]. Copyright 2001, American Physical Society.

the electronic states of Zn-DNA prepared with tris-HCl buffer at pH $=$ 8 have not been resolved yet. It will be shown in Sections. 7.2.3 and 7.3 that the metal ions in *M*-DNA prepared with the ethanol precipitation method are located between the bases of a base pair with hydrating water molecules and form ionic coupling with PO_4^- ions in DNA backbones, as shown in Fig. 7.1.

7.2.2.2 Suppression of the inter-π-band absorption

We cannot find any meaningful difference in the spectral forms of *M*-DNAs from those of B-DNA, except for Fe-DNA. In contrast, the quantitative study of the absorption intensity reveals interesting behavior depending on the kind of base pairing, guanine-*M*-cytosine (G-*M*-C) or adenine-*M*-thymine (A-*M*-T). The absorption intensity of A-*M*-T is definitely suppressed from that of A-T, but nothing of change is found in the intensity of G-*M*-C from that of G-C.

Quantitative change in the absorption spectrum of an aqueous DNA solution before and after the *in situ* addition of MCl_2 powder (approximately 0.5 mM of MCl_2, 10 times that of DNA concentration) to the solution has been investigated for B-DNA, SS 30mer-DNAs, and DS 30mer-DNAs. Figure 7.4 shows the suppression of the absorption spectrum in the Mn-DNA solution of approximately 0.05 mM. The dashed curve of B-DNA was taken before adding MCl_2, and the solid curve describes the spectrum of *M*-complexes after the

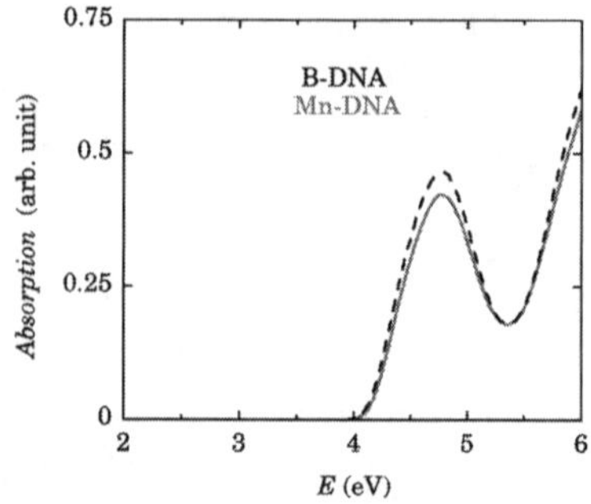

Figure 7.4 Absorption spectra in an aqueous solution of B-DNA (dashed curve) and Mn-DNA+unreacted $MnCl_2$ (solid curve) prepared by *in situ* addition of $MnCl_2$ in the B-DNA. Reprinted with permission from Ref. [27], Copyright 2014 by the American Physical Society.

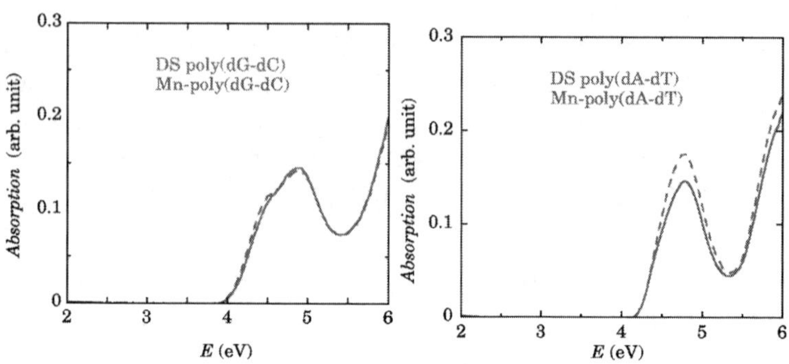

Figure 7.5 Absorption spectra in aqueous solutions of DS poly(dG-dC) and DS poly(dA-dT) (dashed curves). The solid curves represent the spectra of Mn-poly(dG-dC) and Mn-poly(dA-dT) after *in situ* $MnCl_2$ addition. In Mn-poly(dA-dT), the definite suppression of the absorption area from DS poly(dA-dT) by 15 % below 5.5 eV was observed, but no change in Mn-poly(dG-dC) was found from that of DS poly(dG-dC). Reprinted with permission from Ref. [27], Copyright 2014 by the American Physical Society.

in situ addition of MCl_2. Since $MnCl_2$ solution is almost transparent below 6 eV, the solid curve represents only the spectra of Mn-DNA.

With 30mer-DNAs, the difference between G-M-C and A-M-T is revealed, as shown in Figs. 7.5 and 7.6. The spectral area below 5.4 eV in Mn-poly(dA-dT) is reduced to $\approx$85 % of the spectrum before adding $MnCl_2$. The reduction reaches $\approx$30 % in Fe-poly(dA-dT), in which one electron has transferred from Fe^{2+} ion to the base pair,

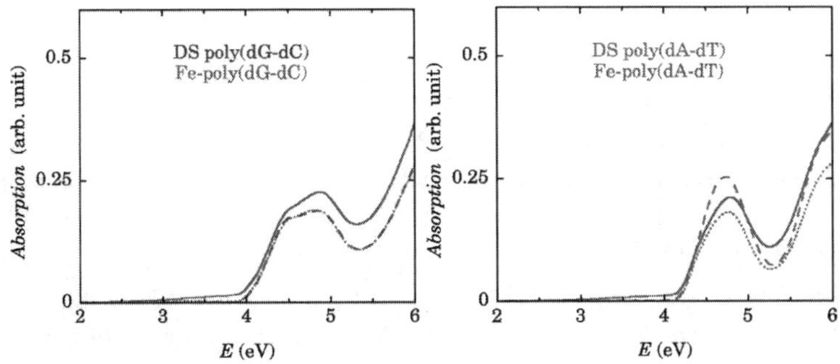

Figure 7.6 Absorption spectra in aqueous solutions of DS Fe-poly(dG-dC) and DS Fe-poly(dA-dT), which are measured 30 min after adding $FeCl_2$ to DS poly(dG-dC) and DS poly(dA-dT), shown by solid curves. Dashed curves represent the spectra before adding $FeCl_2$. Dotted curves represent the difference between the spectra shown by solid curves and the spectra of unreacted $FeCl_2$, and $FeCl_3$ as a reference spectrum for Fe^{3+}, which corresponds to the intra-gap absorption of Fe-DNA below 4 eV (refer to Section 7.2.3 for detail), which is suppressed by $\approx$30 % in Fe-poly(dA-dT). In Fe-poly(dG-dC), the dashed curve overlaps entirely with the dotted curve. Reprinted with permission from Ref. [27], Copyright 2014 by the American Physical Society.

resulting in Fe^{3+}, which gives the intra-gap absorption below 4 eV, as will be described in Section 7.2.3. It is interesting to know that SS (single stranded) 30mer-DNA solutions are also studied with the *in situ* addition of $FeCl_2$ and that the complex of Fe and SS 30mer-DNA also provides the intra-gap absorption, but no suppression of the absorption of 30mer-DNA [27]. The findings are summarized as follows:

(1) The suppression occurs only in the A-M-T complex, but not in the G-M-C.
(2) The suppression was observed both in M = Mn and Fe, under the presence of A-T base pairs.
(3) SS 30mer-DNAs with $FeCl_2$ produce the intra-gap absorption caused by Fe^{3+} ions, which means the formation of A-Fe-A, T-Fe-T, etc., but no suppression of the inter-π-band absorption was observed.

Here, two important points should be stressed: (1) The required condition for the suppression of inter-π-band absorption is the presence of the A-T base pair. (2) Fe^{3+} and generally M^{2+} are able to bind arbitrary pairs of bases, A-M-A, A-M-G, A-M-C, etc. This supports the model that the metal ions are located between the bases of a base pair and bind PO_4^- ions on the DNA backbones by the Coulomb attraction force to form a duplex, as shown in Fig. 7.1. In natural DNA, the stabilizing energy of the hydrogen bonds keeps the duplex structure of DNA against the repulsive Coulomb interaction between the PO_4^- ions on the opposite DNA backbones. The metal ions centered in the DNA duplex strongly stabilize the two DNA backbones via the Coulomb attractive interaction along $PO_4^- - M^{2+} - PO_4^-$.

The mechanism of suppression has no connection with the charge injection to the LUMO band from the metal ions, because no charge injection occurs in Mn-DNA, in which the suppression occurs. As a possibility, the suppression of the inter-π-band absorption is concerned with the decoupling or enhancement of the intra-A-T-base pair interaction by the insertion of a large hydrated metal ion. A related fact is that the peak energies of the inter-π-band absorption of poly(dA) and poly(dT) in Fig. 2.3 overlap each other, but those of poly(dG) and poly(dC) are separated by 0.5 eV, comparable with the full width of the absorption band, 0.5 eV, of SS 30mer-DNAs. Suppression of absorbance in poly(dA-dT) is known as hypochromicity [255], which means diminuation of absorbance when single-stranded DNAs form double-stranded DNAs with the help of hydrogen bonds; the contrary phenomenon is known as hyperchromicity. It is also known that poly(dA-dT) shows higher hypochromicity than poly(dG-dC). Thus, the observation of hypochromicity in M-poly(dA-dT) suggests the strong contribution of the metal ion located between the bases of a base pair to the electronic states of the A-T base pair. The mechanism of the suppression only in the A-M-T complex would be a very interesting issue and there should be some interesting mechanism in it.

7.2.3 *Fe-DNA Solution*

We found that the metal ions in M-DNA kept their valence $+2$ and that there was no charge transfer from the metal ions to DNA, except for Fe-DNA. Several evidences of charge transfer from Fe^{2+} to DNA have been found: color change of the mixture of DNA and $FeCl_2$ from light green to ocher characteristic of Fe^{3+} ion, ESR signals around $g \approx 2$ characteristic of the S-state ($L = 0$) electronic state such as $S = \frac{5}{2}$ and $\frac{1}{2}$, and the magnetization curve ascribable also to the mixture of $S = \frac{5}{2}$ and $\frac{1}{2}$ (refer to Section 7.3 for detail). To investigate the relation between the emergence of Fe^{3+} and the presence of DNA (30mer homonucleotide polymers), the time evolution of the optical absorption in the following three aqueous solutions was measured.

(1) $FeCl_2$ solution to check the possibility of oxidation in air
(2) The mixture solution of $FeCl_2$ and DS poly(dG-dC)
(3) The mixture solution of $FeCl_2$ and DS poly(dA-dT)

Figure 7.7 demonstrates the time evolution of the solutions with the following typical features:

(1) No transformation of Fe^{2+} to Fe^{3+} occurred in the $FeCl_2$ solution at room temperature for 30 min, that is, $FeCl_2$ was not oxidized in this condition.

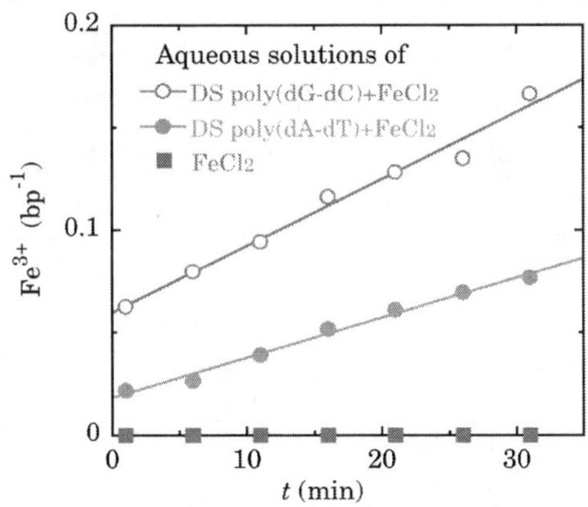

Figure 7.7 Time evolution of Fe^{3+} concentration per base pair (bp) in aqueous solutions of $FeCl_2$, DS poly(dG-dC)+$FeCl_2$, and DS poly(dA-dT)+$FeCl_2$. $FeCl_2$ is principally transformed to FeO(OH) at elevated temperatures but does not transform at room temperature for 30 min. The incremental spectra of DS 30mer-DNAs with $FeCl_2$ are well reproduced by the spectrum of $FeCl_3$, suggesting that the electronic states of Fe-30mer-DNA are ionic in character. Reprinted with permission from Ref. [27], Copyright 2014 by the American Physical Society.

(2) The number of Fe^{3+} ions increases in proportion to the time up to 30 min under the coexistence of Fe^{2+} ions and the G-C base pairs.

(3) The similar increase in Fe^{3+} to the G-C case was found under the coexistence of Fe^{2+} and A-T base pairs, but half of that of the G-C case in the increasing rate.

These findings inform us that Fe^{3+} emerges only under *the coexistence of the base pairs, any of G-C and A-T, and even the homo bases, such as poly(dA), poly(dT), poly(dG), or poly(dC), with $FeCl_2$*. Thus, when Fe^{2+} forms complexes with the base pairs, such as G-Fe-C and A-Fe-T, Fe^{2+} transforms to Fe^{3+} by transferring one electron to the base pair. The double rate of complex formation with G-C than with A-T suggests larger stabilization energy in G-Fe-C than in A-Fe-T, although the required energy to break a base pair is large in G-C compared to that in A-T, as suggested by the number of hydrogen bonds—three in G-C and two in A-T.

Figure 7.8 represents the absorption spectrum of Fe-DNA with the salmon sperm DNA. A characteristic point is the intra-gap

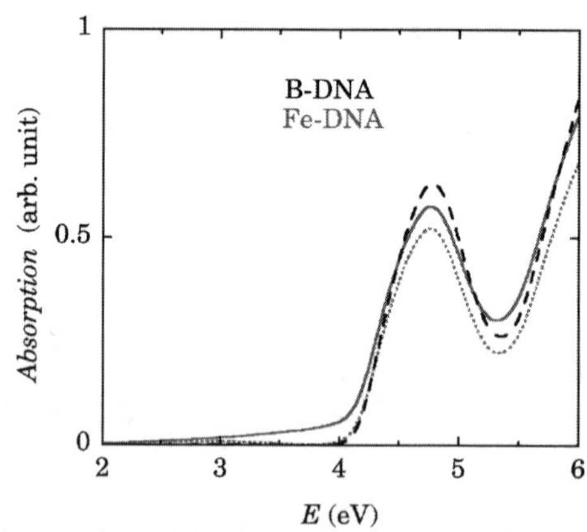

Figure 7.8 Absorption spectra in an aqueous solution of Fe-DNA prepared with the dialysis method from B-DNA and $FeCl_2$ (solid curve). Dashed curve represents the spectra of B-DNA. Dotted curve represents the difference between the spectra of Fe-DNA and $Fe^{3+}Cl_3^-$, which suggests that the new intra-gap absorption below 4 eV has the same electronic states as $FeCl_3$, that is, ionic nature. Reprinted with permission from Ref. [27], Copyright 2014 by the American Physical Society.

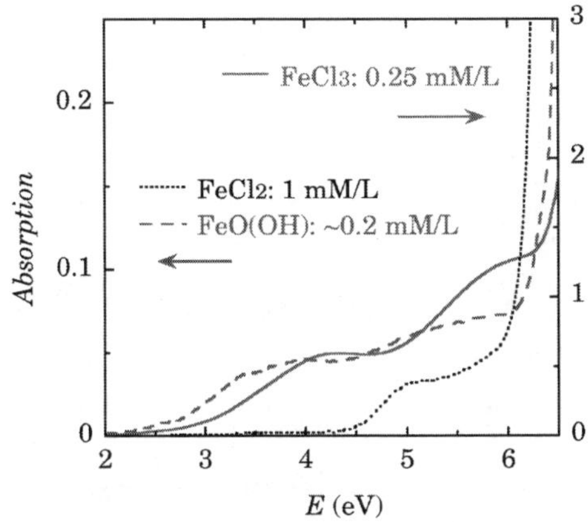

Figure 7.9 Absorption spectra in aqueous solutions of $FeCl_3$, $FeCl_2$, and FeO(OH). The absorption intensity of $FeCl_3$ is stronger than the others more than 10 times. Although the valence of Fe is +3 in both $FeCl_3$ and FeO(OH), the spectra differ qualitatively from each other, so that we can easily discriminate them. Reprinted with permission from Ref. [27], Copyright 2014 by the American Physical Society.

absorption below 4 eV. To know the electronic states of the intra-gap absorption, the absorption spectra of several materials containing Fe ions are measured in Fig. 7.9: $FeCl_2$, $FeCl_3$, and FeO(OH). $FeCl_2$ gives only a weak spectrum above 4.5 eV, which contradicts the intra-gap absorption of Fe-DNA. Both $FeCl_3$ and FeO(OH) contain Fe^{3+} ion, but the spectra are different from each other in spectral shape and intensity. $FeCl_3$ is an ionic salt, but FeO(OH) possesses more covalent nature, formed in a hot water solution of $FeCl_2$ with oxygen molecules in air. Figure 7.8 demonstrates that the spectral shape of $FeCl_3$ is consistent with the intra-gap absorption of Fe-DNA. This fact suggests that the electronic states of Fe^{3+} in Fe-DNA are ionic, which is the same as that of Fe^{3+} in $FeCl_3$. Since metal ions generally prefer to form covalent bonds with nitrogen atoms in a base pair, the metal ions are surrounded by hydrating water molecules, as modeled in Fig. 7.1. An ESR study of Mn^{2+} ions in Mn-DNA in Section 7.3.3.3 for Mn^{2+} gives the same conclusion.

The electron transfer from Fe^{2+} to the bases of a base pair requires the promotion of an electron from the HOMO d band of Fe to the LUMO π band of the bases. The HOMO band of Fe ions is

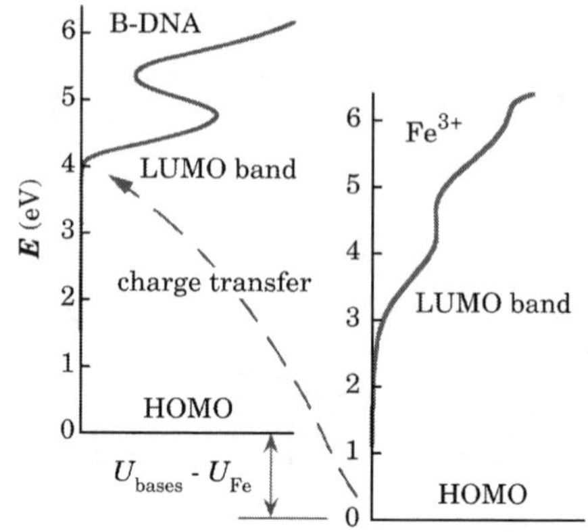

Figure 7.10 Schematic description of the promotion energy from the HOMO of Fe^{3+} to the LUMO of DNA bases. Reprinted with permission from Ref. [27], Copyright 2014 by the American Physical Society.

triply degenerated, t_{2g}, under the octahedral H_2O configuration and filled by six electrons in Fe^{2+} and five in Fe^{3+}. To roughly estimate the promotion energy, the relative potential difference between the HOMO bands of Fe and the bases should be known. We have no information about it, but we can estimate a limit of the maximum difference of 2 eV as $U_{\text{base}} - U_{\text{Fe}}$, since the LUMO band of Fe (2 eV), above HOMO in Fig. 7.9, should be higher than the HOMO level of the bases, as described in Fig. 7.10. If not, the charge will flow from the HOMO band of bases to the LUMO of Fe. Thus, the promotion energy is less than 6 eV (2 eV + 4 eV for the HOMO–LUMO gap of the bases). This energy should be supplied by the Coulomb attractive potential $U_{\text{C},\pi-\text{Fe}}$ between the LUMO π electron of the bases and Fe^{3+} in eV,

$$\frac{U_{\text{C},\pi-\text{Fe}}}{-e} = -\frac{3e}{4\pi\epsilon_0 d} \tag{7.1}$$

where $e = 1.6 \times 10^{-19}$ C is the electron charge and d is the distance between the π electron and the Fe^{3+} ion. Assuming the distance $d = 0.6 \times 10^{-9}$ m between the centers of Fe^{3+} and the hexagon of bases, we obtain $U_{\text{C},\pi-\text{Fe}}/(-e) = -7.2$ eV as the stabilizing energy for the charge transfer, which is large enough to compensate the estimated promotion energy of 6 eV. Thus, we conclude that the electron transfer reasonably occurs from Fe^{2+} to the bases and $U_{\text{base}} - U_{\text{Fe}}$ should be smaller than 2 eV. The transferred π electrons form a half-filled conduction band, which would provide metallic conduction in Fe-DNA. Unfortunately, the observed conductivity of Fe-DNA is smaller than that of B-DNA, which is caused by the strong electron–electron correlation along the base stacking.

7.2.4 M-*DNA Film*

As discussed in Section 2.2.6, DNA strands form solids and interact with each other in the film samples compared with the case of solutions. Thus, some changes from the solution spectra are expected to be induced in the film spectra. The spectrum of B-DNA in the film changed only a little, as shown in Fig. 2.7. However, a much larger effect than that of the B-DNA case is expected in the divalent metal ion–incorporated DNA, *M*-DNA. The possible mechanisms by which the band broadens in the film spectra are as follows. One is the electron transfer between the neighboring DNA strands. The other is the repulsive Coulomb interaction between the ions of the neighboring DNA strands, such as Na^+ and PO_4^-, which tends to expand the separation between the ions of the neighboring strands along the helical axis in the crystalline form of a film. Concerning the separation of the neighboring strands, the electric dipole–dipole interaction of Na^+–PO_4^- or $2PO_4^-$–M^{2+} attracts the neighboring strands. Thus, the helicity of the double strands loosens, giving rise to tiny increase in the base stacking interaction within a double-helical DNA.

In B-DNA, since the charge of PO_4^- ion is almost cancelled by the counter ion of Na^+, the expected effect is small. The observed value is as small as $\approx$0.01 eV for B-DNA, as shown in Section 2.2.6. Figures 7.11, 7.12, and 7.13 show the UV/Vis absorption spectra for six complexes.

The characteristic features of the film spectra of *M*-DNAs are as follows:

(1) Redshifts of the first absorption peak
(2) Broadening of the first absorption bands
(3) Tails into the band gaps

The redshifts of the peaks are from 0.06 eV to 0.10 eV, depending on the ion species, as listed in Table 7.1, which are much larger than

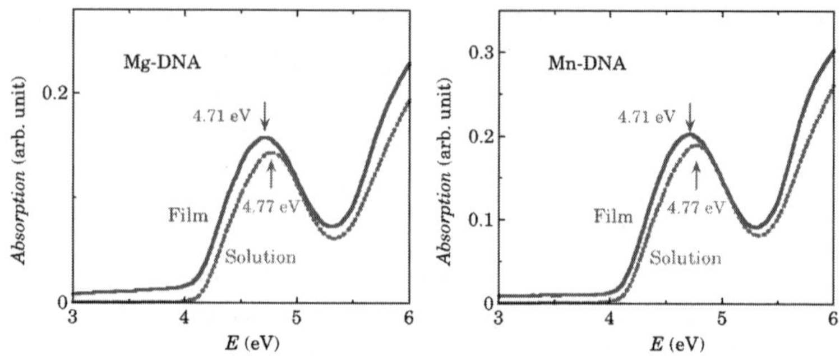

Figure 7.11 Absorption spectra of Mg-DNA (left) and Mn-DNA (right) in a film (solid curves) and in a solution (dotted curves). The π–π^* absorption peak of the bases in the film forms redshifts from that in the solutions.

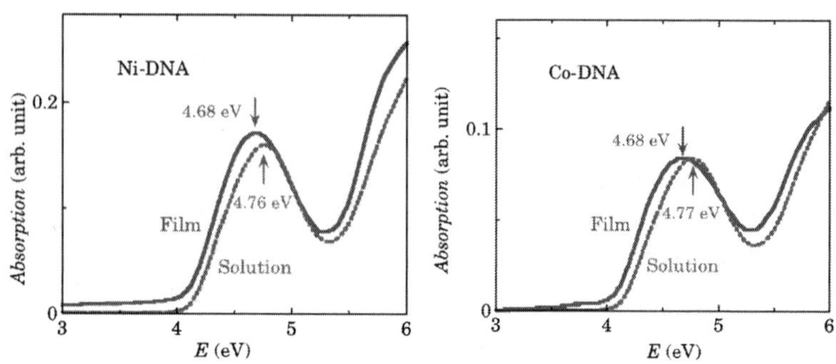

Figure 7.12 Absorption spectra of Ni-DNA (left) and Co-DNA (right) in a film (solid curves) and in a solution (dotted curves).

0.01 eV in B-DNA. These peak shifts accompany the broadening of the absorption band. That is, the increased overlap of the π-electron orbitals of the neighboring base pairs caused by the looseness of the torsion angle increases the bandwidth by the increased transfer integrals and decreases the depth of the periodic potentials of the base pairs along the double-helical axis, which generates the redshift of the absorption peaks. The quantitative analysis of the band broadening is not simple, but the rounding of the absorption peaks of the film samples is pronounced, especially in Co-DNA and Zn-DNA. These findings are rather consistent with the possible mechanisms discussed earlier.

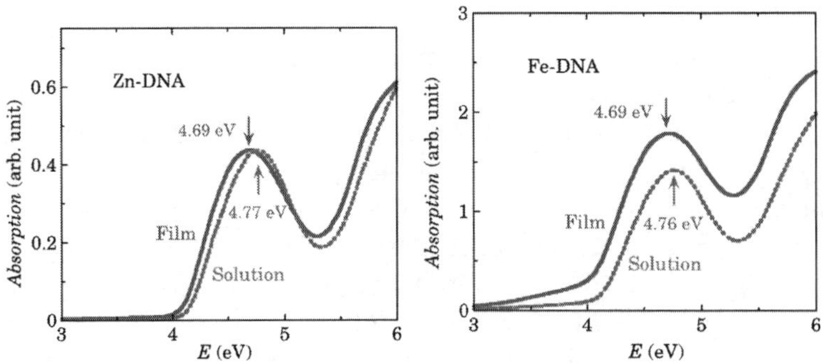

Figure 7.13 Absorption spectra of Zn-DNA (left) and Fe-DNA (right) in a film (solid curves) and in a solution (dotted curves).

Table 7.1 Peak position of the first UV/Vis absorption band of several *M*-DNA complexes in a solution and in a film, and the redshift from the solution spectra in the film. The scattering of the data is of the order of 0.01 eV

eV	Mg-DNA	Mn-DNA	Ni-DNA	Co-DNA	Zn-DNA	Fe-DNA
Solution	4.77	4.77	4.76	4.77	4.77	4.76
Film	4.71	4.71	4.68	4.67	4.69	4.69
Redshift	0.06	0.06	0.08	0.10	0.08	0.07

The tails in the band gaps are observed in several *M*-DNAs. The magnitude of the tails depends on the film thickness: the thicker, the larger. Thus, the tails, except for Fe-DNA film, originate mainly by the degree of transparency of films, which is probably dominated by the presence of nanoscale air bubbles. In the case of Fe-DNA, the absorption of Fe^{3+} starts at 3 eV, levels of around 4 eV, increases again around 5 eV, and then saturates around 6 eV, as demonstrated in Fig. 7.9.

7.2.5 *Summary of Absorption Spectra in* M-*DNA*

The UV/Vis absorption spectra of the metal ion–incorporated DNA contain basic information on the electronic states of *M*-DNA. Except for Fe-DNA, the absorption spectra of all the *M*-DNAs have qualitatively features similar to that of B-DNA. This finding is consistent with the valence of +2 for the metal ions, which suggests that no electron transfer between the bases and the metal ions has occurred in these *M*-DNAs.

In contrast, the suppression of absorbance has been observed only in A-T base pairs and not in G-C base pairs. The mechanism of suppression is not clear now, but a possibility of the suppression of the inter-π-band absorption band is concerned with the decoupling or enhancement of the intra-A-T-base pair interaction by the insertion of a large hydrated metal ion (see next section), which is in connection with the hypochromicity of the A-T base pair [254]. The mechanism of suppression only in the A-*M*-T complex is an interesting issue, and there might be some interesting mechanisms in it.

Fe-DNA shows peculiar physical properties. Fe^{2+} transforms into Fe^{3+} in Fe-DNA, as demonstrated by the emergence of the intra-gap absorption, which is reproducible from that of $FeCl_3$. Such a transformation occurs only in the presence of the complementary base pairs, such as A-T and G-C, but nothing happens without base pairs in air. The electronic states of Fe-DNA have been proposed from the absorption spectra of DNA and Fe^{3+}. These observations suggest that Fe^{3+} ion is hydrated by several water molecules, which prevent Fe^{3+} ion from forming preferable covalent bond with the nitrogen atoms of the bases of a base pair.

The absorption spectra of *M*-DNA films show some solid state effects compared with that of *M*-DNA in a solution. A larger redshift and the broadening of the absorption band compared with that of B-DNA have been observed. The mechanism of the solid state effects was discussed.

7.3 Magnetic Property of *M*-DNA

In the previous section, we found that the absorption spectrum of *M*-DNA (except for Fe-DNA) shows qualitative similarity with that of B-DNA. Mn-DNA, one of such *M*-DNAs, has an advantage over others in the magnetic study of the electronic states of *M*-DNA. The characteristic nature of the Mn ions in Mn-DNA—their location at the center of double helix and their alignment as a quasi-one-dimensional array—provides useful information about M-DNA. The magnetic property of Fe-DNA also provides useful and interesting information. Both Mn^{2+} and Fe^{3+} are good magnetic probes of S-state ion with the spin $S = \frac{5}{2}$ or $S = \frac{1}{2}$ without orbital freedom. Thus, the ESR spectrum of these M-DNAs is simple and easy to analyze for obtaining information about the structure and electronic states of M-DNAs.

7.3.1 *Sample Preparation for Magnetic Study*

M-DNA (M = Mn or Fe) is prepared from an aqueous solution of DNA (salmon or 30mer-DNA), provided by Wako Pure Chemical Ind., Ltd., Hokkaido System Science Co., Ltd., and the Ogata Materials Science Lab., with MCl_2 at the molar ratio of DNA to MCl_2 from 1:5 to 1:10. Here, note that MCl_3 with M^{3+} does not form an M-DNA composite. Excess cold ethanol at $-20°C$ is poured into the transparent DNA-MCl_2 solution, resulting in the formation of a transparent precipitate of M-DNA. The residual MCl_2 is washed out thoroughly from the obtained precipitate with pure ethanol, in which DNA is insoluble. Circular dichroism (CD) spectra of the Mn-DNA solution are examined to confirm the B-form of the double helical structure, as shown in Fig. 7.14 [94]. X-ray fluorescence analysis indicated that the ratio of phosphorus to metal is approximately 2:1 as expected for the proposed structure in Fig. 7.1. The divalent metal ion is located at the center of a base pair in place of the protons and compensates the charges of the two phosphoric anions in the two DNA backbones of a double helix, in place of two Na cations. Thus, the synthesized M-DNA contains only a trace of Na atoms. ESR spectra are taken mainly at X-band ($\approx$9.5 GHz) and Q-band ($\approx$35 GHz). The half-width at the half-height of the absorption spectrum is used as the linewidth of ESR spectra. All the samples studied in this report are of polycrystalline form.

7.3.2 M-*DNA*

The results of the previous section suggest that M-DNAs, except Fe-DNA, have electronic states similar to that of B-DNA. Magnetic properties also provide information about the electronic states.

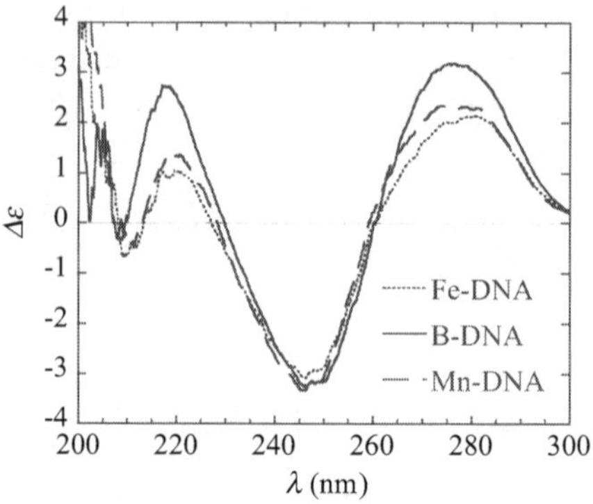

Figure 7.14 Spectra of circular dichroism (CD) in aqueous solutions of B-DNA (solid line), Mn-DNA (dashed line), and Fe-DNA (dotted line). All the three spectra represent typical double helix DNA with the large positive values at 200 nm, in contrast with that around zero for denatured, single-stranded DNA. Reprinted with permission from Ref. [94], Copyright 2007, The Physical Society of Japan.

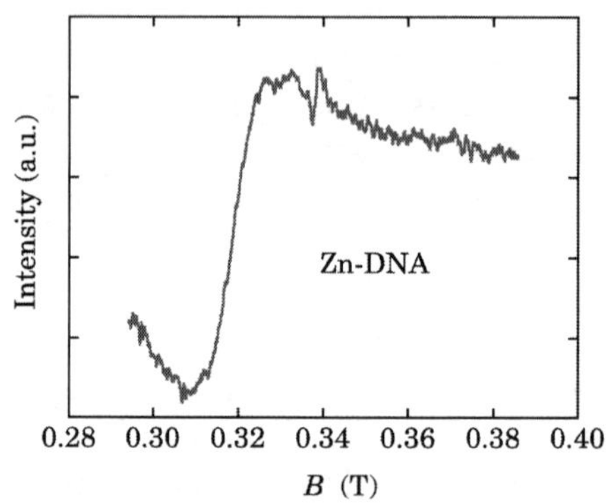

Figure 7.15 ESR derivative spectra in a dry state of Zn-DNA at X-band. At the beginning, we assigned this signal to the Cu^{2+} impurity spins because of the fewer number of spins ($\approx$0.2%/bp). However, new interpretation has been proposed, as discussed in Chapter 9. The sharp signal at $g \approx 2$ is unclear in origin.

Figure 7.15 shows the ESR spectra of Zn-DNA in a dry state. As supposed by the low signal-to-noise (S/N) ratio, the number of spins is estimated to be $\approx$0.2%/bp, which is markedly small, so that the origin of the signal is mainly impurities such as Fe and Cu.

This finding is consistent with the conclusion in Section 7.2 that the divalent metal ions, except Fe, exchange simply for two Na^+ counter ions to the two PO_4^- ions in the two DNA backbones. The findings and discussions in Chapter 9 suggest that the Fe impurity of 0.1%–0.2%/bp in B-DNA is the most probable candidate for the ESR signals in Zn-DNA, especially under dried conditions. Thus, ESR studies are not useful for us in investigating the electronic states of *M*-DNAs. Exceptions are Mn-DNA and Fe-DNA with Fe^{3+} ions, which are *S*-state ions with $S = \frac{5}{2}$ or $S = \frac{1}{2}$ spins depending on the strength of the crystal field of octahedral symmetry. In the later sections, the magnetic properties, especially ESR, of Mn-DNA and Fe-DNA are described in detail.

7.3.3 *Mn-DNA*

Figure 7.16 shows a typical spectrum of a B-form Mn-DNA at X-band (≈9.45 GHz) along with those of ruby standard. The Mn ions of the B-form Mn-DNA form a quasi-one-dimensional (Q1D) linear chain structure along the double helical axis of DNA as proposed by Lee

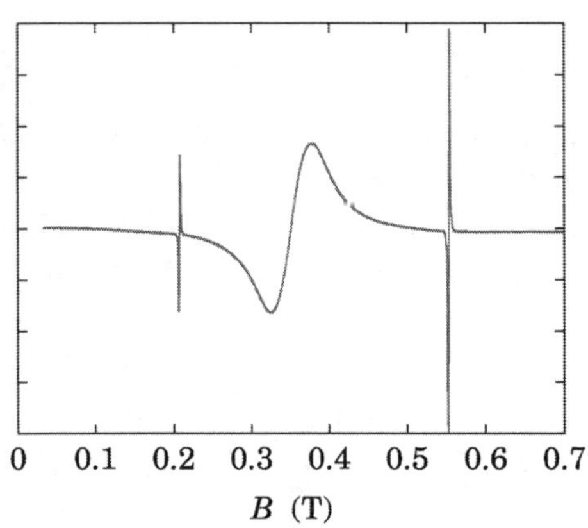

Figure 7.16 EPR derivative spectra of the B-form Mn-DNA stable under wet condition taken at room temperature at X-band. Two sharp peaks above and below the Mn spectrum are the signals from the ruby standard. The resonance appears at $g \approx 2.00$ as expected for the S state ions without angular magnetic moment.

et al. [16] (Fig. 7.1). In this section, we describe the following points in detail:

(1) The ESR linewidth of Mn-DNA is dominated by the magnetic dipolar interaction and the exchange interaction between Mn $S = \frac{5}{2}$ spins, which provides information about the structure of Mn-DNA. We can predict the ESR linewidth of the Mn ESR and compare it with the observed half-width of $\Delta H_{1/2} = 380$ G (0.038 T) at the half-height of the absorption spectra (integrated spectra of that in Fig. 7.16) [157].
(2) Two isomeric structures of DNA, the A-form and the B-form, have characteristic structures, as shown in Fig. 7.20, which reflect also the electronic states and ESR lineshapes [94].
(3) Mn ion carries $I = \frac{5}{2}$ nuclear spin, which couples with the electron spin. As a result, in some cases, the ESR spectrum of Mn ions splits into six hyperfine signals. Since the separation of such signals depends on the spatial distribution of the 5d wavefunction, that is, the bonding nature such as ionic or covalent, the separation of the hyperfine structure of Mn ions provides information about the electronic states of the metal ions and DNA [56].

7.3.3.1 ESR linewidth of Mn-ESR [157]

The magnetic dipole moment $\boldsymbol{\mu} = g\mu_B\mathbf{S}$ of an electron spin carries the magnetic fields, as schematically shown by the solid curves in Fig. 7.17. In a crystal, a magnetic moment feels the sum of the local magnetic fields produced by the neighboring magnetic moments.

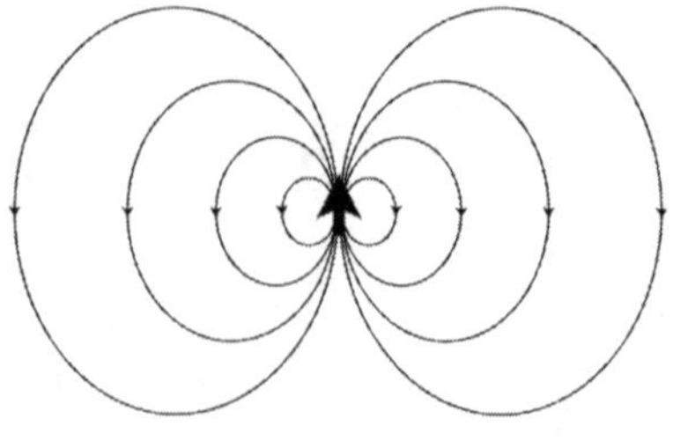

Figure 7.17 Schematic view of the magnetic field around a magnetic moment.

If the crystal is in paramagnetic state at room temperature, each magnetic moment directs at arbitrary directions. Thus, the sum of the magnetic fields by the neighboring moments is expected to be zero on the average, but the local magnetic field at the position of each moment will deviate by ΔH symmetrically around the average. For a powder sample, the angular averaged second moment $\langle \Delta H^2 \rangle$ is given in c.g.s unit by [255, 256]

$$\langle \Delta H^2 \rangle = \frac{3}{5}\gamma^2 \hbar^2 S(S+1) \sum \frac{1}{r^6}. \tag{7.2}$$

where $\gamma = 1.76 \times 10^7\ \mathrm{s}^{-1}$ is the gyromagnetic ratio for the electron, r is the nearest neighbor distance between Mn^{2+} ions, the summation is taken over the nearest neighbors, and the other symbols have their usual meanings. The second moment $\langle \Delta H^2 \rangle$ of the local field is substantially reduced by the exchange interaction between the neighboring electron spins, such as motional narrowing. The exchange narrowed half-width $\Delta H_{1/2,cal}$ at half-height is given by [257]

$$\Delta H_{1/2,cal} = \frac{\langle \Delta H^2 \rangle}{H_{ex}}, \tag{7.3}$$

where H_{ex} is the exchange field defined by $E_{ex} = g\mu_B H_{ex} \sum \mathbf{S}_i \cdot \mathbf{S}_j$ [258]. With the average distance of 3.4 Å between the Mn ions, $S = \frac{5}{2}$, and the exchange field $H_{ex} = 3k_B|\theta|/g\mu_B\sqrt{S(S+1)} = (6.0 \pm 0.9) \times 10^3$ G from a Curie–Weiss temperature of $\theta = -0.8 \pm 0.1$ K determined by the Curie–Weiss susceptibility of Mn-DNA in Fig. 7.18, we obtain $\langle \Delta H^2 \rangle = 2.34 \times 10^6\ \mathrm{G}^2$ and $\Delta H_{1/2,cal} = 390 \pm 50$ G. In the estimation of the second moment, the summation was taken only for the two nearest-neighbor Mn^{2+} ions in the Q1D linear chain, because the second nearest neighbor contributes to the second moment by only 1% to 2%. The inter-helix interaction between the Mn ions through 2 nm of the helix diameter is negligibly small, which is less than 10^{-4} of the estimated value. Thus, the good consistency of $\Delta H_{1/2,cal} = 390 \pm 50$ G with the observed ESR linewidth of $\Delta H_{1/2} = 380$ G supports the proposed structure of the B-form *M*-DNA in Fig. 7.1. In addition to this consistency, as discussed in the next section, the spectral shape of the B-form Mn-DNA is characteristic of the Q1D array of the Mn ions. In the other configurations of Mn ions such as on the surface of the double-helical DNA, the mean separation between the Mn ions inevitably

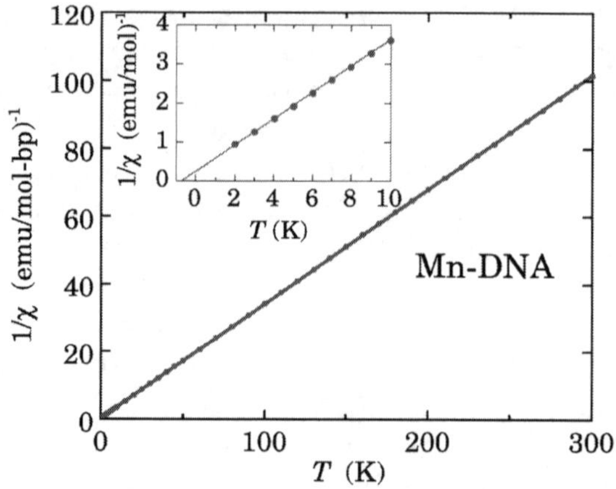

Figure 7.18 Magnetic susceptibility of Mn-DNA measured with a SQUID susceptometer at 1 T. The estimated spin concentration is about one $S = \frac{5}{2}$ spin per base pair. The inset shows the expanded view at low temperatures demonstrating $\theta \approx -0.8$ K. Reprinted with permission from Ref. [157], Copyright 2005 by the American Physical Society.

increases more than 3.4 Å, and thus the ESR linewidth of 380 G can be accounted for by only the proposed structure of *M*-DNA in Fig. 7.20.

7.3.3.2 ESR lineshape of Mn-ESR [94]

Interestingly, the ESR lineshape strongly depends on the humidity conditions of Mn-DNA, which control the isomeric structure of DNA: A-form stable under dried conditions and B-form under humid conditions. Figure 7.19 demonstrates the ESR lineshapes of the two isomeric structures. This lineshape difference is ascribable to the structural difference of the A-form and the B-form Mn-DNAs, as shown in Fig. 7.20. The B-form Mn-DNA carries a Q1D array of the Mn ions along the double-helical axis, in which the Mn ions are isolated by the double helix from the neighboring Mn arrays, whereas the A-form does a coil-like arrangement winding around a rod, in which the Mn ions appear in the surface of the Mn-DNA double helix. In a solid of Mn-DNA, the Mn ion arrays interact with the neighboring Mn-DNAs. In the B-form Mn-DNA, the Mn ions interact with only the Mn ions of the array. In contrast, the Mn ions

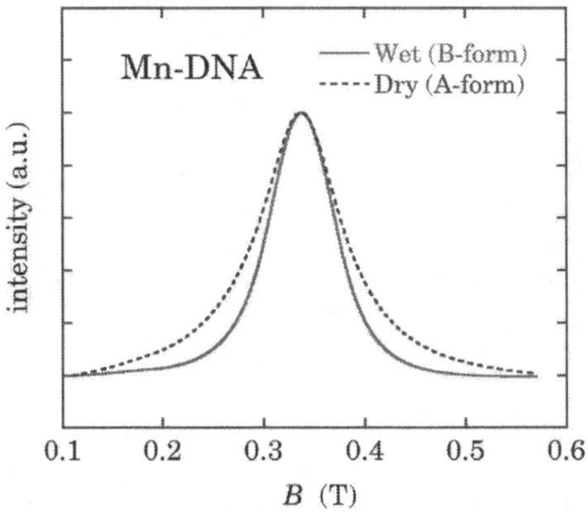

Figure 7.19 The ESR absorption spectra of Mn-DNA under both wet and dry conditions taken at X-band around 9.4 GHz. Note the definite difference in lineshape caused by the hydration conditions.

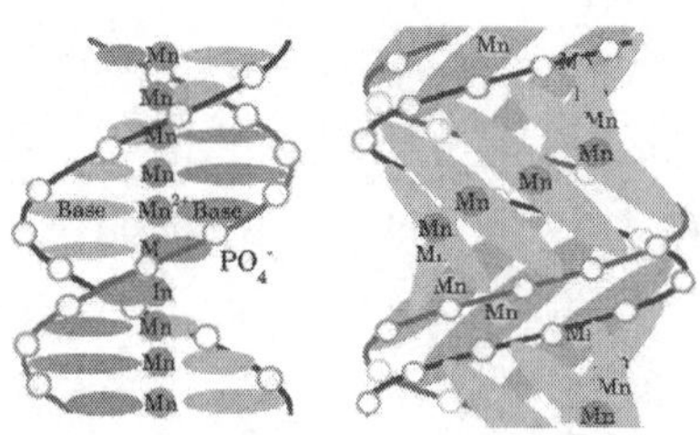

Figure 7.20 Model structures for the B-form (left) and the A-form (right) of Mn-DNA. Ellipsoids represent bases or base pairs. In the B-form, Mn ions occupy the center axis of the double helix, while they form a coil in the A-form DNA such as a rope ladder.

of the A-form Mn-DNA interact not only with the Mn ions of the array but also with the Mn ions of the neighboring Mn-DNAs. That is, the interaction of the Mn ions in the B-form Mn-DNA is restricted within the one-dimensional array, but widely spreads over three dimensions in the A-form Mn-DNA.

The aforementioned discussion helps us to analyze the ESR lineshapes. The spatial range of the interaction between the Mn ions is also applied to the exchange interaction between the Mn ions, which dominates the ESR lineshape through exchange narrowing, as explained in Figs. 7.21 and 7.22. Thus, we analyze the ESR spectra

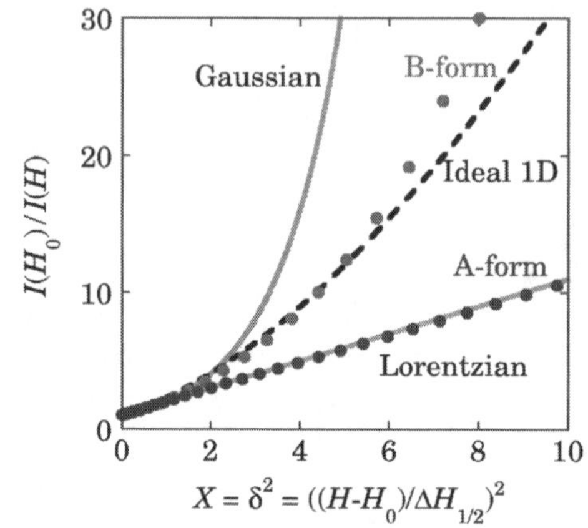

Figure 7.21 Characteristic plot for ESR lineshape analysis. The abscissa represents the square of the magnetic field deviation from the resonance center H_0 normalized by the half-width at the half-height $\Delta H_{1/2}$ of the absorption spectrum. The ordinate is the inverse signal intensity $I(H)$ normalized by the center height $I(H_0)$. The spectral shape for the A-form Mn-DNA is reproduced well with the Lorentzian lineshape, but that for the B-form Mn-DNA with the characteristic lineshape for the one-dimensional exchange narrowing case. Reprinted with permission from Ref. [259], Copyright 1973, by the American Physical Society.

with a special plot, which provides an easy way to analyze visually the ESR lineshapes. If the lineshape is the normalized Lorentzian

$$\frac{I(H)}{I(H_0)} = \frac{1}{1 + \left(\frac{H - H_0}{\Delta H_{1/2}}\right)^2} = \frac{1}{1 + X}, \tag{7.4}$$

where $X = \left(\frac{H-H_0}{\Delta H_{1/2}}\right)^2$, the inverse of the normalized intensity holds the relation

$$\frac{I(H_0)}{I(H)} = 1 + X, \tag{7.5}$$

which is a straight line with a y-intercept of unity, as actually demonstrated with the label "Lorentzian" in Fig. 7.21, in which the inverse of the absorption intensity normalized by the center height is plotted against the square of the normalized deviation field from the center of the resonance. With the normalized Gaussian lineshape

$$\frac{I(H)}{I(H_0)} = e^{-\left(\frac{H-H_0}{\Delta H_{1/2}}\right)^2} = e^{-X}, \tag{7.6}$$

its inverse is given by

$$\frac{I(H_0)}{I(H)} = e^X, \tag{7.7}$$

which is a simple exponential function in the plot of Fig. 7.21 (labelled "Gaussian"). In the case of one-dimensional systems, Hennessy *et al.* have shown that the ESR spectrum has a characteristic lineshape, as labelled "Ideal 1D" [259]. The ESR spectra of Fig. 7.19 are plotted in Fig. 7.21 by the dots with the labels "A-form" for the A-form Mn-DNA and "B-form" for the B-form Mn-DNA. The ESR lineshape of the A-form has the Lorentzian shape and the B-form has approximately the ideal one-dimensional shape, which is consistent with the predicted dimensionality of the exchange interaction from the structures in Fig. 7.20. A further explanation of exchange narrowing is shown in Fig. 7.22.

Since the B-form Mn-DNA has the one-dimensional Mn chain in the double helical axis of DNA, the occurrence of any long-range magnetic ordering is not expected. Figure 7.23 suggests such a situation in the ESR linewidth, which is independent of temperature down to 2 K. In contrast, the ESR linewidth of the A-form Mn-DNA rapidly increases with decreasing temperature below 50 K. This enhancement of the ESR linewidth is ascribed to the growth of the short-range antiferromagnetic (AF) ordering near the antiferromagnetic transition temperature in three-dimensional spin systems [260]. Actually, the A-form Mn-DNA shows a peak of specific heat at $\approx$0.4 K, as shown in Fig. 7.24, which corresponds to the antiferromagnetic ordering temperature, and thus it is a clear evidence of the three-dimensional network of the exchange interaction in the A-form Mn-DNA.

Interesting irreversible effects of humidification have been observed in the dried A-form Mn-DNA. Figure 7.25 shows the ESR spectrum of the dried Mn-DNA and that of the same Mn-DNA after being humidified again. Both the spectra are reproduced by Lorentzian lineshape very well, which is clearly different from the initially synthesized B-form DNA with the lineshape typical of one-dimensional system under humidified conditions, as described in Fig. 7.21. This observation suggests that the A-form Mn-DNA does not transform back to the B-form Mn-DNA under humidified conditions. This is also consistent with the characteristic structure

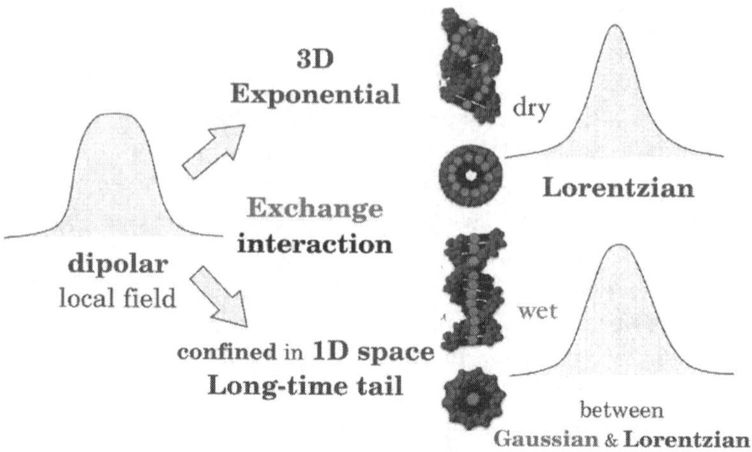

Figure 7.22 Schematic description of the exchange narrowing in the B-form and the A-form Mn-DNAs. The ESR spectrum is produced by the distribution of the dipolar local field at each magnetic moment with the lineshape of approximately Gaussian or more box-type form depending on the lattice type of the magnetic moments [255]. The exchange interaction substantially averages the distribution of the local fields out with the rule determined by the dimensionality of the exchange interaction. In one-dimensional systems, the average is incompletely carried out only within the one-dimensional array, which gives the intermediate lineshape between Gaussian and Lorentzian [259], whereas it is homogeneously done over the three-dimensional exchange systems, resulting in a single exponential correlation with the Lorentzian lineshape.

of *M*-DNAs in Fig. 7.1. The divalent metal ions strongly bind the PO_4^- anions of the DNA backbones, which prevents Mn-DNA from transforming the A-form into the B-form structure. The linewidth of the rehumidified A-form Mn-DNA is $\Delta H = 46$ mT, which is narrower than 51.7 mT for the dried A-form Mn-DNA. This narrowing due to humidification is ascribed to the increased water content not only in the double helix but also between the double helices, which increases the distance between the Mn ions of the neighboring double helices.

To conclude, ESR linewidth described in the previous section and the ESR lineshape are strictly consistent with the structure in Fig. 7.1 and cannot be accounted for without the structure shown in Figs. 7.1 and 7.20.

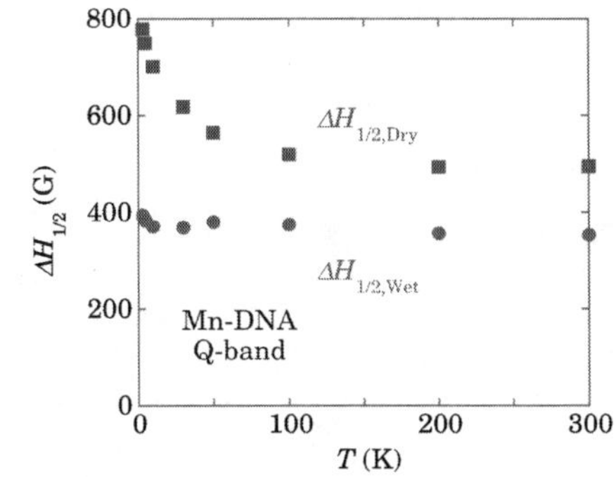

Figure 7.23 Temperature dependence of ESR linewidth for Mn-DNA under wet (B-form: circles) and dry (A-form: squares) conditions. The linewidth is dominantly governed by the electronic–magnetic dipolar interaction between Mn ions at room temperature [94]. Note the low temperature increase in linewidth under the dry condition, typical of antiferromagnetic interaction. Reprinted with permission from Ref. [94], Copyright 2007, The Physical Society of Japan.

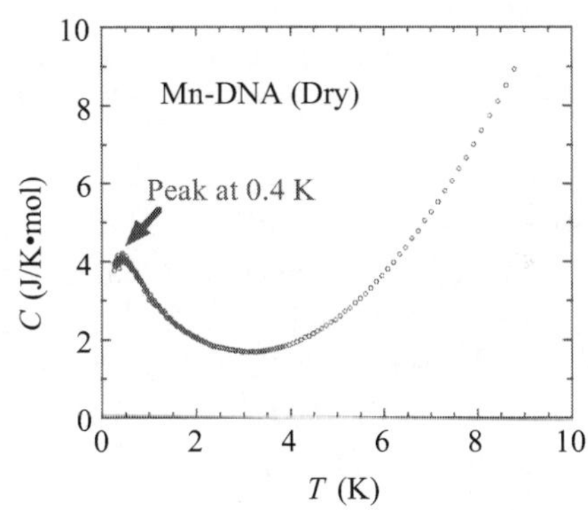

Figure 7.24 The low-temperature specific heat of Mn-DNA in the dry condition (A-form). A peak is found at 0.4 K below the disappearance of the phonon contribution. It is suggested that the peak originates from some AF-like magnetic ordering of Mn ions, which corresponds to the Curie–Weiss temperature of ≈ -2 K. Reprinted with permission from Ref. [94], Copyright 2007, The Physical Society of Japan.

7.3.3.3 Hyperfine structure of Mn-ESR

Mn atom carries the $S = \frac{5}{2}$ electron spin along with the $I = \frac{5}{2}$ nuclear spin, which provides information about the wavefunction of the electron spin from the hyperfine structure of Mn-ESR [56].

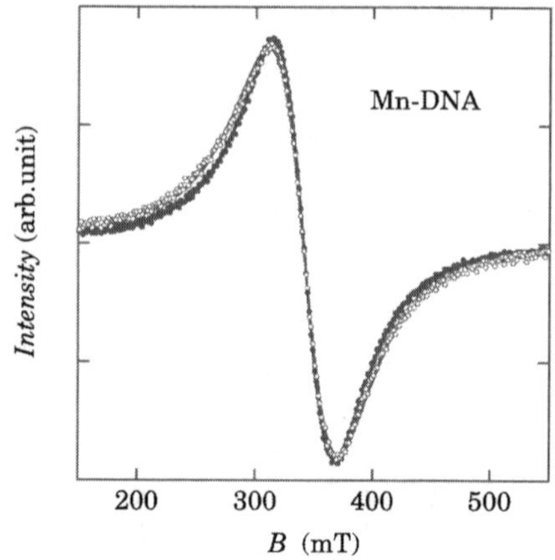

Figure 7.25 ESR derivative spectra of Mn-DNA: open circles represent the dried A-form and closed circles denote the same Mn-DNA but humidified again. The solid curves represent Lorentzian lineshapes fitted to both data sets, which reproduce the spectra well. The initial B-form Mn-DNA has spectral shape typical of one-dimensional exchange narrowing between Gaussian and Lorentzian, as described in Figs. 7.21 and 7.22. Thus, the A-form Mn-DNA keeps the A-form structure even under humidified condition.

In the film samples, as shown in Figs. 7.19, 7.26, and 7.27, the six hyperfine peaks predicted from the $I = \frac{5}{2}$ nuclear spin are absent both in the A-form and the B-form Mn-DNAs because of the exchange interaction between the Mn electron spins. The magnitude of the exchange interaction in Mn-DNAs is of the order of 2 T (2 K) for the A-form and less than 1 T (1 K) for the B-form, which are much larger than $\approx$0.1 T of the hyperfine splitting for Mn ESR and is large enough to smear out the hyperfine splittings. Thus, we utilized Mn-DNA solutions to analyze the hyperfine splittings.

Figure 7.26 shows the ESR spectra of Mn-DNA with different water contents. The bottom two spectra correspond to the solution spectra with the six hyperfine splittings. Through the intermediate shapes of the next two spectra, the B-form spectral shape appears with the lineshape typical of the one-dimensional exchange interaction [259]. The topmost three spectra are well reproduced by the Lorentzian lineshape characteristic of the A-form Mn-DNA. Here, note that the emergence of the hyperfine split spectra even in the

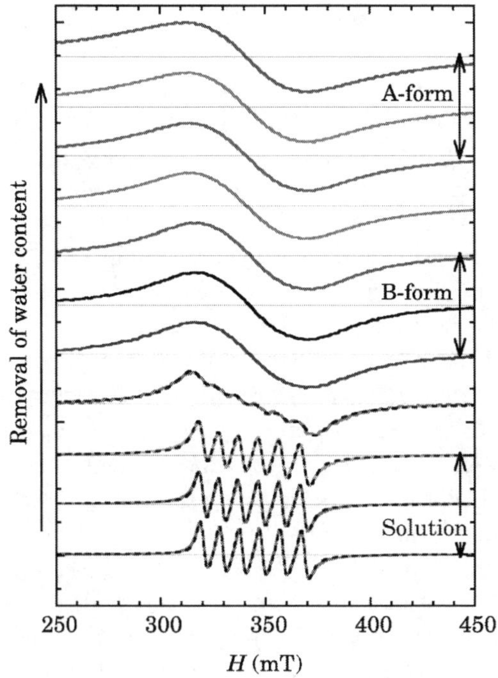

Figure 7.26 Time evolution of the ESR derivative spectra in Mn-DNA with water at X-band. By evaporating water, the spectral shape gradually changes from six hyperfine split peaks to a broad single Lorentzian shape (solid A-form state, top three spectra, $W_B = 50 \pm 3$ mT), via intermediate spectra of a one-dimensional shape typical of the B-form Mn-DNA (fifth to seventh spectra), as discussed in Ref. [94]. The dashed curves represent the results of the simulation with HFS + Lorentzian for the bottom four spectra. Reprinted with permission from Ref. [56], Copyright 2011, The Physical Society of Japan.

solution of Mn-DNA is not self-evident. The dipolar broadening in the B-form Mn-DNA completely disappears in the solution spectra. The random reorientational motion of the double helical DNA is thermally excited in the solution, which averaged the dipolar interaction out angularly, that is, motional narrowing occurs. As expected from the dipolar interaction in Fig. 7.17, the dipolar interaction depends on not only the distance between the spins but also the angle between the vector connecting the interacting spins and the external magnetic field. In contrast, the isotropic exchange interaction depends on only the overlap integrals between the wavefunctions of the neighboring Mn ions. Since, as shown in the CD spectra of Fig. 7.14, *M*-DNAs keep double-helical structure in solution, almost all of the exchange interactions between the neighboring Mn ions should be held regardless of thermal vibrations, which ensures the exchange narrowing mechanism for the hyperfine interaction even in the Mn-DNA solution. Thus, the hyperfine spectra should not be observed if the exchange interaction still survives in solution. We have to investigate the mechanism that invalidates the isotropic exchange interaction.

A possible mechanism is the role of water molecules in Mn-DNA. Figure 7.26 demonstrates that the hyperfine structure gradually weakens with decreasing water content in Mn-DNA. Even in the films, the water content of the B-form Mn-DNA is larger than that of the A-form Mn-DNA, as described in Chapter 3. The exchange interaction is related to the Curie–Weiss temperature of the magnetic susceptibility and is 2 K in the A-form Mn-DNA, which is twice (1 K) that of the B-form isomer [94]. If we suppose that the water molecule mediates or perturbs the exchange interaction in any way, the difference in the exchange interaction in these isomers could be accounted for as the additional mechanism to that of the structural origin. In the solutions, a large number of water molecules dynamically enter and come out of the Mn-DNA, which strongly disturb and suppress the exchange interaction, resulting in the emergence of the hyperfine structure of ESR spectra.

The hyperfine interaction between the electrons and the nucleus of an Mn ion (in T) is given by [255, 261, 262]

$$\frac{\mathcal{H}_{hf}}{g\mu_{\mathrm{B}}} = \mathbf{I}\cdot\mathbf{A}\cdot\mathbf{S} = A_0\mathbf{I}\cdot\mathbf{S} + \mathbf{I}\cdot\mathbf{A}_{\mathbf{ani}}\cdot\mathbf{S}, \tag{7.8}$$

where $\mathbf{I}$ is the nuclear spin, $\mathbf{S}$ is the electron spin, and $\mathbf{A}$ is the hyperfine coupling tensor, which is the sum of the isotropic part A_0 and the traceless anisotropic part $\mathbf{A}_{\mathbf{ani}}$. In the Mn-DNA solution, DNA double helices rapidly tumble, which averages the anisotropic part $\mathbf{A}_{\mathbf{ani}}$ out at zero, as in the case of the dipolar interaction, giving rise to $\mathbf{A} \approx A_0$. Neglecting the small nuclear Zeeman energy relative to that of the electron spin, the spin Hamiltonian for the Mn-DNA solution

$$\frac{\mathcal{H}}{g\mu_{\rm B}} \approx B_0 S_z + A_0 S_z I_z + \frac{A_0}{2}(S_+ I_- + S_- I_+), \tag{7.9}$$

gives the Zeeman energy for each nuclear spin multiplet m_I. The resonance condition of the HFS spectra with the Zeeman energy splittings $\Delta E(m_I)$ is given by [261, 263]

$$\frac{\Delta E(m_I)}{g\mu_{\rm B}} = B_0 + A_0 m_I + \frac{1}{2}\left(\frac{A_0^2}{B_0}\right)(I(I+1) - m_I^2). \tag{7.10}$$

The second term ($\propto m_I$) predicts equally spaced $2I+1 = 6$ peaks for $I = 5/2$, which correspond to each m_I value. The third term ($\propto m_I^2$) provides a linear deviation from the equally spaced peak separation on m_I as a higher-order effect of the hyperfine interaction.

The hyperfine split spectra in Figs. 7.26 and 7.27 were simulated with Eq. (7.10). The parameters deduced from the simulation with Eq. (7.10) for ($Ca_{1-x}Mn_x$)-DNA ($x = 1, 0.1, 0.01$) solutions are summarized in Table 7.2, along with those for the reference

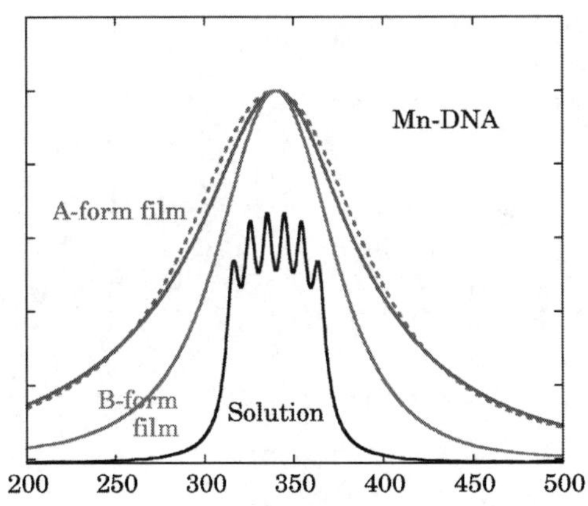

Figure 7.27 ESR absorption spectra at X-band for the Mn-DNA with B-form in solution, and B-form and A-form Mn-DNA films. The spectra in solution agree with each other within the uncertainty. The simulated spectrum for the A-form Mn-DNA in terms of the six Lorentzian spectra with the same separation as those in solution Mn-DNA is also shown by the broken curve, which fails to reproduce the Lorentzian line shape in the A-form Mn-DNA. Reprinted with permission from Ref. [56], Copyright 2011, The Physical Society of Japan.

Table 7.2 Average value of isotropic hyperfine parameters A_0 and A_0^2/B_0 [94], which is a degree of the deviation from the equal spacing caused by the third term of Eq. (7.10), in the units of mT. "expt." is deduced from the simulations of the hyperfine split spectra to Eq. (7.10), and "calc." is deduced from A_0 and B_0. Experimental errors at the least significant digit are indicated in parentheses

Unit (mT)	A_0	A_0^2/B_0	
		expt.	calc.
Mn-DNA solutions	9.62(1)	0.27(1)	0.27
$Ca(Mn)Cl_2$	9.11(1)	0.24(1)	0.25
Mg(Mn)O	8.74(1)	0.22(1)	0.22

materials, Mg(Mn)O and $Ca(Mn)Cl_2$ as shown in Fig. 7.28. Several characteristic features are found in Table 7.2

(1) The isotropic hyperfine parameter A_0 depends markedly on the host atoms: 8.74 mT in Mg(Mn)O, 9.11 mT in $Ca(Mn)Cl_2$, and 9.62 mT in Mn-DNA

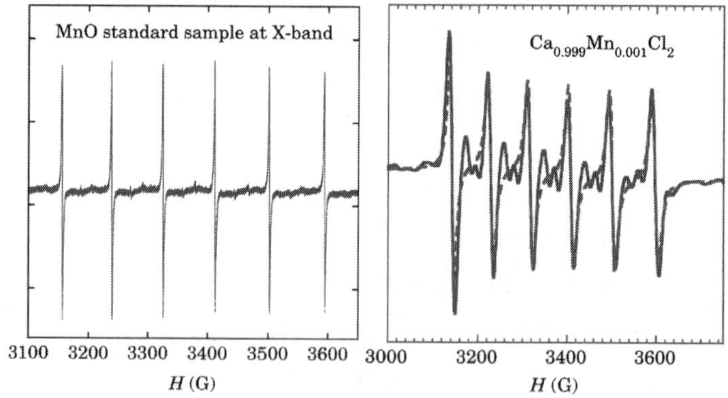

Figure 7.28 Hyperfine split ESR derivative spectra at X-band of Mn ion dilutely substituted for Mg in Mg(Mn)O (left) and $Ca(Mn)Cl_2$ (right) powder samples. The separation between the neighboring peaks is not uniform because of the second-order effect of the isotropic hyperfine interaction, the third term of Eq. (7.10). Forbidden transition with $\Delta m = \pm 1$ in $Ca(Mn)Cl_2$ is much larger than that of Mg(Mn)O because of the lower symmetry of crystalline field, local uniaxial symmetry in this system than the cubic in Mg(Mn)O [261, 262].

(2) The effect of the third term of Eq. (7.10) reproduces well the experimental parameters.

In Fig. 7.29, the isotropic hyperfine parameters A_0 deduced from ESR hyperfine spectra of dilutely substituted Mn ions are shown against various host compounds with combinations of cations and anions, which reflects the ionicity of Mn ions [263]. Table 7.3 shows

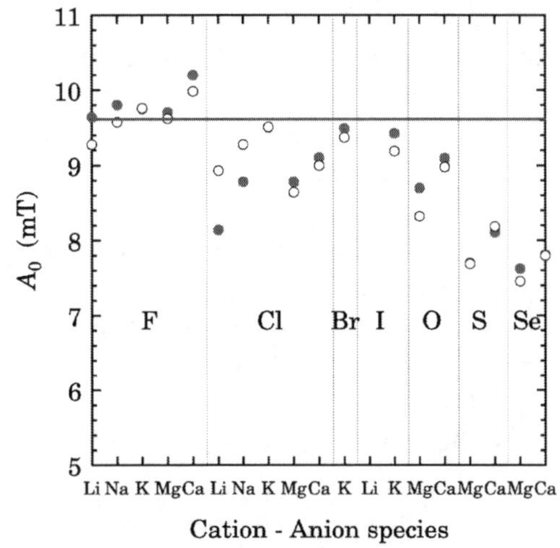

Figure 7.29 Isotropic hyperfine parameters determined by ESR hyperfine spectra in various compounds with dilutely substituted Mn ions (closed circles). The open circles represent the calculated values. The horizontal bar shows 9.62 mT in Mn-DNA. Reproduced with permission from Ref. [263], Copyright 1968, Consultants Bureau.

Table 7.3 Hyperfine coupling constants of Mn^{2+} ions embedded in calcium halides and chalcogenides, in the units of mT [263]. The linear relationship $g\mu_B A_0 = 19.4 + 83.3\ i\ [10^{-4}\ \text{cm}^{-1}] = 2.08 + 8.92\ i$ [mT] has been estimated from the calculated result based on the experimental data set, where i is the degree of ionicity [263]. Note that the slope depends on the host ion species. To convert mT to cm^{-1}, divide it by 1.071×10^3 [mT/cm^{-1}]

Host	CaF_2	CaO	$CaCl_2$	CaS
A_0 (mT) for Mn^{2+}	10.1	9.18	9.11	8.11
Degree of ionicity	0.90	0.80	0.79	0.68

Source: Ref. 56.

A_0 of Mn hyperfine spectra and the degree of ionicity in a series of calcium compounds. CaF has the largest hyperfine parameter of 10.1 mT with the ionicity of 0.90, which demonstrates strong ionic bonds in this compound. In contrast, CaS has the smallest value of 8.11 mT with a low ionicity of 0.68, which suggests a comparable part of the covalent bond in addition to the ionic one.

Since the *p* or *d* electron has no spin density at the Mn nucleus, the isotropic interaction is caused by the interaction with the inner *s*-electron cores. Thus, the isotropic hyperfine interaction A_0 results from the polarization of the inner *s*-electron cores induced by the *d*-electron polarization (the core polarization effect) [263]. The wavefunctions of Mn ions depend on the bonding nature; the wavefunction of the ionic bond has a spherical symmetry typical of the isolated atoms with closed shells. In contrast, the wavefunction of the covalent bond extends to the direction of the bonding atoms, which reduces the overlap integral between the *s*-core wavefunctions, having contact interaction with the Mn nucleus, and the directionally extended *d*-wavefunctions.

Thus, the large A_0 in Mn-DNA suggests the ionic nature of Mn ions, as schematically described in Fig. 7.1. The Mn ions in Mn-DNA should be hydrated with several water molecules, which prevents Mn ions from forming covalent bonds preferably with the nitrogens of the surrounding bases. The same conclusion has been obtained in Fe-DNA on the basis of UV/Vis absorption spectra in Section 7.2.3. If the hydrating water molecules are evacuated completely, Zn ions of Zn-DNA actually form covalent bonds with the nitrogen atoms of bases, as discussed in Chapter 9.

7.3.4 *Fe-DNA*

The UV/Vis absorption study in Section 7.2 demonstrates that Fe-DNA is an exceptional system in the *M*-DNAs. Fe-DNA is synthesized in a mixture solution of DNA and $FeCl_2$, in which Fe^{2+} transforms to Fe^{3+} by donating an electron to the bases of a base pair. Fe^{3+} ion is hydrated by water molecules, which prevents Fe ion from forming covalent bonds with the nitrogen atoms of the bases, as in Fig. 7.1. To unveil the electronic states of Fe-DNA, magnetic properties are examined in this section. ESR of Fe-DNA shows characteristic feature of Fe^{3+} with three components corresponding to the high-spin state of $S = \frac{5}{2}$, the low-spin state of $S = \frac{1}{2}$ for Fe^{3+}, and $S = \frac{1}{2}$ for the π-electron of the base pair, which has been transferred from Fe^{2+}. The spin states of Fe^{3+} would be dominated by the local hydration structure. It will be demonstrated that such an assignment of the ESR spectra is consistent with the magnetization curve of Fe-DNA at 2 K.

7.3.4.1 Electronic states of Fe

Figure 7.30 shows the ESR spectrum of Fe-DNA along with that of Mn-DNA measured at room temperature and Q-band around 35 GHz [252]. The resonance position of Mn ESR is at the free electron value, $g \approx 2$, characteristic of S-state ion with the spin of $S = \frac{5}{2}$ and without orbital angular momentum. The ESR spectrum of Fe also appears at $g \approx 2$, which is a definite evidence for the valence

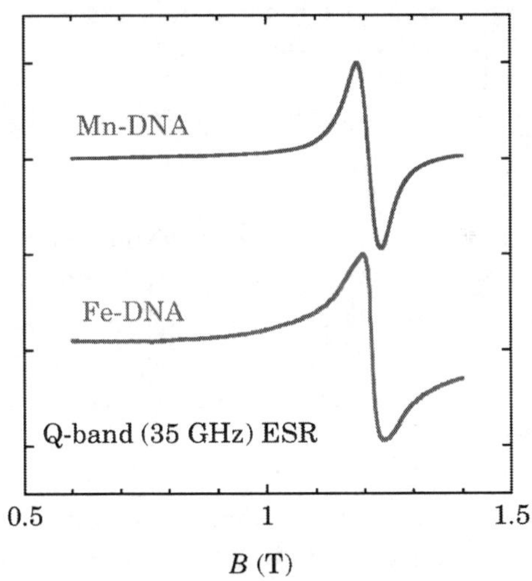

Figure 7.30 ESR spectrum of Fe-DNA taken at Q-band ($\approx$35 GHz, 1.21 T corresponds to $g = 2$), together with Mn-DNA spectrum. From g-factor nearly equal to 2 in Fe-DNA, it is suggested that the electronic state of Fe is Fe^{3+} with $S = \frac{5}{2}$ for the high-spin state, and $S = \frac{1}{2}$ for the low-spin state, which depends on the crystalline field, in which metal ions are located. Reprinted with permission from Ref. [252], Copyright 2015, MDPI AG (Basel, Switzerland).

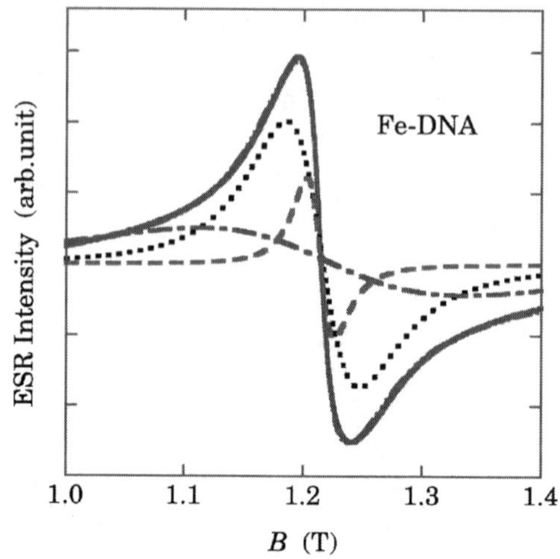

Figure 7.31 Simulation of Fe ESR at room temperature by three Lorentzians with linewidths of 189 mT (dash-dotted line), 54 mT (dotted line), and 20 mT (dashed line). The broad and the medium signals are assigned to $S = \frac{5}{2}$ and $S = \frac{1}{2}$ of Fe^{3+}, respectively. Since the π-electrons are delocalized along the base stacking of the double helix, ESR spectrum should be narrower than those of Fe^{3+}. Reprinted with permission from Ref. [252], Copyright 2015, MDPI AG (Basel, Switzerland).

state of Fe ion to be Fe^{3+}. This fact is consistent with the conclusion of the XAFS (X-ray absorption fine structure) study on Fe-DNA in Section 5.3. Note that the difference of the Fe spectrum from the Mn spectrum is shoulder structures and a steep transition from a peak to a valley at the center of spectrum. Thus, we tried to simulate the Fe spectrum by three Lorentzians at $g \approx 2$ with different linewidths, as shown in Fig. 7.31 [252]. The ESR spectrum of Fe-DNA is reasonably reproduced by the assumption of three Lorentzians: a broad signal of 189 mT for the high-spin state, a medium one of 54 mT for the low-spin state of Fe^{3+}, and a sharp one of 20 mT for the π-electron spins. At 5 K, a similar simulation of the Fe spectrum is applied, and the obtained parameters are summarized in Table 7.4 together with the parameters deduced from the SQUID magnetization data in Fig. 7.32 by applying multiple Brillouin functions fittings

Figure 7.32 shows the magnetization curve of Fe-DNA at 2 K. The magnetization at 7 T is markedly smaller than that under the simple

Table 7.4 Parameters deduced from Fe ESR spectra at 5 K and 300 K, and SQUID magnetizations at 2 K of Fe-DNA. Half-width at the half-height $\Delta H_{\frac{1}{2}}$ and relative absorption heights are presented for three ESR components. Percentages of the number of spins are derived from ESR analysis with three components and from SQUID magnetization, assuming two or three Brillouin functions (BF). In the three BF and ESR analyses at 5 K, the Curie–Weiss temperature of −45 K is assumed to estimate the number of π-electrons.

		Fe^{3+}		π
		$S = \frac{5}{2}$	$S = \frac{1}{2}$	$S = \frac{1}{2}$
ESR	$\Delta H_{\frac{1}{2}}$ (mT)	220	44	15
5 K	Relative height	1.62	1.64	0.94
	$\frac{N_i}{N_{Fe,\frac{5}{2}}+N_{Fe,\frac{1}{2}}}$ (%)	22	78	100 ($\theta = -45$ K)
ESR	$\Delta H_{\frac{1}{2}}$ (mT)	189	54	20
300 K	Relative height	6.83	7.87	1.76
	$\frac{N_i}{N_{Fe,\frac{5}{2}}+N_{Fe,\frac{1}{2}}}$ (%)	21	79	6.6
SQUID	two BF (%)	23	77	-
2 K	three BF (%)	26	74	100 ($\theta = -45$ K)

assumption of the $S = \frac{5}{2}$ high-spin state for all the Fe^{3+} ions, as in Mn-DNA, which suggests a dominant contribution of the low-spin states with $S = \frac{1}{2}$. In Section 7.2.3, we concluded that in a DNA and $FeCl_2$ solution, Fe^{2+} ions react as follows:

$$Fe^{2+} + 2Cl^- + DNA \Rightarrow Fe^{3+}\text{-}DNA^{2-}(\pi^-) + 2Na^+ + 2Cl^-, \quad (7.11)$$

where DNAs(π^-) are located in the bases of a base pair, and for brevity, the Na ions of "DNA" are not shown explicitly. Following this conclusion, the number of π-electrons produced by Eq. (7.11) is equal to that of Fe ions and they carry $S = \frac{1}{2}$. The three Brillouin functions (BF) fitting in Fig. 7.32 (right) takes into account the third spin species of the π-electrons. The Curie–Weiss temperature of −45 K was chosen to be consistent with the simulation parameters of Fe ESR at 5 K. Both the two and three BF fittings yield consistent ratios of the number of spins with each other for Fe^{3+} ions. Since the experimental ratios of the number of spins for Fe^{3+} depend on the humidity in air, the scattering of the ratios in Table 7.4 should not be taken seriously. On the other hand, the ratio of the π-electrons

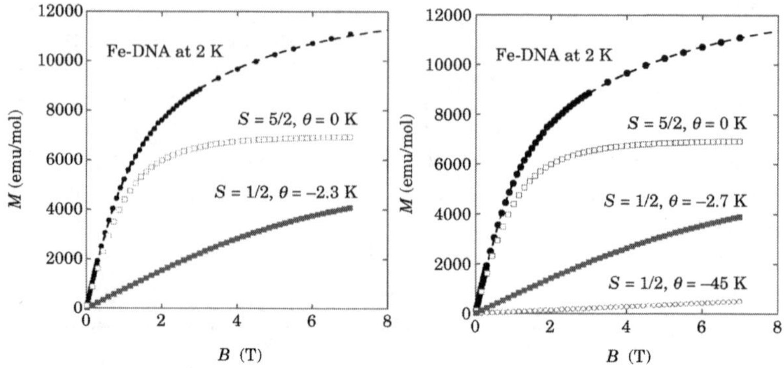

Figure 7.32 Magnetization curve of Fe-DNA at 2 K. The dashed curve represents trial fittings: (left) double Brillouin functions fitting (BFF) of the high-spin, $S = \frac{5}{2}$ with $\theta = 0$ K and the low-spin, $S = \frac{1}{2}$ with $\theta = -2.3$ K of Fe^{3+} ions, and (right) three Brillouin functions of the π-electrons with $\theta = -45$ K transferred from Fe^{2+} into the bases, in addition to $S = \frac{5}{2}$ with $\theta = 0$ K and $S = \frac{1}{2}$ with $\theta = -2.7$ K of Fe^{3+} ions. The obtained ratio of the number of spins is summarized in Table 7.4. Reprinted with permission from Ref. [252], Copyright 2015, MDPI AG (Basel, Switzerland).

estimated at 300 K markedly deviates from the expected 100%. Thus, the assignment of the sharp component of Fe ESR requires further consideration, such as incorrect assignment of the sharp component and/or an incorrect interpretation of the electronic states of the π-electrons. As discussed in Chapter 9, the π-electrons induced by freeze drying (FD) to remove water molecules have nonmagnetic ground state in the dehydrated FD-Zn-DNA, in which any ESR and SQUID signals corresponding to the π-electrons are not observed.

In conclusion, the valence of Fe ions in Fe-DNA is +3 with five d-electrons and $S = \frac{5}{2}$. The electronic states of the π-electrons transferred from Fe^{2+} are not simple Curie–Weiss behavior but would be nonmagnetic states, as in the dehydrated FD-Zn-DNA discussed in Chapter 9. The electronic states of Fe^{3+} are a mixture of the high-spin $S = \frac{5}{2}$ and the low-spin $S = \frac{1}{2}$, and the ratio of the number of spins $N_{\frac{5}{2}}:N_{\frac{1}{2}}$ is from 1:3 to 1:4. A mechanism for the appearance of two spin species of Fe^{3+} surely relates to the presence of two Fe^{3+} sites with different local symmetries of crystal field.

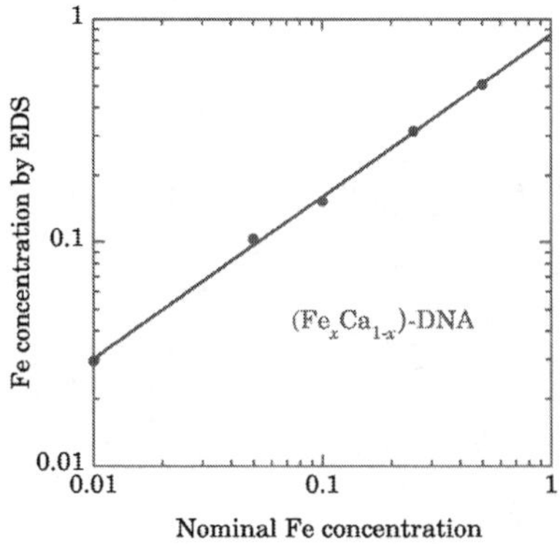

Figure 7.33 Fe concentration deduced by energy dispersive X-ray spectroscopy (EDS) is plotted against nominal ratio of Fe to Ca in sample preparations.

7.3.4.2 Concentration dependence of ($Fe_{1-x}Ca_x$)-DNA

In this section, we study the Fe concentration dependence to obtain further information on the electronic states of Fe^{3+}. Fe concentration is studied by preparing alloys of Fe-DNA and Ca-DNA, as (Fe_xCa_{1-x})-DNA. Since Ca-DNA is a nonmagnetic system, the observed magnetic properties reflect directly the concentration dependence of Fe. Figure 7.33 shows the Fe concentration determined by energy dispersive X-ray spectroscopy (EDS) against the nominal Fe concentration in aqueous solutions of $x$$FeCl_2$+(1-$x$)$CaCl_2$+DNA. The tendency is that the actual concentration of Fe is enhanced at the low concentration region less than $x = 0.1$.

Figure 7.34 shows the concentration dependence of the magnetic susceptibility [left] with Curie–Weiss behavior $\chi = \frac{C}{T+\theta}$, $\theta \approx -1.6$ K, and shows the magnetization curves [right] of (Fe_xCa_{1-x})-DNA at 2 K. With decreasing x, a functional form of the magnetization curve gradually changes. Figure 7.35 shows the concentration dependence of the ESR spectra at Q-band. The spectral tails of the broad component are normalized for comparison of the medium component. The concentration of the $\frac{5}{2}$ spin deduced from the analyses of magnetization curve and ESR spectra is summarized in Fig. 7.36, along with the data deduced from the samples prepared under oxygen-free argon atmosphere. All the data treated in air

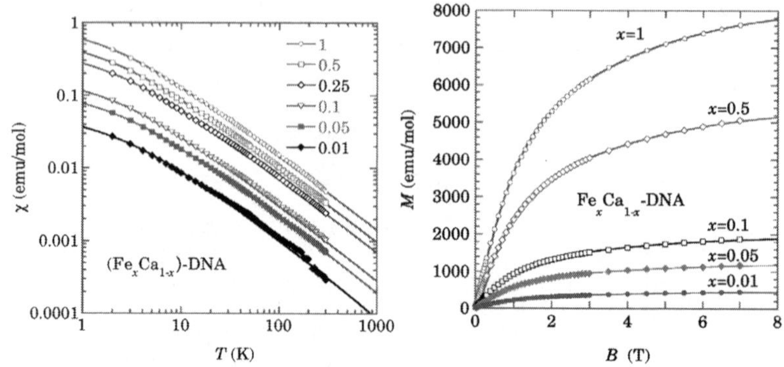

Figure 7.34 [Left] Fe concentration dependence of the magnetic susceptibility of (Fe_xCa_{1-x})-DNA at 1 T. The solid curves represent the Curie–Weiss fitting with the Weiss temperature of -1.6 K. [Right] Magnetization curves at 2 K for a variety of x. The solid curves represent the double Brillouin functions fittings.

are consistent with each other within the scattering of the data, which suggests a weak increase in the $S = \frac{5}{2}$ spin concentration with a decrease in the Fe concentration x. ESR data show the larger scattering because of the three Lorentzian fittings with nine parameters. The increase in the $S = \frac{5}{2}$ spin with decreasing x

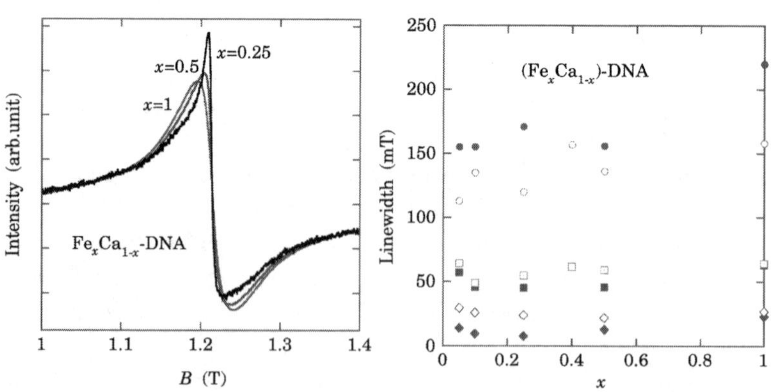

Figure 7.35 [Left] Concentration dependence of Q-band ESR spectra of (Fe_xCa_{1-x})-DNA at room temperature. The magnitude of the spectra is normalized in the tail of the broad component. [Right] Linewidth of three components.

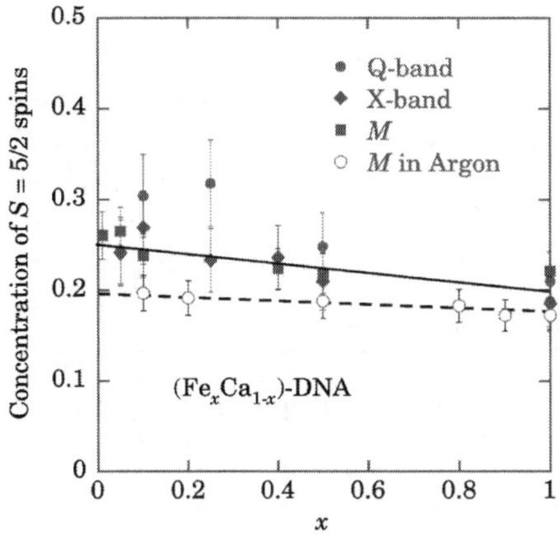

Figure 7.36 $N_{S=\frac{5}{2}}$ versus the concentration x of (Fe$_x$Ca$_{1-x}$)-DNA, deduced from ESR lineshape analysis in Fig. 7.31 at X-band (≈9.5 GHz) and Q-band (≈35 GHz), and from magnetization analysis in Fig. 7.32 at 2 K. The open circles represent the data of (Fe$_x$Ca$_{1-x}$)-DNA prepared under argon atmosphere.

means the increase in the Fe sites with higher local symmetry of the crystal field, which suggests that the higher symmetry for Fe^{3+} ions realizes preferably with the calcium neighbors. In contrast, the data of (Fe$_x$Ca$_{1-x}$)-DNA synthesized under argon atmosphere and dried under vacuum show the suppression of the $S = \frac{5}{2}$ spin concentration from that of the samples in air and depend only a little on the Fe concentration, as shown by the broken line in Fig. 7.36. This finding suggests that some factors, oxygen and/or water molecules, which are removed from the air atmosphere, enhance the $S = \frac{5}{2}$ spin concentration. Then, we studied the magnetization at 2 K under the controlled atmosphere.

7.3.4.3 Fe-DNA under controlled condition

The magnetization curve at 2 K is useful to investigate the concentration of the $S = \frac{5}{2}$ spins in a mixture of the high-spin $S = \frac{5}{2}$ and the low-spin $S = \frac{1}{2}$ states of Fe^{3+} ions in Fe-DNA. It was suggested in the previous section that the oxygen and/or water molecules are the determining factor of the concentration of the $S = \frac{5}{2}$ spins. Thus, we proceeded the following experiments:

Table 7.5 Typical concentrations of $S = \frac{5}{2}$ and $S = \frac{1}{2}$ spins in Fe-DNA prepared by several procedures, which are estimated by double Brillouin functions fittings with the assumption that the magnetism of the π-electrons transferred from the Fe ions is nonmagnetic

Fe-DNA	Synthesis	Dry	Fe^{3+}	
			$S = \frac{5}{2}$	$S = \frac{1}{2}$
(1)	Air	Air	0.22	0.78
(2)	Argon	Vacuum	0.17	0.83
(3)	Air	Evacuate completely	0.15	0.85

(1) Natural drying in air
(2) Synthesis under argon and dry in vacuum
(3) Synthesis in air and completely dry in vacuum

Table 7.5 shows the $S = \frac{5}{2}$ spin concentration of Fe-DNA prepared under several different conditions. The $S = \frac{5}{2}$ spin concentration of (1) shows the maximum concentration of 22% as a typical value. Fe-DNA (1) prepared in air contains more water and oxygen molecules than Fe-DNA (2) and (3). Especially, Fe-DNA (2) with the intermediate value of 17% is free from oxygen molecules in air but contains some water molecules. Fe-DNA (3) with the least value of 15% contains only a trace of oxygen and water molecule. Thus, Fe-DNA (2) and (3) contain minimum number of oxygen molecules, but different quantity of water molecules with the different quantity of the $\frac{5}{2}$ spin concentration, which suggests that water molecules are determining factors of the spin concentration in Fe-DNA. The other factor is a kind of the neighboring metal ions, as derived from the Fe concentration dependence of (Fe_xCa_{1-x})-DNA in Fig. 7.36, which would act as a stabilizer for the hydrating water molecules between Fe and Ca ions. A possible model for the coexistence of the $\frac{5}{2}$ and $\frac{1}{2}$ spins is that the six water molecules hydrating Fe ion provide a weak octahedral crystal field, which gives rise to the high-spin state with $S = \frac{5}{2}$, but the four water molecules form relatively strong square planer crystal field, which yields the low-spin state with $S = \frac{1}{2}$ [261].

In conclusion, we summarize the electronic states of Fe-DNA as follows:

(1) The valence of Fe ions is +3.
(2) The high-spin state and the low-spin state of Fe^{3+} coexist.
(3) The ratio of the $\frac{5}{2}$ spin to $\frac{1}{2}$ is ≈1:4, which depends on the hydrating water molecules at Fe ions, that is, an octahedral crystal field with $6H_2O$ and a square planer field with $4H_2O$.

7.3.5 *Summary of Magnetic Property in* M-*DNA*

The electronic states of *M*-DNAs have been clarified by the magnetic properties in addition to the UV/Vis absorption study. The *M*-DNAs prepared by the alcohol precipitation method contain the divalent metal ions M^{2+} between the bases of a base pair, on the basis of the existence of ESR signal relevant to the inserted metal ions, ESR linewidth, and lineshape analyses of Mn-DNA, except for Fe-DNA. Fe-DNA gives ESR signal at $g \approx 2$, as the strong evidence for the Fe valence of +3 in Fe-DNA. Interestingly, the magnetization and ESR spectrum analyses revealed that both the high-spin and low-spin states of Fe^{3+} coexist with the approximate ratio of one to four. The suggested mechanism of coexistence is that the six hydration water molecules of Fe ions provide an octahedral crystal field, which corresponds to the high-spin state, and the four water molecules give a square planer crystal field, giving rise to the low-spin state of Fe^{3+} ions. These results are consistent with the conclusions deduced by the UV/Vis absorption study in Section 7.2.

Chapter 8

IR Spectral Studies on *M*-DNA

Hiroshi Matsui
Department of Physics, Tohoku University, Sendai, Miyagi, 980-8578 Japan
matsui_ldp@m.tohoku.ac.jp

8.1 Introduction

A conventional Na^+ counterion in dry natural DNA can be replaced by other metallic ions from alkali metals, alkaline earth metals, and transition metals [264]. A natural DNA sample with a metallic ion (M) is written as M-DNA. Using divalent metallic ions (Co^{2+}, Ni^{2+}, and Zn^{2+}) incorporated inside the helix [16, 57], metallic conductions were suggested by the I–V characteristics [17, 29]. Furthermore, the reduction of π–π^* in Zn-DNA was pointed out in ultraviolet spectroscopy [46], and strong electronic correlation was reported [24]. The ESR signal due to π electrons, however, was never detected in M-DNA ($M = Zn^{2+}$, Mn^{2+}, Ca^{2+}, and Mg^{2+}) [27, 94, 157]. From the viewpoint of ESR studies, the incorporation of divalent metallic ions is not so easy to inject a carrier in M-DNA. On the other hand, the valence of Fe ions incorporated in Fe-DNA is found to be 3+ [94], and the charge transfer from Fe^{2+} to DNA is indicated by the intragap absorption [27].

DNA Engineering: Properties and Applications
Edited by Kenji Mizoguchi and Hirokazu Sakamoto

ISBN 978-981-4669-46-7 (Hardcover), 978-981-4669-47-4 (eBook)
www.panstanford.com

A monovalent Na^+ is situated near negatively charged PO_2^- as illustrated in Fig. 3.10. How do divalent and trivalent metallic ions compensate a charge of phosphate group? To balance the charge in the whole system, the incorporation effect of divalent and trivalent metallic ions may spread not only to the phosphate group, but also to the base and sugar molecules. To resolve this issue, we systematically studied the arrangement of metallic ions in the double helix and the local structural change in terms of the infrared spectra of dry *M*-DNA (M = Li, Na, Mg, Ca, Mn, Fe, and Zn) at room temperature [253].

8.2 Infrared Spectra in *M*-DNA

The present *M*-DNA samples were based on a salmon sperm DNA (Wako Pure Chemicals). First, the aqueous solution of original salmon sperm DNA (1 mmol/L) was mixed with 5–10 mmol/L MCl_2 solution. Second, impurities were removed by ethanol precipitation, and excess metallic ions were excluded by dialyzation. The thick filmy sample made on an Si plate was mounted on our optical cryostat to control relative humidity.

Figure 8.1 shows the infrared spectra for all the *M*-DNA samples at 90% RH (a) and 0% RH (b) in the range 2500–3800 cm^{-1}. At 90% RH, the water content in Na-DNA is approximately 20 wpn, so that the primary hydration shell is almost filled with water molecules. Other *M*-DNA samples may probably possess similar water content to Na-DNA. If the configuration of divalent and trivalent metallic ions differs from that of the phosphate group, the hydration structure in *M*-DNA is not always the same with Fig. 3.10 obtained for Na-DNA.

For 90% RH in Fig. 8.1(a), broad bands around 3300 cm^{-1} are dominated by the OH stretching vibration of water molecules hydrated to dry *M*-DNA. Peaks at 3372 cm^{-1} and kinks at 3180 cm^{-1} are related to the stretching modes of NH_2, CH, and CH_3. Small peaks at 2960 cm^{-1} are associated with the NH or CH_3 stretching modes. The wavenumber of these modes depends a little on the type of metallic ions.

For 0% RH in Fig. 8.1(b), the absorption around 3300 cm^{-1} is hugely suppressed owing to dehydration. The water content is 1 wpn

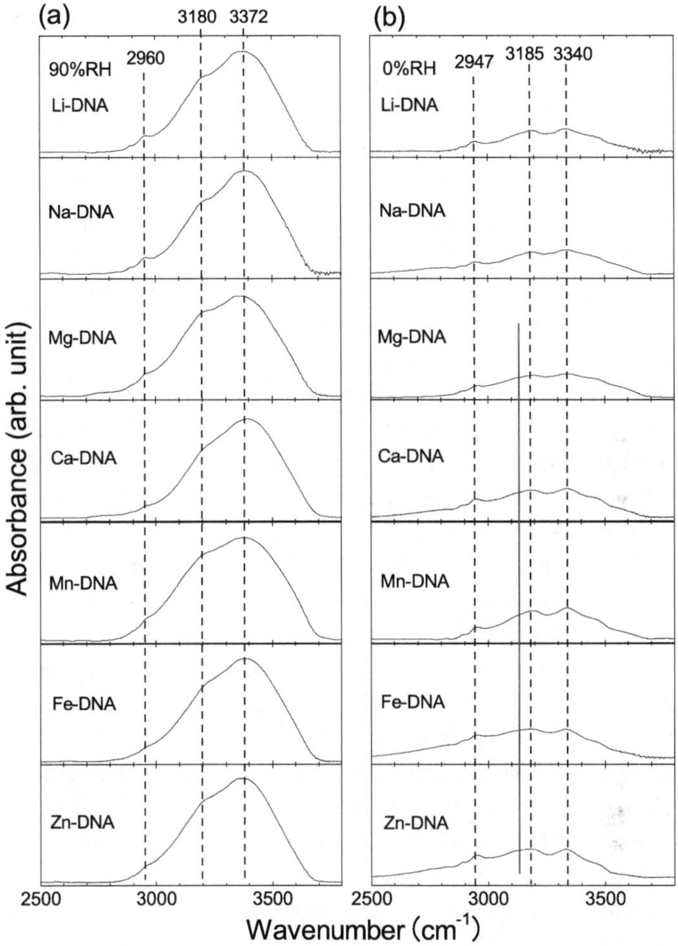

Figure 8.1 Infrared spectra at 2500∼3800 cm^{-1} for 90% RH (a) and 0% RH (b) in *M*-DNA.

at most. Then the spectra around 3300 cm^{-1} mainly stem from the molecular vibrations due to DNA itself. There appears a peak at 3185 cm^{-1}, which shifts to higher wavenumber than the kink observed at 3180 cm^{-1} for 90% RH. On the other hand, the stretching bands at 2947 and 3340 cm^{-1} for 0% RH shift to lower wavenumber compared to one at 2960 and 3372 cm^{-1} for 90% RH. These shifts are responsible for the conformational transformation from B-form to A-form.

As indicated by a red solid line in Fig. 8.1(b), a kink slightly emerges around 3100 cm^{-1} only in *M*-DNA with divalent and trivalent metallic ions for 0% RH [253]. The kink structure seems to be remarkable in Fe-DNA and Zn-DNA in comparison with Mn-DNA, Ca-DNA, and Mg-DNA. Nevertheless, the spectra in Li-DNA and Na-DNA show monotonic change around 3100 cm^{-1}. The kink indicates that the divalent and trivalent metallic ions have some effects on the imino group (NH) contributing to the hydrogen bond between base molecules. We shall return to this point later.

The spectra for 90% RH at 700~1800 cm^{-1} are shown in Fig. 8.2(a). The small absorption peak at 838 cm^{-1} is a B-form marker band corresponding to a puckering of C2′-endo in the sugar molecule. Other sugar vibration is observed at 1052 cm^{-1}. The symmetric stretching band of PO_2^- is observed at 1092 cm^{-1} in Li-DNA and Na-DNA, while the bands in *M*-DNA with divalent and trivalent metallic ions exhibit constant red shift at 1087 cm^{-1}, as shown by a red solid line.

The absorption band at 1232 cm^{-1} is attributed to the asymmetric stretching mode of PO_2^- reflecting B-form. Red shifts emerge in *M*-DNA with divalent and trivalent metallic ions as well as in the symmetric stretching mode just mentioned above. The amount of red shift, however, tends to increase gradually from 1232 cm^{-1} (Li-DNA and Na-DNA) to 1218 cm^{-1} (Zn-DNA), as denoted by a blue broken line. The red shift implies that the environment around PO_2^- changes depending on the metallic ions.

The absorption band at 1600~1750 cm^{-1} is composed of various stretching modes (C=C, C=N, C=O) and NH_2-bending vibrations due to the base molecule, and additionally the HOH-bending vibration of water molecules is superposed. Two peaks at 1664 and 1709 cm^{-1} depend little on the type of counterions. Among them, the peak at 1709 cm^{-1} is classified as a B-form marker band.

As demonstrated in Fig. 8.2(b), the infrared spectra for 0% RH are distinctly changed by dehydration. Weak absorptions identified as an A-form marker band due to the sugar molecule are detected around 885 cm^{-1}, which shifts to higher wavenumber compared to

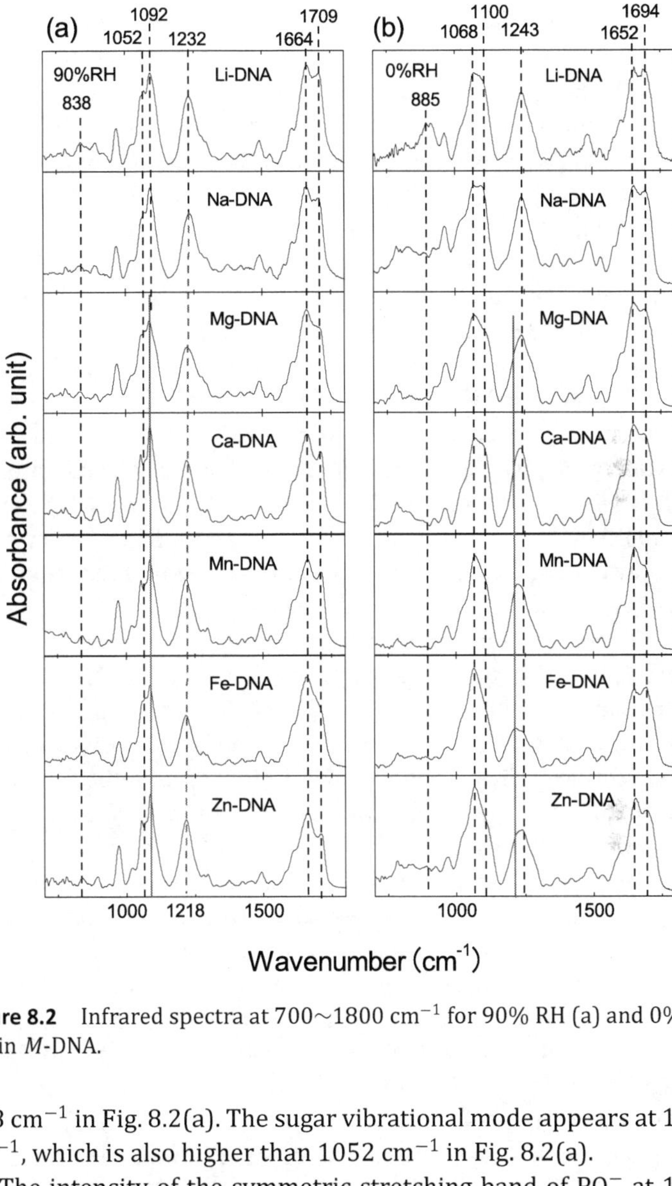

Figure 8.2 Infrared spectra at 700~1800 cm^{-1} for 90% RH (a) and 0% RH (b) in *M*-DNA.

838 cm^{-1} in Fig. 8.2(a). The sugar vibrational mode appears at 1068 cm^{-1}, which is also higher than 1052 cm^{-1} in Fig. 8.2(a).

The intensity of the symmetric stretching band of PO_2^- at 1100 cm^{-1} in Fig. 8.2(b) has a tendency to reduce when seen from top (Li-DNA) to bottom (Zn-DNA). The reduction reveals that the bandwidth is broadened with an increase in the atomic number of metallic ions. Since the asymmetric stretching band of PO_2^- appears at 1243 cm^{-1}, the conformation is identified as A-form. In Li-DNA and Na-DNA, the absorption band is well fitted by a single Lorentzian curve. To reproduce the absorption bands for other *M*-DNA, however, two Lorentzian curves are required, and then a splitting is found to occur in the asymmetric stretching band through the incorporation of divalent and trivalent metallic ions. The high-frequency component is positioned at around 1243 cm^{-1}, which coincides with the peak frequency in Li-DNA and Na-DNA. As indicated by a red solid line, the low-frequency component emerges at 1214 cm^{-1}, which is irrelevant to the kinds of metallic ions. The wavenumber of 1214 cm^{-1} is indeed close to one for a B-form marker band, though the whole structure of *M*-DNA never takes B-form at low humid condition. Therefore, the incorporation of divalent and trivalent metallic ions conducts local structural change around PO_2^-. These experimental facts suggest that the configuration of the divalent and trivalent metallic ions may be different from that of the neighborhood of the phosphate group.

We should remind here that the asymmetric stretching band for 90% RH (Fig. 8.2(a)) exhibits the red shift in *M*-DNA with divalent and trivalent metallic ions. Taking these experimental facts into consideration, the stretching modes are very sensitively affected by the valence and configuration of counterions in *M*-DNA. Furthermore, the hydration structure, especially around PO_2^-, considerably differs from Fig. 3.10.

A symmetric stretching mode is associated with the amplitude change of an electric dipole due to vibrational motions against electric fields of infrared light. In contrast, an asymmetric stretching mode corresponds to orientational change of an electric dipole in addition to amplitude change. If the phosphate group is influenced by additional electric fields inducing spatial anisotropy, the asymmetric stretching frequency should exhibit a shift in comparison with nearly isotropic environments. The anisotropy seen from PO_2^- originates from the anisotropic charge distribution corresponding to the configuration of divalent and trivalent metallic ions in *M*-DNA. In addition, the electric dipole of hydrated water molecules may also contribute to the anisotropy.

A large amount of water molecules hydrate and screen the electric dipole of PO_2^- for 90% RH. The screening suppresses the anisotropy around the phosphate group. For 0% RH, however, the anisotropy is hardly screened by the water dipole, and then we are allowed to observe the distinct spectral splitting and variation of magnitude.

In Li-DNA and Na-DNA, the charge neutrality is achieved by a monovalent ion, which contacts to single phosphate group. If a divalent (trivalent) ion is hydrated to two (three) phosphate groups, considerable anisotropies should exist even in high relative humidity, and a band splitting must be observed in the asymmetric stretching modes of PO_2^- for 90% RH. As far as the present infrared spectroscopy is concerned, it is difficult to consider that the divalent and trivalent metallic ions contact to the phosphate group. This result conducts the presence of a free phosphate group. Then large amounts of water molecules have to hydrate and screen the negative charge in the phosphate group.

The molecular vibrations themselves due to sugar and base molecules, except for the imino group, are independent of the type of metallic ions introduced. The original structures are fundamentally conserved at the portions of these molecules. On the other hand, the incorporation of divalent and trivalent metallic ions has a profound effect on the phosphate group and imino group. Such effect reveals that the divalent and trivalent metallic ions are located close to the base molecules.

According to the far-infrared measurement reported so far, extremely broad absorptions with weak intensity are observed around 310 cm^{-1} for *M*-DNA with divalent and trivalent metallic ions [253]. The results of ESR, NMR, and fluorescence experiments also show that imino protons are possible to replace a metallic ion [16, 28]. Taking account of these results, the broad band around 310 cm^{-1} is attributed to novel N–*M* stretching vibrations [265]. Because of the formation of N–*M* bond, a free phosphate group without counterions emerges in *M*-DNA. As a consequence, the hydration structure especially around PO_2^- in *M*-DNA with divalent and trivalent metallic ions is hugely different from that in Li-DNA and Na-DNA.

8.3 Summary of the Infrared Spectra in *M*-DNA

In terms of the infrared spectral difference in *M*-DNA samples, the intercalation of divalent and trivalent metallic ions yields the novel N–*M* bonding and the distinct variation of hydration structure especially around the phosphate group without counterions. Infrared spectroscopic studies on DNA systems are quite helpful not only to examine the basic properties, but also to develop further device applications and engineering.

Chapter 9

Charge Doping in Zn-DNA

Kenji Mizoguchi and Hirokazu Sakamoto
Department of Physics, Tokyo Metropolitan University, Hachioji, Tokyo, 192-0397 Japan
mizoguchi@phys.se.tmu.ac.jp, sakamoto@phys.se.tmu.ac.jp

9.1 Introduction

In most *M*-DNA systems, the charge transfer from metal ions to DNA does not occur, except for Fe-DNA, where Fe^{2+} transforms into Fe^{3+} by releasing an electron to the bases of a base pair, as described in Chapter 7. UV/Vis absorption spectra revealed that Fe^{3+} absorption typical of $FeCl_3$ emerged in addition to that of the base π bands. The magnetism of Fe^{3+} was observed as a mixture of the $S = \frac{5}{2}$ high spins and the $S = \frac{1}{2}$ low spins. However, the magnetism of the π electron spins was not observed probably because of nonmagnetic ground state of the π spin system, in which the localized nature of the π band is concerned. In this chapter, the other example of the charge-transferred system, Zn-DNA, is described in detail, which shows an interesting role of water molecules in governing the magnetic properties of Zn-DNA. The electronic states of Zn-DNA

DNA Engineering: Properties and Applications
Edited by Kenji Mizoguchi and Hirokazu Sakamoto

ISBN 978-981-4669-46-7 (Hardcover), 978-981-4669-47-4 (eBook)
www.panstanford.com

depend on the sample preparation procedure [28]. The conventional ethanol precipitation procedure produces usual *M*-DNA systems with M^{2+} ions surrounded by hydrating water molecules, which prevent the metal ions from forming covalent bonds with the nitrogen atoms of bases. In contrast, Omerzu *et al.* reported that a freeze-drying procedure leads to different electronic states from that by the ethanol precipitation procedure, as shown in Figs. 9.1 and 9.2 [24]. The electronic states of the new freeze-dried Zn-DNA (FD-Zn-DNA) system have been uncovered by the other approach than that by Omerzu *et al.* in sample preparation [28]. The FD procedure removes almost all the water molecules from Zn-DNA and induces

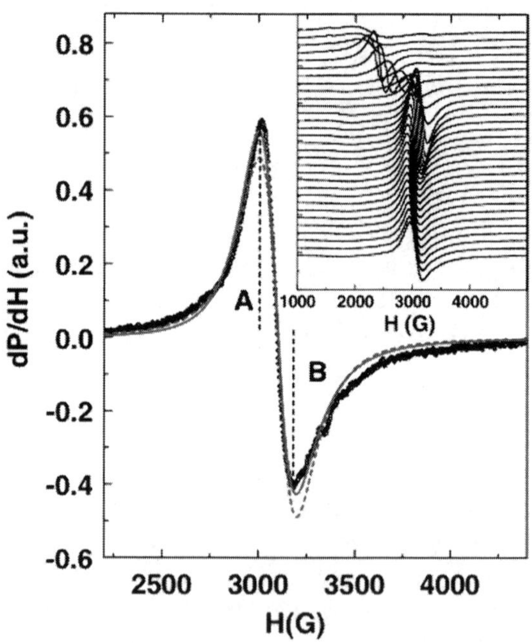

Figure 9.1 The room-temperature ESR spectrum of Zn-DNA (black circles) fitted with an asymmetric Lorentzian (Dysonian) (solid red line) and a Lorentzian (dashed blue line) curve. The inset shows the temperature development of the Zn-DNA ESR resonance measured on cooling from the room temperature (the bottom curve) down to 10 K (the top curve) measured at every 10 K. Reprinted with permission from Ref. [24]. Copyright 2010, American Physical Society.

the metal ions to form covalent bonds with the nitrogen atoms of bases, in place of imino hydrogen of thymine or guanine. The electronic states of the induced π electron system are dominated by the circumstance of ambient humidity. The dehydrated FD-Zn-DNA has a nonmagnetic π spin system because of antiferromagnetic interaction between the π spins. In contrast, the hydrated FD-Zn-DNA shows Pauli-like temperature-independent paramagnetism with the π-band width of 0.24 eV.

9.2 Freeze-Dried Zn-DNA

Omerzu *et al.* have reported a new type of Zn-DNA synthesis. They applied the freeze-drying procedure to a mixture of Zn-DNA, $ZnCl_2$, and tris-HCl buffer solution [24]. Figure 9.1 shows the ESR spectra at room temperature and X-band, along with the temperature dependence of the spectra in the inset. The ESR lineshape was fitted by an asymmetric Lorentzian, which shows a better reproduction than that with a Lorentzian. The temperature dependence of the ESR intensity together with the ESR linewidth and g value, reprinted in Fig. 9.2, looks like Pauli-like temperature-independent paramagnetism, with which they concluded that the degenerated π electron system had been realized with DNA. They proposed that the charge transfer from the tris-HCl buffer to DNA is the mechanism underlying the observation.

One definite thing resulted from their report is that the electronic states of Zn-DNA are not unique but depend on the sample preparation method. Unfortunately, it would be difficult to make the mechanism unambiguously clear with their mixture sample, because the mixture sample is not necessarily suitable for studying the physical properties other than ESR. On the basis of our experience so far, we assumed that the removal of water molecules from Zn-DNA by the freeze-drying procedure would be an intrinsic mechanism underlying these findings. Thus, we started this investigation of the mechanism using high-purity Zn-DNA prepared by ethanol precipitation. Finally, a new model for describing all the present findings is proposed.

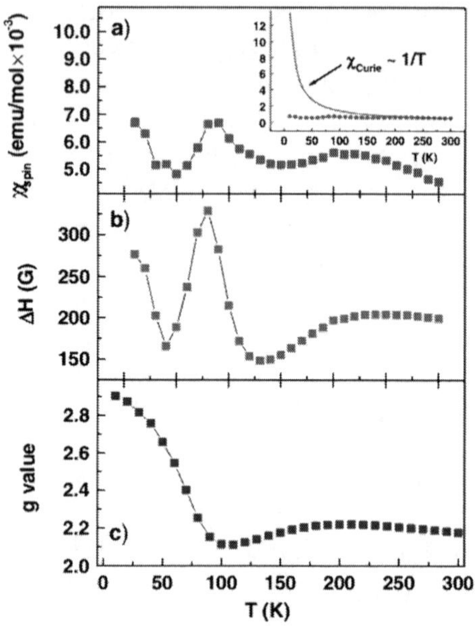

Figure 9.2 Temperature dependences of (a) spin susceptibility χ_{spin}, (b) a resonance linewidth ΔH, and (c) a g value. The inset of (a) shows comparison of the temperature dependence of the spin susceptibility of Zn-DNA with the Curie law on an expanded scale. The susceptibility is expressed in moles of DNA base pairs. Reprinted with permission from Ref. [24]. Copyright 2010, American Physical Society.

9.2.1 *Preparation of High-Purity Freeze-Dried Zn-DNA*

Zn-DNA is prepared with an aqueous solution of 2 mM salmon sperm DNA supplied by the Ogata Materials Science Lab. and Sigma-Aldrich and 20 mM $ZnCl_2$ purchased from Wako Pure Chemicals and Sigma-Aldrich. Excess cold ethanol at $-20°C$ is poured into the transparent DNA-$ZnCl_2$ aqueous solution resulting in a transparent precipitate of Zn-DNA. The residual $ZnCl_2$ is washed out thoroughly from the obtained precipitate with an excess amount of pure ethanol, in which DNA is insoluble. Thus, we obtained pure Zn-DNA without $ZnCl_2$ or any other buffer materials. For the preparation of FD-Zn-DNA, an aqueous solution of Zn-DNA was dripped into a flask at liquid nitrogen temperature and was frozen instantly. The flask was immersed in a low-temperature bath down to −15°C and evacuated with a cold trap for several days to a week. The final product, completely dehydrated FD-Zn-DNA, has a white-colored polystyrene form. Samples for a superconducting quantum interference device (SQUID) susceptometer and ESR were sealed in quartz tubes with He gas for thermal exchange.

9.2.2 *Magnetic Susceptibility of Freeze-Dried Zn-DNA*

Figure 9.3 shows SQUID magnetization curves of B-DNA and Zn-DNA at 100 K, prepared by different procedures. The open diamonds with the solid line show the diamagnetism of B-DNA. The magnetization of Zn-DNA dried in air ("Air") is represented by the closed circles, which shows the large DNA diamagnetism with tiny paramagnetism, which was assigned to magnetic impurities [157]. This result is consistent with the previously reported conclusion on the ethanol-precipitated *M*-DNAs, that is, the metal ions maintain divalency with ionic bonds in *M*-DNA and thus no charge transfer from M to DNA has occurred [25, 93–95, 157]. An effect of freeze drying applied to Zn-DNA is demonstrated typically in FD-Zn-A and FD-Zn-B (Fig. 9.3). FD-Zn-A shows markedly large nonlinear paramagnetism against the magnetic field B with the diamagnetism of Zn-DNA. In contrast, FD-Zn-B has smaller nonlinear paramagnetism, i.e., less than one-third that of FD-Zn-A, and linear paramagnetism in addition to the diamagnetism. Here we have two characteristic features to discuss concerning the FD-Zn-DNA:

(1) Nonlinear paramagnetism in all Zn-DNAs
(2) Linear paramagnetism in FD-Zn-B

Concerning the first point, in the 1960s, strong nonlinear paramagnetism of the FD-Zn-DNA was observed and extensively studied. Blois Jr. *et al.* concluded on the basis of their experiments and consideration that the submicron ferromagnetic particles tightly attached to the nucleic acid yielded strong nonlinear paramagnetism, as described in Section 2.3.1 [84]. In recently available commercial DNA samples, such strong contamination is eliminated. Recently, however, Lee *et al.* have observed a nonlinear paramagnetism at 300

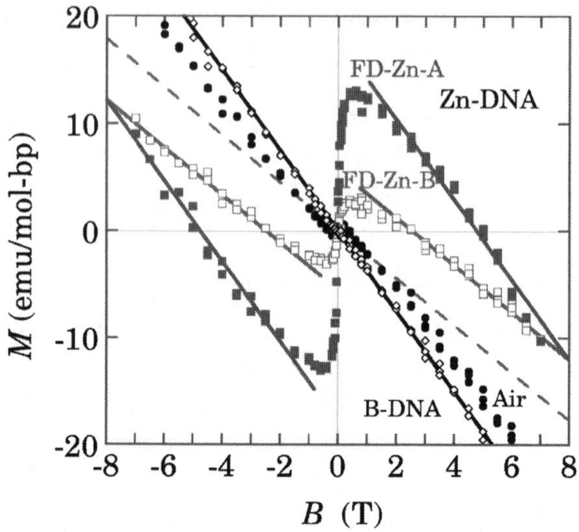

Figure 9.3 Magnetization per unit containing one base pair of B-DNA and various Zn-DNAs at 100 K, at which oxygen molecules contaminating the data have only a small contribution, as demonstrated in Figs. 2.12 and 2.13 [94]. Open diamonds represent the diamagnetism of as-received salmon DNA (B-DNA). Closed circles show the magnetization of Zn-DNA dried in air ("Air") with large diamagnetism of Zn-DNA and small paramagnetism. FD-Zn-A and FD-Zn-B represent the magnetization of two different batches of the FD-Zn-DNA with much larger nonlinear paramagnetism than that of the Zn-DNA dried in air. The solid straight line on FD-Zn-A is almost parallel to the diamagnetism of B-DNA. In contrast, the baseline of FD-Zn-B represented by the dashed line clearly contains the additional linearly increasing paramagnetism with the magnetic field B over the diamagnetism of B-DNA. Reprinted with permission from Ref. [28], Copyright 2014, The Physical Society of Japan.

K, especially in B-DNA evacuated for a long term (*without Zn* ions), but with very small amplitude compared with that of FD-Zn-A, as described in Section 2.3.4 [34]. The present data shown in Fig. 9.3 are, however, reasonably understood from the simple saturating, nonlinear paramagnetism up to 7 T within the uncertainty, as suggested by the straight lines for FD-Zn-A and FD-Zn-B. By a long-term evacuation of B-DNA at room temperature, we also reproduced a similar magnitude of nonlinear paramagnetism with one-tenth of M_0 of FD-Zn-A. Furthermore, we confirmed by the freeze-drying procedure that B-DNA without Zn ions shows large nonlinear paramagnetism comparable to that of FD-Zn-A, as shown in Fig. 2.12. These facts suggest that nonlinear paramagnetism emerges in the completely dehydrated DNA as a subsidiary effect of the freeze-drying procedure but is not related to the presence of Zn ions.

One key feature for interpreting the origin of the nonlinear paramagnetism is its magnitude, which is described as $M_0 = Ng\mu_B S$ in the ferromagnetic case, corresponds to as small as 0.3% of the $S = \frac{1}{2}$ spin or 0.01% of the $S = \frac{5}{2}$ spin per base pair, suggesting an impurity effect. The Fe impurity concentration of 0.1%–0.2% per base pair obtained by X-ray fluorescence analysis is consistent with that of a mixture of the high spin states ($S = \frac{5}{2}$) and the low spin states ($S = \frac{1}{2}$). The second point will be discussed in Section 9.3 with other new data and with the model for interpreting all the data consistently.

9.2.3 *ESR of Freeze-Dried Zn-DNA*

The ESR spectra of FD-Zn-A are shown in Fig. 9.4. The intensity of the main broad ESR signal at $\approx$0.326 T ($g \approx 2.1$) is approximately temperature independent. In contrast, the Curie law behavior of the intensity of the six small hyperfine split signals for isolated Mn impurities at 0.3–0.36 T is characteristic. The temperature independence of the main ESR signal is apparently consistent with both the nonlinear paramagnetism (Fig. 2.12) and the Pauli-like temperature-independent paramagnetism. Thus, it is noteworthy that when one interprets the temperature-independent ESR intensity, such as in Fig. 9.4 and in the report by Omerzu *et al.* reproduced in Figs. 9.1 and 9.2 [24], it is impossible to differentiate the Pauli-like paramagnetism from the nonlinear paramagnetism without the magnetization data. In contrast, this fact is incompatible with the reported conclusion based on the orbital magnetization of the persistent ring current, as discussed in Section 2.3.4 [34], which does not give ESR signals in general. The details of the ESR spectra in Fig. 9.4 do not necessarily agree with Figs. 9.1 and 9.2; the ESR

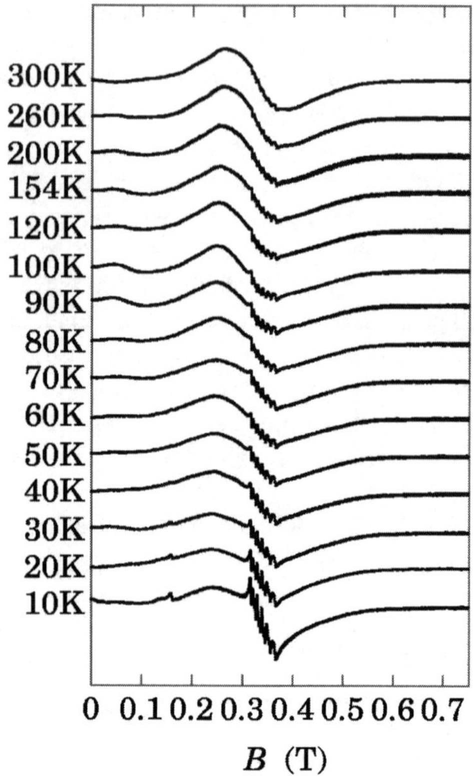

Figure 9.4 ESR spectra of FD-Zn-A at X-band from 10 to 300 K. The main signal at $\approx$0.326 T ($g \approx 2.1$) corresponds to the nonlinear paramagnetism shown in Fig. 9.3. The six hyperfine split signals at 0.34 T originate from Mn impurities, which are isolated from each other. Note that the intensity of the isolated Mn signals increases with decreasing temperature following the Curie law, but the broad main signal shows no marked change in its area with temperature. The linewidth of the main signal is $\approx$0.1 T at 300 K and gradually increases with decreasing temperature. Reprinted with permission from Ref. [28], Copyright 2014, The Physical Society of Japan.

linewidth in Fig. 9.4 is several times broader than that in Figs. 9.1 and 9.2 and the g value also shows some difference, especially in the low temperature range below 100 K. These differences suggest a naive nature of these ESR signals, as discussed throughout this chapter.

9.3 Nature of Freeze-Dried Zn-DNA

9.3.1 *Effect of Moisture on Magnetic Properties*

The observation of the unique magnetic behavior in the course of sample treatments gave us a crucial insight into the electronic states of FD-Zn-DNA. Freeze-dried samples appear to have a polystyrene-like tiny closed cell structure with nanoscale holes for water molecules to exit from the cells. Just after the freeze-drying procedure, the closed cells of a sample are almost empty because of the long-term evacuation in the procedure. After that, the sample was transferred to a quartz tube in air, and He gas for thermal exchange was introduced after a short-term evacuation. This procedure in air introduces air contamination into the sample, and the short-term evacuation is insufficient to remove all the oxygen molecules via the nanoscale holes, as confirmed in the χ–T data shown in Fig. 9.5, which reveal considerable oxygen

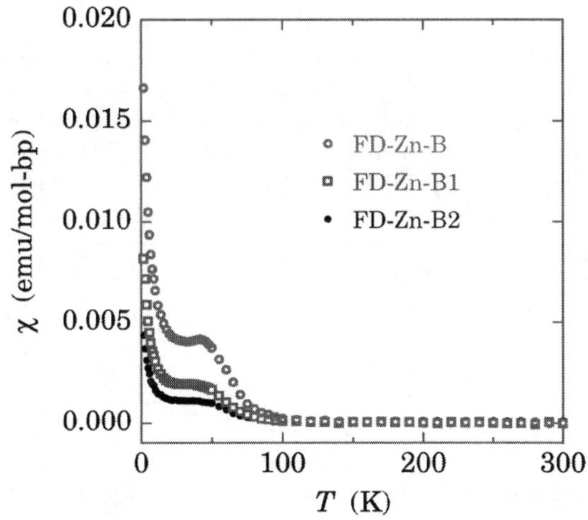

Figure 9.5 Temperature dependence of the magnetic susceptibility χ of FD-Zn-B, FD-Zn-B1, and FD-Zn-B2 at 1 T. The large bump below 100 K and the Curie–Weiss behavior below 30 K originate from the oxygen molecules trapped in the cells of the FD-Zn-DNA. The effect of the mechanical treatment is clearly demonstrated as a reduction in the oxygen contribution below 100 K. Reprinted with permission from Ref. [28], Copyright 2014, The Physical Society of Japan.

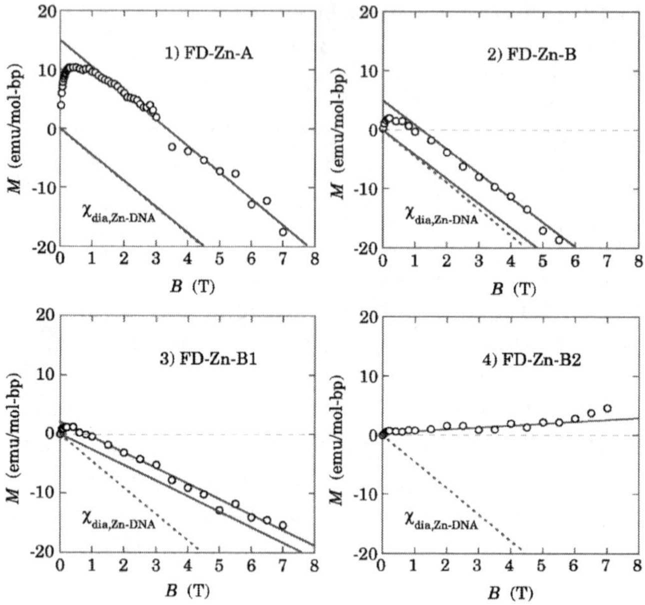

Figure 9.6 Magnetization curves of FD-Zn-DNA at 300 K, prepared by the treatments described in the text. The solid straight lines in the above panels represent the baseline and the saturation lines of nonlinear magnetization. The dotted line describes the diamagnetism of Zn-DNA. Reprinted with permission from Ref. [28], Copyright 2014, The Physical Society of Japan.

contamination below 100 K [33]. Thus, we tried to break the closed cells of the sample mechanically, which assists in removing air from the sample with the short-term evacuation.

The evolution of magnetization in FD-Zn-DNA by mechanical treatments is shown in Figs. 9.5 and 9.6. FD-Zn-A and FD-Zn-B are prepared from the same source materials but belong to different freeze-drying batches. After measurements of FD-Zn-B, the sample was exposed to an "air+humidity" condition for carrying out a treatment to remove the closed cell structure. FD-Zn-B1 is FD-Zn-B treated in "air+humidity," that is, FD-Zn-B1 corresponds to the partially hydrated FD-Zn-B. After measurements of FD-Zn-B1 in a sealed quartz tube, FD-Zn-B1 was treated again in "air+humidity" to obtain FD-Zn-B2 as further hydrated FD-Zn-B and measured in a quartz tube. The oxygen contamination below 100 K in the magnetic susceptibility [33] was markedly reduced by the treatments, as shown in Fig. 9.5.

In addition to the oxygen reduction, this treatment generated interesting reproducible changes in the magnetic properties, as demonstrated in Fig. 9.6:

(1) Suppression of the saturation magnetization M_0, and
(2) Emergence of the temperature-independent linearly increasing paramagnetism χ_{para} with increasing B.

Figure 9.7 shows the linear, nonsaturating paramagnetic susceptibility of the well-hydrated FD-Zn-B2, which is almost temperature independent. As the first step to understanding the electronic states of FD-Zn-B2, we assumed the Curie–Weiss formula with one $S = \frac{1}{2}$

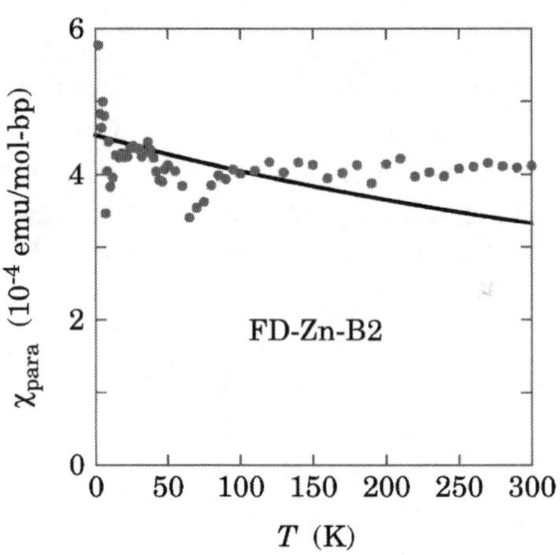

Figure 9.7 Temperature dependence of the paramagnetic susceptibility χ_{para} of FD-Zn-B2, which was obtained by subtracting the diamagnetism of Zn-DNA, the oxygen contamination, and the small nonlinear paramagnetism shown in Fig. 9.6 - 4) from the SQUID susceptibility data. The ambiguity of the estimated χ_{para} depends on the temperature; below 100 K, the subtraction of the oxygen contribution gives a large ambiguity, but that above 100 K is represented by the data scattering. The solid curve represents the least-squares fit to the Curie–Weiss formula, Eq. (9.1), which is in poor agreement with the data. Reprinted with permission from Ref. [28], Copyright 2014, The Physical Society of Japan.

spin per base pair:

$$\chi_{\mathrm{CW}} = \frac{N_{\mathrm{A}} g_e^2 \mu_{\mathrm{B}}^2 S(S+1)}{3k_{\mathrm{B}}(T+\theta)}, \tag{9.1}$$

where N_{A} is the Avogadro constant, g_e is the g-factor for a free electron, μ_{B} is the Bohr magneton, and k_{B} is the Boltzmann constant. The solid curve in Fig. 9.7, representing χ_{CW} with $\theta \approx 820$ K, is consistent with the magnitude of the observed paramagnetism but fails to reproduce the Pauli-like temperature-independent behavior of FD-Zn-B2.

The next possible interpretation is the Pauli susceptibility with $\chi_{\mathrm{Pauli}} = N(E_{\mathrm{F}})\mu_{\mathrm{B}}^2$, where $N(E_{\mathrm{F}})$ is the density of states per eV at the Fermi energy. With this relation, the observed $\chi_{\mathrm{para}} = 4.0 \times 10^{-4}$ (emu/mol-bp) in Fig. 9.7 gives 12.4 (states/eV), which indicates that FD-Zn-B2 is a strongly correlated π electron system. With the one-dimensional tight-binding approximation, the π-band width is estimated as $\approx$0.05 eV ($E_{\mathrm{F}} = 25$ meV), which is inconsistent with the experimental finding of the temperature-independent susceptibility up to 300 K ($\approx$25 meV). Therefore, we assumed that the electronic structure of the hydrated FD-Zn-DNA with A-form [39, 93, 94] is three dimensional (3D). Actually, the exchange interaction of A-form Mn-DNA is 3D [94]. With the relation $\frac{N(E_{\mathrm{F}})}{N} = \frac{3}{2E_{\mathrm{F}}}$ for the 3D free-electron model, the bandwidth is estimated as 0.24 eV. This order of the π-bandwidth is common in organic systems, such as molecular conductors and fullerides. Thus, it is reasonably concluded that the hydrated FD-Zn-DNA has a highly correlated π-band with a width of $\approx$0.24 eV. This small bandwidth restricts the intraband absorption energy to much less than that of visible light, which is consistent with the transparency of the sample.

In contrast, the completely dehydrated FD-Zn-DNA such as FD-Zn-A carries no χ_{Pauli} but shows only nonlinear paramagnetism as a subsidiary property to the freeze-drying procedure applied to DNA, as discussed in Sections 2.3.4 and 9.2.2. Thus, FD-Zn-A carries no paramagnetism originating from the π spin system. It is, however, not probable that the hydrated water molecules in the mechanical treatments cause DNA to create this π spin system. Therefore, this fact suggests that the freeze-drying procedure generates the π electron spin system, which loses its magnetism in the dehydrated FD-Zn-DNA, probably because of the strong antiferromagnetic coupling between the neighboring π electron spins. Further discussion will appear in Section 9.3.2 with the model for interpreting these observations.

The temperature-independent paramagnetic susceptibility, shown in Fig. 9.7, is consistent with the ESR intensity reported by Omerzu *et al.* [24]. Unfortunately, they could measure the ESR intensity only at the X-band because of the dilute mixture sample of the FD-Zn-DNA with excess $ZnCl_2$ and tris-HCl buffer. It is also difficult to extract the temperature-independent paramagnetism using the SQUID magnetometer with the dilute mixture sample. Thus, it is unclear whether they have observed the nonlinear paramagnetism (Figs. 9.3 and 9.4) or the Pauli paramagnetism (Fig. 9.7). The signal intensity of the nonlinear paramagnetism is very strong because of the ferromagnetic enhancement of the local microwave field, which is consistent with the enough strong ESR signal even with their dilute mixture sample. Furthermore, on the basis of the fact that the dehydrated FD-Zn-DNA carries only nonlinear paramagnetism, it is considered that they observed the nonlinear paramagnetism in the dehydrated FD-Zn-DNA mixture.

9.3.2 *Electronic States of Freeze-Dried Zn-DNA*

We propose a model for interpreting the intrinsic magnetic properties of FD-Zn-DNA: the freeze-drying procedure generates one $S = \frac{1}{2}$ spin at each base pair, which shows Pauli paramagnetism in the hydrated FD-Zn-DNA but no Pauli paramagnetism in the dehydrated state. The divalent metal ions in the conventional *M*-DNA synthesized by ethanol precipitation have usually been hydrated with many water molecules, as shown in Fig. 7.1. These water molecules prevent the metal ions from forming covalent bonds with the neighboring nitrogen atoms of the bases [25, 56]. As a result, the metal ions form ionic bond with PO_4^- anions of the DNA backbones.

The freeze-drying procedure removes water molecules from a frozen aqueous solution, which ideally retains the separation between DNA double helices in the solution. Finally, all the water molecules around the metal ions are removed and then each metal

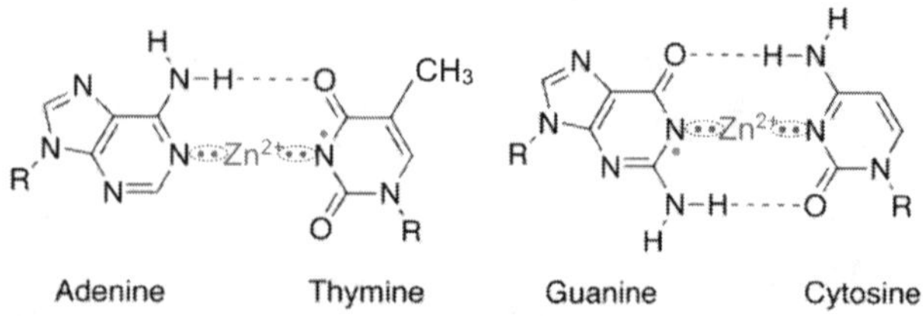

Figure 9.8 Model for the FD-Zn-DNA. All the hydrated water molecules are removed by the freeze-drying procedure. Zn ions prefer to form configurational bonds in place of the imino hydrogen atoms of guanine and thymine bases. As a result, one unpaired electron on the nitrogen atom provides $S = \frac{1}{2}$ spin per base pair. Reprinted with permission from Ref. [28], Copyright 2014, The Physical Society of Japan.

ion is allowed to form preferable configurational bonds with two nitrogen atoms, having a lone electron pair, in place of the imino hydrogen atom of the guanine or thymine base, as shown in Fig. 9.8, similar to metal phthalocyanines. The loss of the imino hydrogen atom in the guanine or thymine base changes the number of electrons from even to odd, that is, the creation of an unpaired π electron spin in each base pair of FD-Zn-DNA takes place.

The generated π electron spins should delocalize all over the base pair with the Zn ion to reduce the kinetic energy. The magnetic interaction between the spins of neighboring base pairs separated by approximately 3.4 Å is expected to be antiferromagnetic. The dehydrated FD-Zn-DNA shows only diamagnetism ascribable to that of Zn-DNA, which suggests that the expected paramagnetism of the π electron spins is completely suppressed by the strong antiferromagnetic coupling between the spins of the neighboring base pairs. Thus, the possible electronic states of the dehydrated FD-Zn-DNA at room temperature are some nonmagnetic ground states with a transition temperature much higher than 300 K.

There are several possible nonmagnetic states in the antiferromagnetic quasi-one-dimensional chain, such as antiferromagnetic (AF) and spin density wave (SDW) states with B parallel to the spontaneous magnetization, spin Peierls (SP), charge density wave (CDW), and charge ordered (CO) states. Since the present sample is in the powdered form, there is no possibility of AF and SDW states for a candidate of the nonmagnetic state. The magnetic susceptibility in the nonmagnetic state disappears exponentially below the transition temperature T_C. Then, T_C should be higher than 600 K, twice the observed temperature of 300 K. It is interesting for SP states that the Hubbard antiferromagnetic exchange energy of a hole carrier in the A-form DNA (A-DNA) is estimated as $J = -2t^2/U = -0.05$ eV with the published values for the transfer energy t on the order of 0.25 eV and the onsite Coulomb repulsion U on the order of 2.5 eV for AT–AT and GC–GC pairs [266]. Thus, the SP state appears to be one of the possible models based on the magnitude of the exchange energy for the dehydrated FD-Zn-DNA at 300 K. It should, however, be noted that (1) the present system of the dehydrated FD-Zn-DNA does not have the same electronic structure as that of the calculated A-DNA; for example, the present estimation of the π-bandwidth is only 0.24 eV; (2) the Peierls instability of the DNA systems concerning on SP and CDW seems to be implausible because of the sizable structural disorder inherent in DNA [267]; and (3) the coherence length of the phonons in DNA is relatively short [268]. The last candidate is the CO state, in which the off-site Coulomb repulsion V is a crucial parameter other than U. From the value of estimated U and V by Starikov [266], V is comparable to or larger than U in AT–AT and GC–GC pairs, which suggests the strong possibility of the CO state in the dehydrated FD-Zn-DNA. A molecular size more than 15 Å for a base pair of Zn-DNA, which is much larger than the separation of ≈3.4 Å between the neighboring base pairs, would cause the condition $V > U$. Thus, the π electrons prefer an alternate double occupancy instead of every single occupancy, by paying a cost of U in place of V, resulting in the CO state with the repetition $(\ldots 0 : -2e : 0 : -2e : 0 \ldots)$.

In the case of the hydrated FD-Zn-DNA, the water molecules play an important role in reducing the off-site Coulomb repulsion of the π electrons, that is, the screening effect by the water molecules surrounding the base pairs, which results under the condition $V < U$ and the Pauli paramagnetic state with the single occupancy in every base pair instead of the double occupancy. Thus, the present model for Zn-DNA treated by the freeze-drying procedure successfully describes the intrinsic part of the anomalous magnetic behavior with the screening effect of the water molecules.

9.4 Reported Theoretical Models for *M*-DNA

Electronic properties of some *M*-DNAs were discussed experimentally in detail up to the former section. *M*-DNA has been first reported as a new structure of DNA by Lee *et al.* [16], as discussed in Section 1.2.2. They also suggested a structural model of *M*-DNA and metallic conductivity due to the metal *d*-band [17, 57]. Experimental results on *M*-DNA in this book can be interpreted on the basis of the structure proposed in Ref. [17], which has metal ions located between the bases of a base pair in a DNA duplex. Following the reports by Lee *et al.*, theoretical researches have been performed on electronic properties of *M*-DNA: origin of a HOMO band and a HOMO–LUMO gap. We review three theoretical papers, in which metal ions are situated between the bases of a base pair. Two of them concern the electronic structure of *M*-DNA including a Zn^{2+} ion [21, 22]. The third one deals with the magnetic property of Mn-DNA [23], which is the topic related to the discussion of Section 7.3. Kino *et al.* also discussed theoretically the insertion of carriers into DNA by metal cations (Mg^{2+}, Zn^{2+}, and Ca^{2+}) and emphasized the importance of dehydrated metal cations [20]. The cations, however, are close to the phosphates of DNA backbones, outside a duplex. Thus, this model is inconsistent with the experimental results shown in this book.

In 2006, Alexandre *et al.* used first principles calculations to investigate the modified forms of DNA with divalent Zn, Co, Fe cations bonded to the two helix strands of poly(dC)-poly(dG) [21]. They first relaxed a single C-Zn^{2+}-G monomer with the sugar-phosphate backbones, in which a Zn^{2+} ion replaced the guanine imino proton and binding guanine and cytosine under three different coordination geometries: octahedral, square-planar, or tetrahedral. In order to get charge neutrality, the divalent Zn^{2+} cation (that replaces a proton as proposed by Rakitin *et al.* [17]) was saturated with one hydroxyl group OH^-, in addition to one or more neutral water molecules. The initial structures of octahedral and square-planar geometries converged to the same final geometry, with the Zn^{2+} ion in a tetrahedral geometry and a single water molecule and the hydroxyl bound to it. Another initial structure, based on the proposal by Rakitin *et al.* [17], converged to a different, distorted tetrahedral geometry, which was unsuitable for periodical stacking. They achieved similar geometries for C-Co^{2+}-G and C-Fe^{2+}-G. The initial geometries of dimmer and trimer fragments were generated by appropriate rotation and translation of the tetrahedral geometry of the relaxed monomer. Using the central base pair of the relaxed trimer, they generated a periodic 11-base pair helix with an OH^--Zn^{2+}-OH^- zigzag chain, for which they studied the electronic properties of *M*-DNA.

Figure 9.9 shows the band structures and the density of states of Zn-DNA, Co-DNA, and Fe-DNA, close to the Fermi level. They found, in all cases, that the incorporation of the metal ions resulted in a large reduction in the original DNA band gap of 2.0 eV to much smaller values of 0.65 eV for Zn-DNA, 0.50 eV for Co-DNA, and 0.27 eV for Fe-DNA. In Zn-DNA, the states with a large weight in the metal ions are far from the Fermi level and the HOMO is strongly localized in guanine, as in pure DNA. The reduction in the HOMO–LUMO gap is because the deformation produced by the Zn ion decreases the distance between consecutive guanine bases and increases their π–π hopping matrix elements. In Co-DNA and Fe-DNA, the states close to the Fermi level have a large weight in the metal cations and the HOMO states are localized around the metal ion, suggesting a large conductivity effect along the metal-hydroxyl chain. As discussed in Chapter 6, Kumeta *et al.* proposed the mechanism of carrier production in FD-Zn-DNA, in which the guanine imino proton is replaced by the Zn^{2+} ion like the model used by Alexandre *et al.* [28]. The Zn^{2+} ion, however, forms configurational bonds with the guanine and thymine bases, which is different from the theoretical result by Alexandre *et al.*

In 2006, Fuentes-Cabrera *et al.* carried out a comprehensive *ab initio* study of eight plausible models for the G-Zn-C base pair to elucidate the structure and electronic properties of Zn-DNA [22]. They studied whether a relaxed model prefers to be planar on the basis of the idea that such a "planar" model is more suitable than a nonplanar one for building an *M*-DNA duplex: favorable stacking of consecutive base pairs is easier when the base pairs are planar. Out of all the studied models, there is only one that preserves its planarity upon full geometry optimization, but this model leads to a parallel Zn-DNA architecture only. Next, to estimate

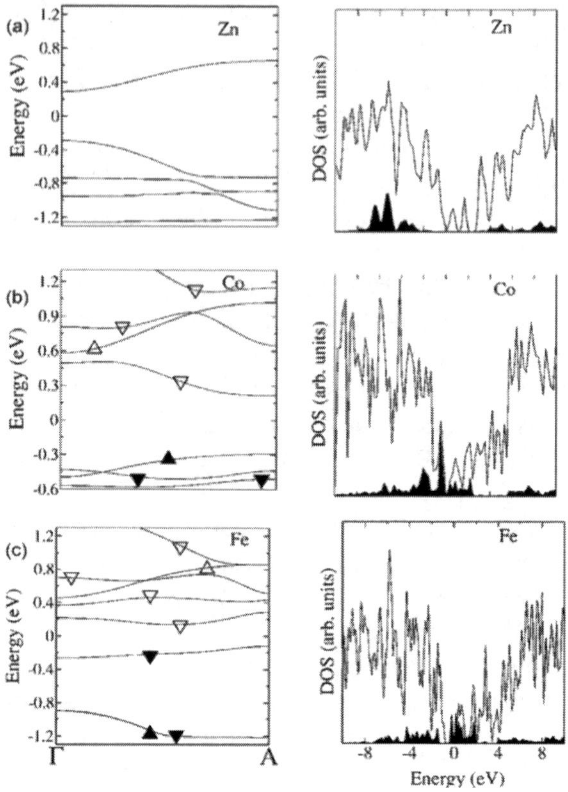

Figure 9.9 Band structure (left) and density of states (DOS, right) of Zn-DNA, Co-DNA, and Fe-DNA. Upward- and downward-pointing triangles indicate up- and down-spin bands, with filled and empty signs indicating filled and empty states. The Fermi level is the zero of the energy scales. The magnetic moment per base pair is 0, 3, and 4 Bohr magnetons for Zn-DNA, Co-DNA, and Fe-DNA, respectively. Empty and filled curves are the total DOS and the projected DOS on the metal atoms. Reprinted with permission from Ref. [21]. Copyright 2006, American Physical Society.

how much energy it takes to flatten a nonplanar model, they have investigated the difference in energy between a nonplanar model and its planar counterpart (obtained by a restricted optimization that enforced planarity). They have found that the particular model, which they call the planar Lee model, requires only 4.5 and 7.1 kcal/mol to be flattened in gas and solvent conditions, respectively, and so suggest that the planar model is a good starting point for creating an antiparallel Zn-DNA duplex. In this model, the imino proton is replaced by Zn that sits in the minor groove of the GC base pair. Thus, this duplex should not show the imino proton in NMR measurements, which would agree with the experimental evidence for Zn-DNA. The metal atom placed in the minor groove has also been suggested by other authors. To study the electronic properties that correlate with Zn-DNA having a metal-like conductivity, they have calculated the nature of the HOMO level and the value of the HOMO–LUMO gap for each of the presented eight G-Zn-C models. They used three different exchange-correlation functionals: the local density approximation (LDA), the generalized gradient approximation (GGA), and hybrid functional B3LYP. The HOMO level of neither of the models studied contains Zn *d*-states, irrespective of the approximation used, indicating that it is unlikely that Zn-DNA has a Zn *d*-band situated at the Fermi level that could facilitate charge transport. The enhanced conductivity of Zn-DNA could be due to a significant decrease in the HOMO–LUMO gap as compared to that of DNA. Indeed, they found that the HOMO–LUMO gap of the G-Zn-C base pairs is, at most, 28% smaller than that of a Watson–Crick GC base pair, but they recognize that more research will be needed to understand whether further reduction in the HOMO–LUMO gap takes place upon stacking base pairs.

In 2007, Mallajosyula and Pati studied the alignment of magnetic ions Cu^{2+} and Mn^{2+} along the modified DNA helix using first principles density functional theory (DFT) calculations to find low-dimensional spin channels [23]. They used a three-base-pair step segment (GC-GC-GC) of the Drew–Dickerson dodecamer as the template for their simulations [269]. The metallobase pairs (X*M*X, where X corresponds to hydroxypyridone) are incorporated into the DNA template by replacing the central base pair step in the structure by the metallobase pair (GC-X*M*X-GC) [18, 270]. They optimized the geometry of this new structure and obtained the initial structures. The incorporation of the metallobase pairs did not deform the helical shape and the pitch of the helix. In the monomer fragments extracted from the optimizing calculations, Cu^{2+} and Mn^{2+} are in distorted square-planar geometry and coordinated to the oxygens of the hydroxypyridone base. They found that the band gap of natural

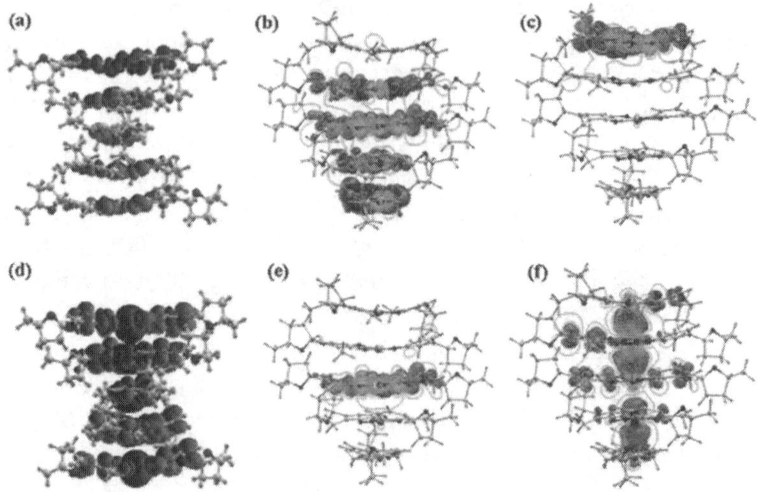

Figure 9.10 Isosurface plot for difference in spin density ($\Delta\rho$) and wave function (state) plots with additional contour plot about the plane containing the metal ions for *M*-DNA. Top: Cu-DNA (a) $\Delta\rho$ Cu-DNA, (b) state, and (c) state. Bottom: Mn-DNA (d) $\Delta\rho$ Mn-DNA, (e) state, and (f) state. The isosurface plots are shown at the same isovalue. Reprinted with permission from Ref. [23]. Copyright 2007, American Physical Society.

DNA calculated to be 1.6 eV reduces to 0.25 eV for trimer and 0.15 eV for pentamer of Cu-DNA and to 0.27 eV for trimer and 0.10 eV for pentamer of Mn-DNA. For the dimer structure of both Cu-DNA and Mn-DNA, they also found that coupling between spins on the magnetic ions is ferromagnetic with the ferromagnetic exchange coupling J ($\mathcal{H} = J\mathbf{S}_i \cdot \mathbf{S}_j$) of -0.075 eV and -0.19 eV for Cu^{2+} and Mn^{2+} ions, respectively. Figure 9.10 shows the spin density $\Delta\rho = \rho_\uparrow - \rho_\downarrow$. For Cu-DNA, the spin state has major contributions from $d_{x^2-y^2}$ orbital of Cu and thereby the in-plane delocalization. On the other hand, for Mn-DNA, the spin state is delocalized through the out-of-plane due to the contribution from the d_{z^2} orbital of Mn. It is worth to note ESR measurements by Mizoguchi *et al.* [157]. They showed that magnetization of Mn-DNA is antiferromagnetic below $T \approx 0.4$ K for A-form, as shown in Fig. 7.24 and Curie-like for B-form, as described in Fig. 7.18. The above-mentioned theoretical calculations are inconsistent with the experimental results.

9.5 Summary of Electronic Properties in Freeze-Dried Zn-DNA

We unambiguously demonstrated that the freeze-drying procedure applied to Zn-DNA generates a π electron spin at each base pair by the formation of the covalent bonds between the Zn ion and the nitrogen atoms of the bases. The magnetic properties of FD-Zn-DNA can be controlled by adjusting the humidity. The intrinsic nature of the π electron spin system generated by the freeze-drying procedure is the nonmagnetic charge-ordered ground state with the off-site Coulomb repulsion $V > U$ in the dehydrated FD-Zn-DNA. In contrast, FD-Zn-DNA hydrated by water molecules shows the Pauli paramagnetism of the π-band with a width of $\approx$0.24 eV. This marked change with hydration is due to the screening effect of the off-site Coulomb repulsion V between the π electrons of the neighboring base pairs by the water molecules.

Omerzu *et al.* have reported that the FD-Zn-DNA has a degenerate electron spin system on the basis of the temperature dependence of ESR intensity [24]. In the present study, we successfully reproduced the temperature-independent paramagnetic susceptibility in the FD-Zn-DNA. However, the magnetization saturates below 0.1 T in the dehydrated state, whereas the Pauli paramagnetism emerges in the hydrated FD-Zn-DNA. Thus, it is uncertain whether they observed the nonlinear paramagnetism or the Pauli paramagnetism, but the nonlinear paramagnetism would be more probable.

The nonlinear paramagnetism appears commonly in the freeze-dried B-DNA and Zn-DNA, which suggests that the magnetic impurities in DNA cause it, that is, the nonlinear paramagnetism is a subsidiary effect of the freeze-drying procedure regardless of the formation of a metal–DNA complex. The saturation magnetization M_0 corresponds to 0.3% of the $S = \frac{1}{2}$ spin or 0.01% of the $S = \frac{5}{2}$ spin per base pair, which is compatible with the possible magnetic impurities in B-DNA. The nonlinear paramagnetism provides a usual ESR signal around $g \approx 2.1$ with the temperature-independent intensity, which contradicts the reported interpretation with the orbital magnetization of the persistent ring current [32, 34]. One possible scenario for nonlinear paramagnetism is super-paramagnetism of magnetic ion clusters: one divalent magnetic ion inserted in a base pair would trigger cluster formation by breaking the hydrogen bonds of the near-neighbor base pairs and by making it preferable for the other divalent magnetic ions to insert into a disconnected base pair and then to form a cluster. It was found that nonlinear paramagnetism disappeared with increasing humidity, suggesting a role of water molecules in reducing the coherence length of the exchange interaction, which is effectively carried out by the water molecules with their inhomogeneous distribution, as suggested by the hyperfine splitting anomaly in Mn-DNA [56].

Theoretical researches on the electronic states of *M*-DNA suggested that the reported enhancement of the electrical conductivity is due to a reduction in the HOMO–LUMO gap. Although the predicted semiconducting band structure is consistent with the experimental results, few of the theoretical researches take account of the effect of water molecules. Since the role of the water molecules is essential for the electrical and magnetic properties of *M*-DNA, as shown in Zn-DNA and Mn-DNA, the reported models are insufficient to explain the experiments. Thus, a new theoretical model is expected, by which we can understand further the experimental data on *M*-DNA.

Chapter 10

Structure of *M*-DNA Studied with STM

Kenji Mizoguchi and Hirokazu Sakamoto
Department of Physics, Tokyo Metropolitan University, Hachioji, Tokyo, 192-0397 Japan
mizoguchi@phys.se.tmu.ac.jp, sakamoto@phys.se.tmu.ac.jp

10.1 Introduction

In Chapters 7 and 8, we have found a lot of indications and evidences that the metal ion inserted into *M*-DNA is located between the bases of a base pair, as shown in Fig. 7.1. Several are listed as follows:

(1) The π–π^* absorption spectrum of A-T base pairs is markedly suppressed by the insertion of metal ions, such as Mn and Fe, as shown in Figs. 7.5 and 7.6, suggesting strong interaction between the bases and the metal ions.
(2) The valence of Fe in Fe-DNA is +3 because of the charge transfer from Fe^{2+} to a base pair, suggesting also the strong interaction between Fe and the bases of the base pair.
(3) The ESR linewidth of B-form Mn-DNA film is well accounted for by the electron spin dipolar interaction of a one-dimensional chain of Mn ions located at the center of double-stranded B-form DNA, as described in Section 7.3.3.1.

DNA Engineering: Properties and Applications
Edited by Kenji Mizoguchi and Hirokazu Sakamoto

ISBN 978-981-4669-46-7 (Hardcover), 978-981-4669-47-4 (eBook)
www.panstanford.com

(4) ESR lineshapes of a B-form Mn-DNA film are qualitatively different from that of an A-form Mn-DNA film. Such a change of the ESR lineshape is uniquely understood by the characteristic dimensionality of Mn ions in the B-form and the A-form Mn-DNA, as discussed in Section 7.3.3.2.
(5) ESR spectra of Mn-DNA films are unchanged after the treatments; dissolve it in an NaCl aqueous solution and recrystallize it. If Mn ions are not located between the bases of a base pair, the Mn ions should be exchanged for abundant ions of Na in the solution, because the Mn ions are hydrated in Mn-DNA, as demonstrated in Section 7.3.3.3.
(6) Metal ions form bonds with the nitrogen atoms of bases, as demonstrated in Chapter 8.
(7) The Zn ions of freeze-dried Zn-DNA form covalent bonds with the nitrogen atoms of bases, as demonstrated in Chapter 9.

Thus, the structure in Fig. 7.1 is highly probable for *M*-DNA. It is very interesting to observe it directly with a scanning tunneling microscope (STM) as a most expectable tool to reach an isolated nanoscale object of *M*-DNA double helix. An atomic force microscope (AFM) applicable to insulating materials cannot provide enough resolution to resolve a single atom.

It has been difficult to visualize DNA in atomic details even with an excellent nanoscale observation technique of STM developed by Binnig and Rohrer in 1982 [271]. Recently, partial sequencing of a single DNA molecule with STM was successfully attained with the use of characteristic electronic states of guanine base molecules [272]. A variety of reports on STM images of DNA with atomic resolution have appeared so far [273, 274]. The most significant issue to be overcome is the insulating nature of DNA, since STM works only with metallic samples, which have finite density of states $N(E_F)$ at the Fermi energy E_F of an STM probe tip. In general, STM cannot be applied to molecules, which have no density of states at E_F. However, it was reported that the bases of a single-stranded (SS) oligo-DNA [275] and an inert gas molecule Xe [276] attached to a metallic substrate could be imaged with STM in atomic details because of their flat molecular or single atom structure, in which only a single atom is located between the metallic substrate and the STM probe tip. In this situation, the interaction between the conduction electrons of the metallic substrate and the electrons in the highest occupied molecular orbital (HOMO) of the molecule produces small but finite density of states $N(E_F)$ through the resonance broadening. Since a Watson–Crick type helical structure of 2 nm in diameter is too thick to produce the finite $N(E_F)$, imaging of a helical DNA in atomic details with STM would be hopeless. In contrast, if there is a novel flat structure of double-stranded DNA, such as the flat SS oligo-DNA, a new possibility to visualize DNA in atomic details would open up. We have examined this possibility to visualize the structure of engineered DNA.

Highly oriented pyrolytic graphite (HOPG) has been frequently used as a substrate for the observation of DNA and other molecules because even under the ambient condition, it is possible to observe the atomic scale images of the carbon atoms of HOPG. Thus, in 2007, we started STM study on the *M*-DNA structure with a substrate of HOPG for several years and found several candidate images for DNA and *M*-DNA in atomic details. However, a caution has been aroused on the mimic structures similar to the Watson–Crick double helix in size of several nm and in periodicity of 1.8–5.3 nm, which is compared with 3.4 nm for B-form DNA periodicity [277, 278], but is much larger than 0.6–0.7 nm of the DNA backbone periodicity found in this chapter. After that, the other substrates for STM have been reported, for example, single crystals of metal, or insulating glass plates or mica plates covered with a conducting layer of water that carries tunneling current [279]. The other technique of atomic force microscope (AFM) has also been reported, even with lower resolution than STM.

In this section, we demonstrate the results obtained by experimental effort for several years.

10.2 Sample Preparation for STM Study

Salmon sperm DNA was provided by Wako Pure Chemical Ind., Ltd., which has a fibrous morphology. Mn-DNA is prepared from an aqueous solution of salmon sperm DNA, with $MnCl_2$ at the molar ratio of DNA to $MnCl_2$ 1:10. Excess cold ethanol at $-20°C$

is poured into the transparent DNA-$MnCl_2$ solution, resulting in the formation of a transparent precipitate of Mn-DNA. The residual $MnCl_2$ is washed out thoroughly from the obtained precipitate with pure ethanol, in which DNA is insoluble. Thus, the obtained Mn-DNA is stored in a refrigerator and dissolved in pure water for STM measurements.

A freshly cleaved HOPG surface is prepared by removing the surface layers with sticky tape. To check any mimic structures [277, 278] of HOPG supplied by Nanosurf, the freshly prepared surfaces have been carefully surveyed without/with passing through pure water without DNA samples, but any mimic structures similar to the rope–ladder form of DNA, found in this chapter, could not be found, except for HOPG mono- or multi-layer steps. The DNA samples for STM measurements were prepared by lifting a freshly cleaved HOPG substrate from the 10^{-5}–10^{-7} mol-base pair (bp)/l aqueous DNA solution, with keeping the surface up by 10 degree from a vertical position. The STM images are taken under ambient condition by EasyScan 2 (Nanosurf) with the probe tip of a mechanically sharpened $Pt_{0.8}Ir_{0.2}$ wire of 0.25 mmϕ and typically the bias voltage of $+100$ mV and the current set point of 1 nA ($10^8\ \Omega$).

10.3 STM in DNA

Figure 10.1 shows the STM image of salmon sperm DNA on the HOPG surface, in which brightness corresponds to the height of the STM probe tip. This image looks like a loosened double-helical structure with the flat rope–ladder form. The separation of the two bright arrays is $\approx$2 nm, which corresponds to the diameter of a DNA double helix. The sequential repetition of $\approx$0.65 nm agrees well with that of DNA backbones. Thus, this image can be interpreted as follows. DNA double helix in an aqueous solution sticks to the edge of a step in the HOPG surface. Then the helix becomes loose by some reason, for example surface tension of water, and finally adheres to the HOPG conductive surface by the electrostatic attraction, as schematically described in Fig. 10.2.

The flat rope–ladder structure on a conductive substrate is stabilized by the Coulomb attraction between the charges of DNA or

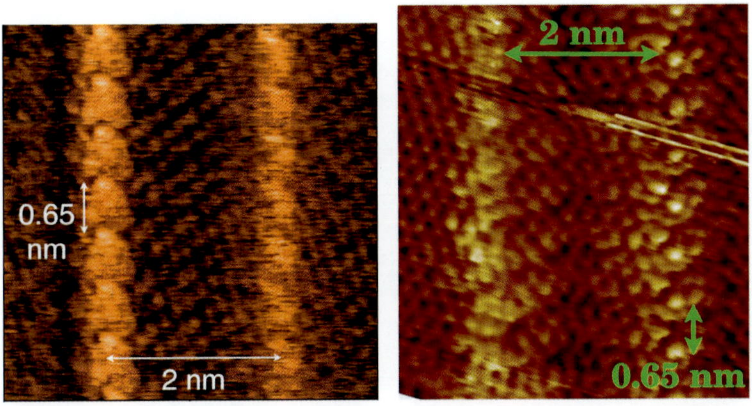

Figure 10.1 Two examples from the STM images of DNA in (left) 4.3 × 4.3 nm^2 and (right) 4.6 × 4.8 nm^2, taken at 100 mV and 1 nA. The separation of two bright arrays is ≈ 2 nm and the sequential repetition is ≈ 0.65 nm for left, which are similar to those of an unwound double-stranded DNA. In right, the sequential repetition is similar, but the separation is somewhat larger than 2 nm, suggesting elongated hydrogen bonds. These scales are consistent with another example of STM image for DNA. The relative height of the brightest part corresponds to ≈0.15 nm. Reprinted with permission from Ref. [25], Copyright 2011, Society of Photo Optical Instrumentation Engineers.

M-DNA in a solution, such as PO_4^- and M^{2+}, and the image charges on the conductive HOPG substrate with signs opposite to the charges of DNA or *M*-DNA, as shown in Fig. 10.2. The image charges on the conductive surface are produced by the electron evacuation (positively charged) in the substrate near the negative charge of PO_4^- or by the electron attraction (negatively charged) in the substrate near the positive charge of M^{2+}. In addition, the bases parallel to the HOPG surface are attracted to the HOPG surface by the van der Waals interaction, as reported in the SS oligo-DNA [275]. As the other example similar to the flat rope–ladder structure, Cluzel *et al.* reported that B-DNA transforms to S-DNA with the length of 1.7 times longer than that of B-DNA by mechanical stretching [280]. The detailed structure of S-DNA is unknown, but Léger *et al.* have proposed [281] that the helical pitch of S-DNA is 22 nm with 38 base pairs per right-handed turn, which means the helical structure is partly unwound, similar to the rope–ladder structure in Figs. 10.1

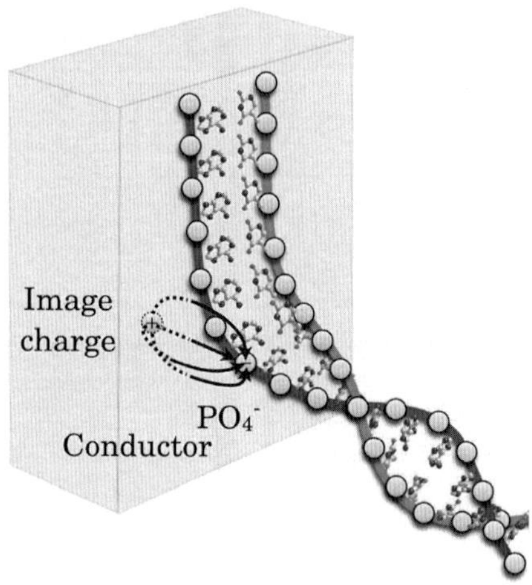

Figure 10.2 A schematic image for the formation process of the flat rope–ladder DNA on a conductive substrate such as HOPG, by the Coulomb attraction between the induced image charge, and the PO_4^- and M^{2+} of *M*-DNA in an aqueous solution. The surface tension of the solution pulls down DNA, which helps to unwound the helical structure of DNA.

and 10.2. Lebrun *et al.* have reported several molecular models of stretched B-DNA [282]. One of the models has a rope–ladder structure similar to the present findings.

The region between the bright arrays in Fig. 10.1 is relatively dark, in which the base pairs of DNA should be located. This is, however, an expected behavior: in principle, the isolated base molecules have no density of states at E_F of HOPG, giving rise to no tunneling current to HOPG and no extra height from the HOPG surface. However, the small but finite interaction between the bases and the conduction electrons of HOPG can generate extra heights, as in the case of the Xe atom on the Ni surface [276]. Concretely, Xe on Ni yielded the observed height of $\approx$0.15 nm from the Ni surface, which is much smaller than the van der Waals diameter of Xe, $\approx$0.42 nm.

There is a further disadvantage in generating the STM height. A mismatch of the atomic positions between HOPG and the bases reduces the interaction between the atoms of bases and HOPG, as described in Fig. 10.3. This argument is also applied to the observed

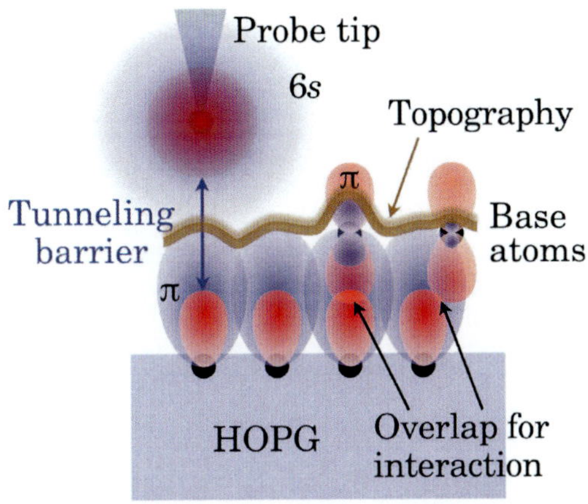

Figure 10.3 Schematic image of wavefunctions in STM measurement. The wavefunctions are schematically represented by the small (red) circle for the probe tip and ellipses for HOPG, which correspond to the classical electron orbitals. The region outside the circle and ellipses, in which the tails of wavefunctions decay exponentially, works as a tunneling barrier, and the overlap of the tails of the probe tip and HOPG gives tunneling current [283]. To keep the tunneling current unchanged, the smaller the local density of states $N(E_F)$ at E_F is, the closer the probe tip should be. Schematically, the large (blue) circle for the probe tip and ellipses for HOPG represent such constant $N(E_F)$ curve, and the resulting scanned trace (called topography) is represented by the curve along them [284, 285]. $N(E_F)$ for the base atoms can be generated only by the interaction between the base electrons and the HOPG conduction electrons [276]. Thus, the effective height for the base atoms depends on "the overlap for the interaction" between the small (red) ellipses of the base and HOPG; in the left atom of the base, the effective height is much larger than that in the right one because of better overlap of electron orbitals.

height of less than 0.1 nm for the bright arrays assigned to the massive DNA backbones, which is much smaller than the structural height of 0.3–0.5 nm, depending on the actual conformations of the DNA backbones. It is also interesting to note that the two bright DNA backbones in Fig. 10.1 show the remarkable difference from each other in structure and brightness. This fact coincides with the opposite running directions of the two backbone chains composed of deoxyribose sugar and phosphate groups. In one of the oppositely directed backbone chains, phosphate groups touch with the HOPG

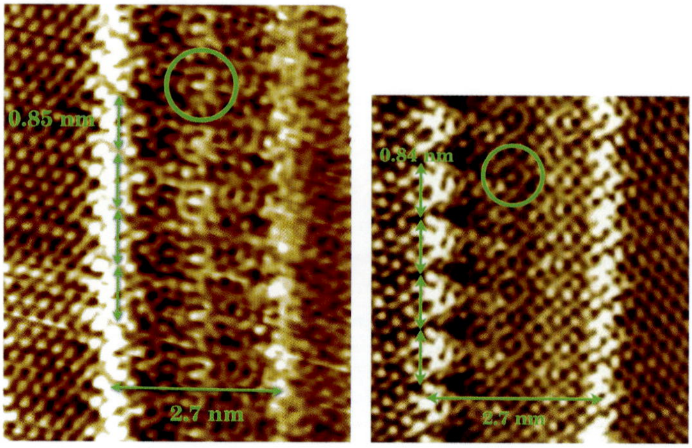

Figure 10.4 STM images of two different Mn-DNA samples on HOPG. The area is 5.3×6.8 nm^2 for left and 4.7×5.1 nm^2 for right. The separation between the bright arrays is $\approx$2.7 nm, which is much wider than 2.0 nm for DNA. The sequential repetition of $\approx$0.85 nm is also longer than 0.65–0.7 nm in Fig. 10.1 and in Ref. [25]. The definite difference from DNA is in the center part between the bright arrays. One bright spot surrounded by a ring-like structure at the center part is found in each sequential unit, as marked by circles. The relative height of the brightest part corresponds to 0.25 nm for left and 0.2 nm for right. Dots outside two bight arrays are images of graphite. Reprinted with permission from Ref. [25], Copyright 2011, Society of Photo Optical Instrumentation Engineers.

surface, but deoxyribose sugar groups do in the opposite chain. They have different heights from each other. Thus, the STM image describes the opposite sides of the backbones.

10.4 STM in *M*-DNA

Figure 10.4 shows the independent STM images of two different Mn-DNA samples measured 2 months apart. The scales are similar to each other; the separation of 2.7 nm between the bright arrays and the sequential repetition of 0.84–0.85 nm, which are much larger than 2.0 nm in separation and 0.65 nm in sequential repetition of DNA in Fig. 10.1. The analysis of the Mn ESR concluded that the nature of the Mn ion in Mn-DNA is purely ionic with hydrating

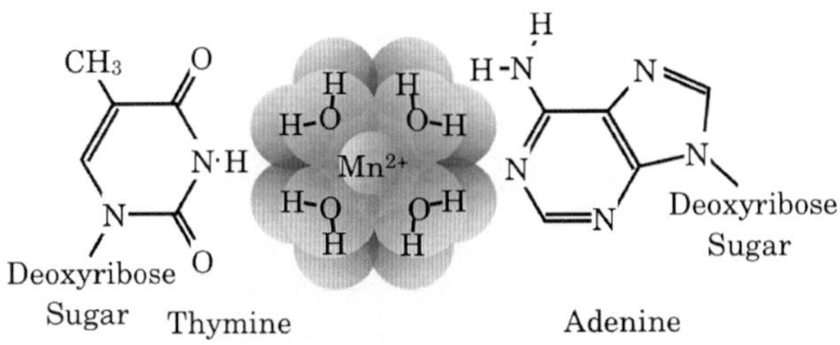

Figure 10.5 Model structure of thymine-Mn-adenine of Mn-DNA prepared with the ethanol precipitation method. Metal ions are hydrated by water molecules and form ionic coupling with two PO_4^- ions in DNA backbones.

water molecules, as described in Section 7.3.3.3 [56]. Thus, these quantitative differences in scale are consistent with the structure of Mn-DNA in Fig. 10.5, in which the hydrated Mn ion with $\approx$0.7 nm of diameter is located between the bases of a base pair. The circles in Fig. 10.4 suggest the corresponding circles of hydrating water molecules with a dot of Mn^{2+} ion. In the right-hand side region of the circles, fused ring-like images are also found, corresponding to the purine bases, and in the left region, single rings of the pyrimidine bases can be found, especially in Fig. 10.4(left).

In contrast, details of the two sets of the bright arrays and the areas between them are rather different from each other. The difference of details looks like natural phenomena, if one takes into account the flexible structure of DNA backbones and the requirement for getting STM spots of the base atoms that sufficient overlap of the wavefunctions with that of HOPG is necessary. As an extreme case, if any of the wavefunctions in the base does not sufficiently overlap with any carbon atoms of HOPG, the image of the base cannot be observed with STM, but that of HOPG instead. Thus, the DNA image on HOPG sensitively depends on a position relative to HOPG.

Figure 10.6 shows the STM image of Co-DNA. We found clear differences in scales and in structures from that of Mn-DNA in Fig. 10.4. The scales would reflect the difference of the center structure of the hydrated Co^{2+}. The semicircle structure of the hydrated Co^{2+} reduces the sequential repetition from 0.85 nm for Mn-DNA to 0.67 nm for Co-DNA. These differences suggest a variety

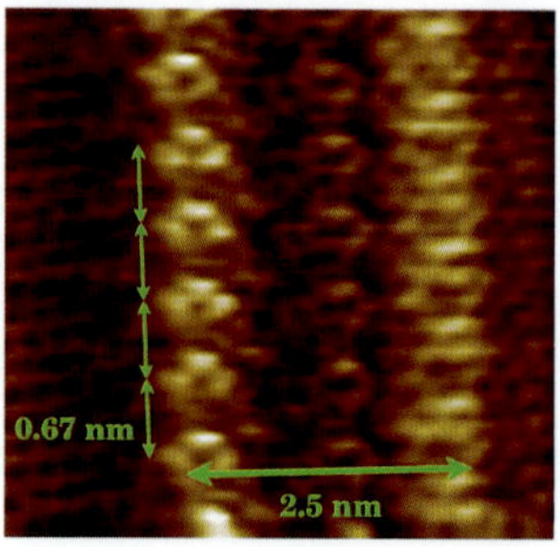

Figure 10.6 An STM image of Co-DNA on HOPG. Note the definite difference from that of Mn-DNA in Fig. 10.4. Main difference is the center structure of the inserted Co^{2+} ions and water molecules, which look like semicircles. The close distance between the semicircles and the right bright array suggests that the right backbone might fold over the bases at the right-hand side of the semicircles. The relative height of the brightest part corresponds to $\approx$0.15 nm.

of electronic states and bonding nature of M^{2+} in *M*-DNA, which is interesting to study comparatively with the other techniques.

10.5 Summary of Structure of *M*-DNA

Finally, we have discovered the novel structure of DNA with the flat rope–ladder form on HOPG. By virtue of this discovery, direct visualization of DNA and *M*-DNA with STM becomes successful in atomic details. This breakthrough provides new possibility to study DNA structure with STM. *M*-DNA is highly stable, that is, it refuses to duplicate the genetic codes and to produce proteins. Thus, the present result might be helpful to understand the role of the heavy metal ions to the lives. However, several issues remain to be overcome for the further development of the present chapter. The most significant issue to expand this work is the sample preparation technique. Since HOPG is a highly stable substrate, it is rare for *M*-DNA in aqueous solution to attach to the HOPG surface and is further rare to find the flat rope–ladder structure on HOPG. The second is the localized nature of the π-electron wavefunctions in HOPG, which prevents us from obtaining the complete image of

M-DNA on HOPG, because of the mismatching in positions of the wavefunctions of bases and HOPG. If one could use plain-wave-like, spatially extended wavefunctions of the conduction electrons of noble metals, the complete STM image of *M*-DNA would be obtained. We hope the present technique develops further with the aid of a variety of research fields to provide a useful way to unveil the DNA structures directly.

Part IV

Applications of DNA

Chapter 11

DNA as a Material

Naoya Ogata[†]
Emeritus Professor of Sophia University and Chitose Institute of Science and Technology, 3-3-7-704 Aoba, Chitose, Hokkaido, 066-0015 Japan
n-ogata@photon.chitose.ac.jp

11.1 Introduction

It is well known that deoxyribonucleic acid (DNA) is a biopolymer with a helical structure, carrying genetic information for all the lives on the earth, and contained in a nucleus of the cell, which constitutes the lives. DNA can be expressed as the root of the life. The DNA molecule is a polyphosphate of super molecular weight from 10^7 to 3×10^9 base pair (bp) with a double-helical structure composed of A-T and G-C base pairs made from the four bases adenine (A), thymine (T), guanine (G), and cytosine (C), which carry the genetic codes.

It is also well known that Watson and Crick proposed the double-helical structure of DNA in 1953 [2], for which they were awarded the 1962 Nobel Prize in Physiology or Medicine, together with Wilkins [37]. Although the sequences and physiological information technology had been globally reported, application of DNA as a

[†] This is his posthumous work.

DNA Engineering: Properties and Applications
Edited by Kenji Mizoguchi and Hirokazu Sakamoto

ISBN 978-981-4669-46-7 (Hardcover), 978-981-4669-47-4 (eBook)
www.panstanford.com

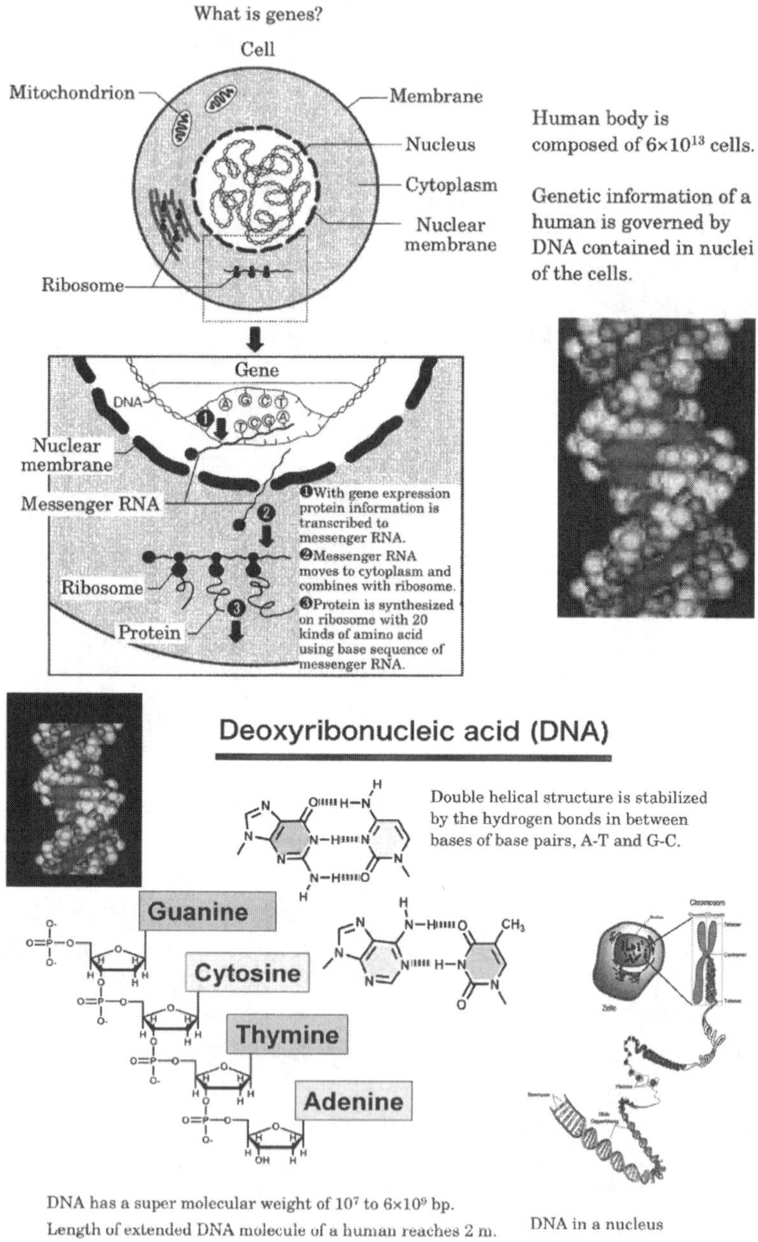

Figure 11.1 DNA in a cell.

material was started only after Okahata, in the Tokyo Institute of Technology, successfully synthesized a DNA–lipid complex film, soluble in alcohol, in 1996 [286].

DNA is soluble only in water, which limits the variety of applications of DNA as a material. However, this weakness has been overcome by exchanging the usual counter-cations of Na for cationic lipids. The obtained DNA–lipid complex is soluble in polar solvents such as alcohol with the double-helical structure preserved. This makes many applications of DNA as a material possible.

In the double-helical structure, the base pairs of A–T and G–C stack up parallel with the molecular planes, between which various aromatic compounds are intercalated by π–π interaction. This peculiar phenomenon of intercalation can strongly enhance the optical and electrical properties of the aromatic compounds. For example, the fluorescence intensity of ethidium bromide intercalated in DNA is enhanced by more than 100 times. Hence, the intercalation of aromatic compounds into DNA opened new possibilities for applications as optical and electrical materials [287–297].

11.2 Production of High Purity DNA

11.2.1 *Production of DNA from Marine Resources*

The molecular weight of DNA folded in the nucleus of a living cell ranges from 10^{10} to 2×10^{12} for a human DNA, whose length reaches as long as 2 m. It is not well known how such a huge DNA molecule can be precisely replicated when a cell divides, which is a secret of life. It is not so simple to extract DNA molecules from the nuclei of cells because the nucleus is a complex of DNA with histone, that is, basic protein. Since DNA molecules occupy only less than 1% of cell weight, it is desirable for mass production to utilize bioresources with higher abundance of DNA in a cell, such as scallops and salmon-milt, as described in Fig. 11.2, with the abundance of up to 10%.

First, salmon-milt is transformed to a gel form by a homogenizer and then a large quantity of protein, such as histone, is decomposed into amino acid by adding a lot of enzymes, such as protease. Next, a massive scale of alcohol is added to a purified aqueous solution

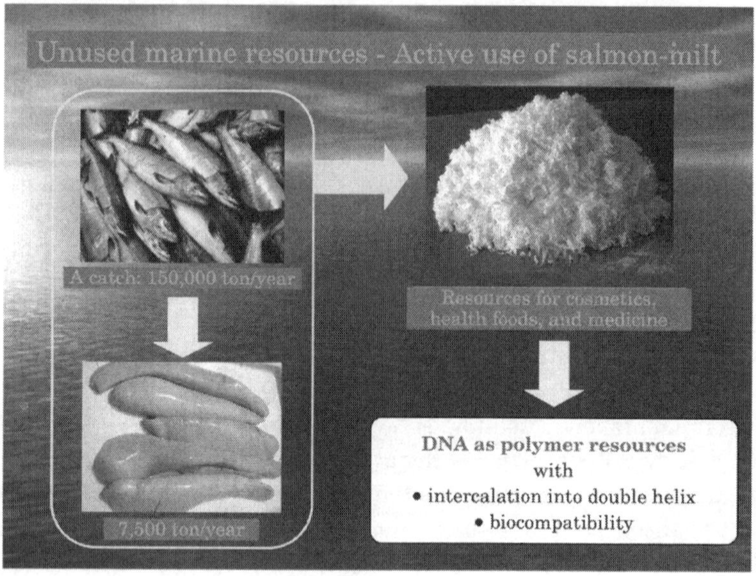

Figure 11.2 Active use of DNA from unused marine resources.

of DNA after the filtration of unreacted materials, and then the purified DNA is collected as a precipitate. After that, the precipitate is dissolved again in water to repeat the procedure until the purity of DNA reaches 98%. The procedure of separation and refinement of DNA from bioresources is shown in Fig. 11.3.

The viscosity of an aqueous solution of DNA is very high because of the huge molecular weight of DNA. Thus, an aqueous solution of 1% DNA is still like a gel without fluidity, which means that in a device process with the double-helical structure as a characteristic of DNA molecule, a new technique to control the molecular weight of double-helical DNA is required. In the following section, such techniques are reviewed.

11.2.2 *Molecular Weight Control of DNA*

Several methods for controlling the molecular weight of double-helical DNA are summarized as follows:

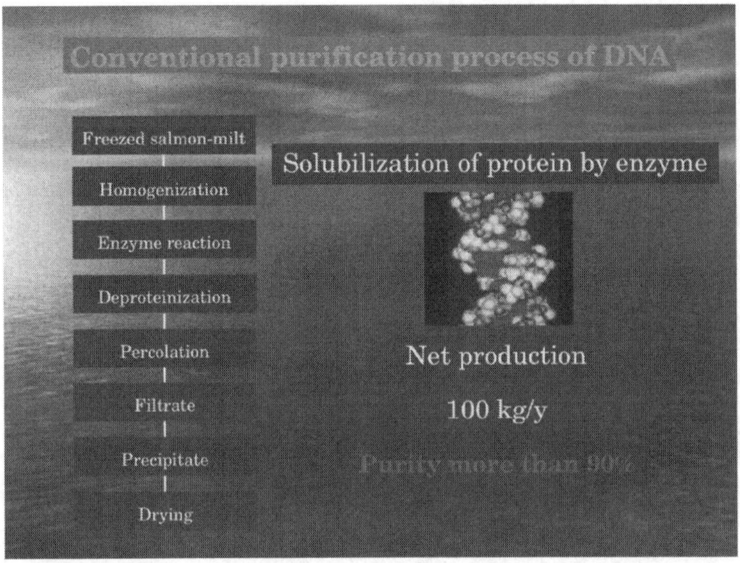

Figure 11.3 Separation and refining of DNA from salmon-milt.

- Thermolysis method: The temperature of an aqueous solution of DNA is raised up to 100°C and then cooled rapidly. Molecular weight and circular dichroism (CD) spectra do not show any change.
- Enzyme method: Double-helical DNA is decomposed with fetal bovine serum as a breakdown enzyme added to an aqueous solution of DNA.
- Sonication method: An aqueous solution of DNA is irradiated with a supersonic wave of 20–100 kHz. The molecular weight can be controlled by the sonication energy applied to the DNA solution.

11.2.2.1 Molecular weight control by sonication

A supersonic wave is generated by a piezoelectric crystal with the piezoelectric effect, such as $PbTiO_3$ and $NbLiO_3$, excited by applying an AC voltage with frequency between 20 kHz and 100 kHz, as shown in Fig. 11.4. The sonication method for mechanically

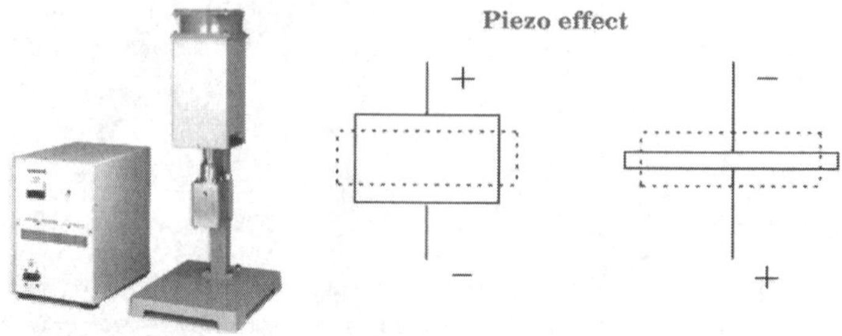

Figure 11.4 Supersonic wave generator with piezoelectric effect. The voltage applied to the piezoelectric crystal along special direction generates a distortion of atomic distances of the crystal.

decomposing DNA while moderately keeping the double-helical structure is preferred by research groups in the United States [292], as shown in Fig. 11.5.

11.2.2.2 Molecular weight control by fetal bovine serum

It is more easy to utilize a breakdown enzyme for controlling the molecular weight of DNA, but the breakdown enzyme is too expensive for industrial applications. Thus, a new method of breaking down DNA with fetal bovine serum was developed, because a serum of blood contains a DNA breakdown enzyme. Details are described as follows.

Six gels of DNA separated from salmon-milt, (1)–(6), with different solvents and DNA concentrations were prepared to investigate the effect of solvent species in reducing the viscosity of DNA gel. DNA gels were located in a six-well plate, as in three-dimensional cultivation experiments, which were incubated under a CO_2 atmosphere. Then the viscosity was measured with the evolution of day. In the meantime, electrophoresis of the parts of gels was also studied.

The prepared DNA gels are:

(1) 2% DNA/H_2O,
(2) 3% DNA/H_2O,
(3) 2% DNA/FBS,
(4) 2% DNA/DMEM (no FBS),

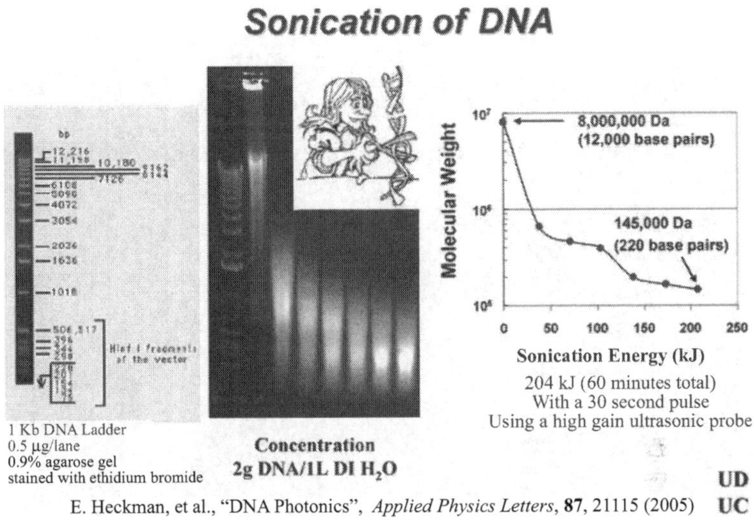

Figure 11.5 Reduction of molecular weight of DNA by irradiation with a supersonic wave. Reprinted with permission from Ref. [291]. Copyright [2005], AIP Publishing LLC.

(5) 4% DNA/DMEM (FBS), and

(6) 2% DNA/10%FBS/H_2O,

where FBS is fetal bovine serum qualified, and DMEM is Dulbecco modified eagle medium. The composition of the culture solution is:

- DMEM: 500 ml,
- FBS qualified: 55 ml,
- Penicillin–streptomycin (AB): 6 ml.

Then the solution is adjusted in a 500 ml bottle to be 10% FBS and 1% AB.

The observed viscosity changes are as follows:

- In (1), (2), and (3), the viscosity did not change.
- In (4), the viscosity did not change for 3 days and then decreased a little after 7 days.
- In (5), the viscosity decreased day by day and became low enough to treat with a pipette 7 days later.

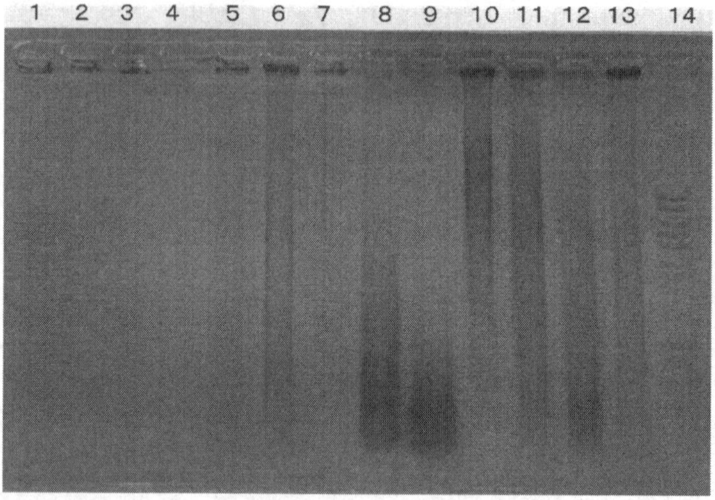

Figure 11.6 Change in molecular weight after the addition of fetal bovine serum to various DNA gels described in text.
Gel (1) — 1: stored at 4°C, 2: 5 days later, 3: 10 days later.
Gel (4) — 4: stored at 4°C, 5: 3 days later, 6: 7 days later.
7: DNA/DMEM 2% with FBS at 4°C, 8: 3 days later, 9: 6 days later.
Gel (5) — 10: stored at 4°C, 11: 3 days later, 12: 7 days later.
Gel (6) — 13: 4 days later.
14: molecular weight markers.

- In (6), the viscosity decreased day by day, but finished 4 days later because of mold.

The molecular weight of DNA controlled by fetal bovine serum is analyzed by electrophoresis, as shown in Fig. 11.6, which suggests that the molecular weight of DNA can be controlled at arbitrary size by the reaction period. The CD spectra of the size-controlled DNA remain unchanged, suggesting the preservation of the double-helical structure of DNA and the possibility for industrial applications.

11.2.3 *Production of DNA from Onions*

Generally, plants are not suitable for the mass production of DNA because the presence of DNA is less than 1%. As an exception, onion cells contain considerable percentage of DNA, suitable for

DNA production. The procedure for separating and refining DNA from onions is described as follows. The obtained DNA shows the same properties as the DNA from salmon-milt. A DNA–lipid complex film is also produced from the onion DNA, which is insoluble in water.

The procedure for the separation and refinement of DNA from onions and applications in LED (light-emitting diode) are described following the US Air Force Research Laboratory.

- Grind onions and percolate it.
- Keep it at 50°C after adding amylase (an enzyme for carbohydrate decomposition) to the filtrate.
- Add protease (an enzyme for protein decomposition) and keep it at 37°C.
- Precipitate and separate 98% pure DNA by adding ethanol.
- Form DNA–CTMA complex.
- Add fluorescent dye into an ethanol solution of DNA–CTMA complex and form a film with it.
- Form a layer-built LED.

Thus, the obtained LED made from onion DNA has a performance comparable or superior to that made from salmon-milt.

11.3 DNA as Optical and Electrical Materials

The optical and electrical functions of aromatic compounds, such as dye, intercalated into the double-helical structure of DNA are markedly enhanced. DNA can be transformed into a DNA–lipid complex with double helix, which is soluble in alcohol but insoluble in water, by exchanging counter ions of Na in DNA for lipids of ammonium cations with long alkyl chains, such as cetyltrimethylammonium cation (CTMA) [287, 288, 290, 291, 293–295, 297]. The procedure is simple; the DNA–CTMA complex is immediately precipitated by stirring an aqueous solution of somewhat excess 1% CTMA and 1% DNA. Then it is percolated and washed thoroughly and dried in vacuum to obtain a cotton-form DNA–CTMA complex. The film of DNA–CTMA complex is obtained by

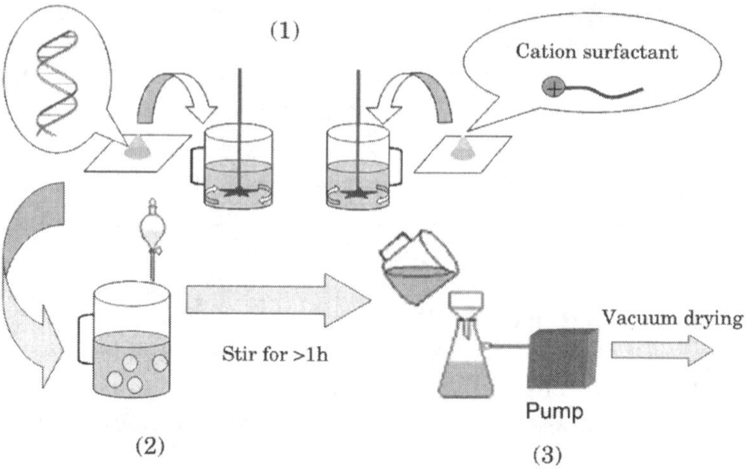

Figure 11.7 Preparation of DNA–lipid complex. (1) Prepare two 1 L aqueous solutions: One has 6.5 g of DNA, and the other has 6.5 g of cation surfactant. (2) Drop the DNA-Na solution slowly into the cation surfactant. (3) Filter the DNA–lipid complex, insoluble in water, with suction.

drying a Petri dish filled with a 5% ethanol solution of DNA–CTMA complex. The procedure is schematically shown in Fig. 11.7.

11.3.1 *Optical Switch*

The DNA–CTMA film is immersed in an aqueous solution of nonlinear fluorescence dye, 4-[4-(dimethylamino)stylyl]-1-dococyl-pyridinium bromide (Stilbene-type) (DMASDPB) to form a dye-intercalated DNA–CTMA film [287]. The intensity of fluorescence in the DNA–CTMA film intercalated by DMASDPB is compared with that of the concentration of DMASDPB with single-stranded DNA, as shown in Fig. 11.8. A remarkably strong enhancement of the fluorescence intensity was found in the dye and the double-stranded DNA–CTMA complex system, but the dye with the single-stranded DNA–CTMA complex system reduced the enhancement less than half, which suggests that the double-helical structure of DNA would play an important role in the intercalation of dye.

This large enhancement of fluorescence intensity is supposed to result from the electron transfer between the bases, mainly A-T,

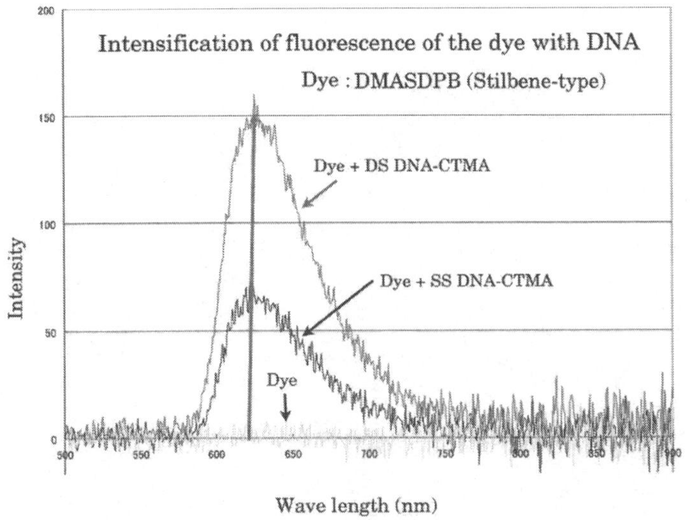

Figure 11.8 Effect of the helical structures of DNA–CTMA complexes on the fluorescence intensity of the dye DMASDPB. The double-stranded DNA–CTMA (DS DNA–CTMA) complex led to the maximum enhancement of the fluorescence intensity.

and the dye intercalated into the layer structure of base pairs in the double-helical structure of DNA.

An optical switch by light irradiation was fabricated to utilize the optical enhancement of the dye intercalated into DNA–CTMA [290, 295]. As shown in Fig. 11.9, a spiropyran compound shows the open–close structural transformation by the irradiation of ultraviolet (UV) light for open and visible (Vis) light for close, in which the refractive index is also modified. Thus, the spiropyran intercalated DNA–CTMA film can be utilized as the optical switch by

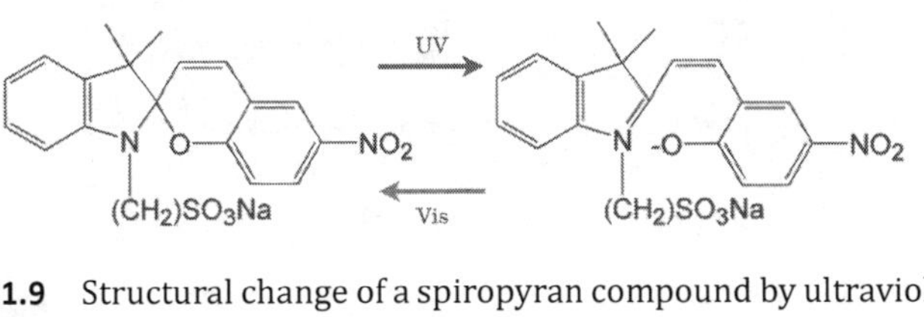

Figure 11.9 Structural change of a spiropyran compound by ultraviolet and visible light irradiations.

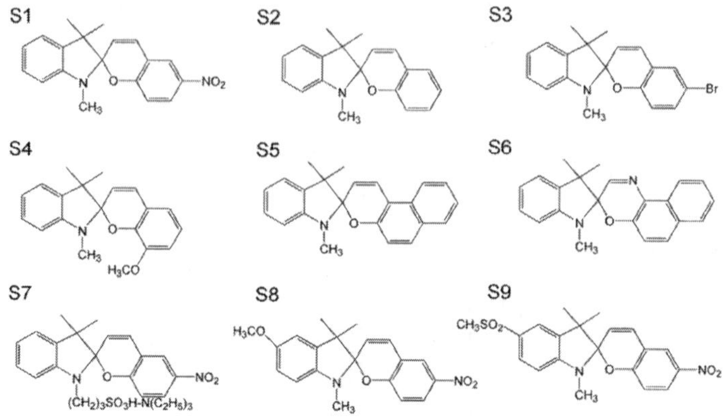

Figure 11.10 Chemical structures of spiropyran compounds for the optical switch.

UV and Vis light irradiation [290]. Several spiropyran compounds were investigated for the optical switch, as shown in Fig. 11.10.

The characteristics of the fluorescence of spiropyran compounds intercalated into DNA–CTMA films are summarized in Table 11.1 [295]. Figure 11.11 and Ref. [295] suggest that better systems for the optical switch are a spiropyran with a nitro group, S9 in Fig. 11.10 and S6, which depend on the switching light power and show relatively faster response time of the fluorescence as a preferable property for the optical switch [295].

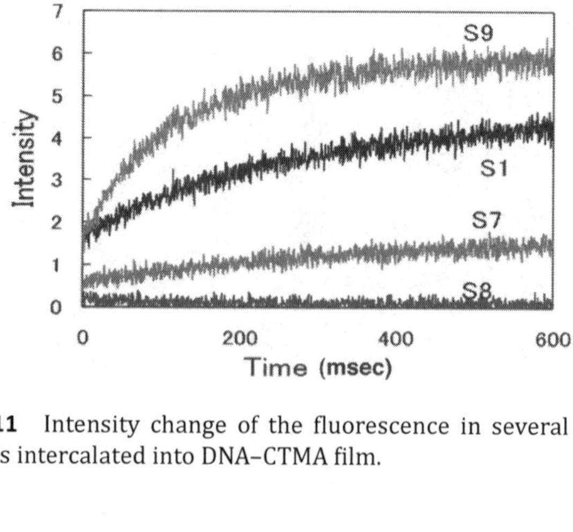

Figure 11.11 Intensity change of the fluorescence in several spiropyran compounds intercalated into DNA–CTMA film.

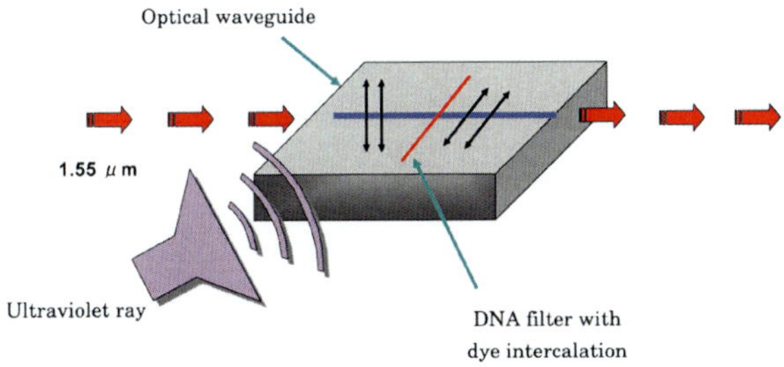

Figure 11.12 A basic structure of the optical switch made of the DNA–CTMA film intercalated by a spiropyran compound.

The basic structure of the optical switch made of DNA is shown in Fig. 11.12. The change in the refractive index of the waveguide by UV light irradiation enables us to optically control the ON and OFF features of the optical switch. A working model of the optical switch was made experimentally, as shown in Fig. 11.13.

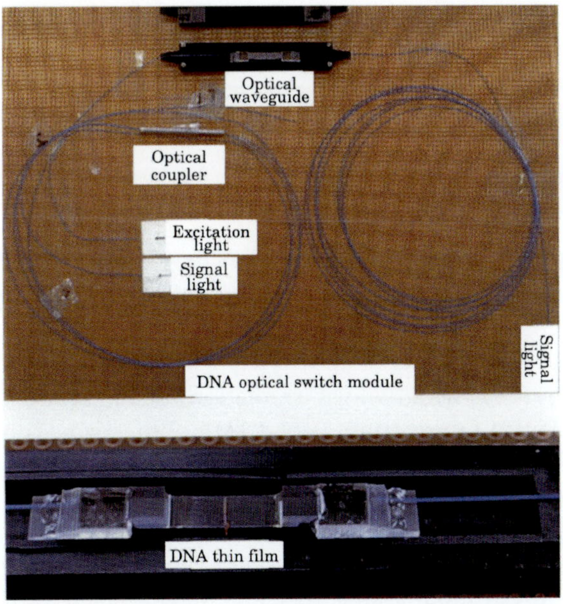

Figure 11.13 Test module of the optical switch made of a DNA–CTMA film intercalated with a spiropyran compound.

Table 11.1 Characteristics of fluorescence of DNA–CTMA films intercalated by a variety of spiropyran compounds [295]. vs: very strong, s: strong, w: weak, and vw: very weak in the intensity column

	Color change by UV irradiation		Fluorescence			
			Fluorescence	Excitation	Intensity	
	Before	After	nm	nm		
S1	yellow	reddish	600	400	w	2→4
		purple	600	560	s	25→80
S2	–	–	356	307	vw	3
			430	345	vw	4.4
S3	–	–	435	355	vw	5.6
S4	–	–	359	307	vw	4.1
			502	474	vw	4.5
S5	light	light	436	377	w	12.5
	pink	pink	595	575	w	8.7
S6	transparent	light blue	426	360	vs	230
S7	light reddish	reddish	600	406	vw	0.6→0.9
	purple	purple	600	560	s	20→30
S8	yellow	blue	–	–		
	green	green				
S9	yellow	purple	600	395	w	1.2→8.6
			615	560	s	39→58

11.3.2 *Laser*

A differential feedback (DFB) laser was developed with an etchless grating of polymethyl methacrylate (PMMA) substrate coated by DNA–CTMA film intercalated by cyanine dyes, such as DMASDPB [288] and 4-[4-(dibutylamino)stylyl]-1-methylpyridinium iomide (DBASMPI) [297] with UV light irradiation. A fluorescence spectrum of a DBASMPI-intercalated DNA–CTMA film coated on a glass substrate under 560 nm excitation light and a corresponding excitation spectrum for the 600 nm fluorescence peak are shown in Fig. 11.14 [Above]. These results were obtained for the DNA–CTMA film with the DNA to laser dye molar ratio of 20 to 1. The intensity of the fluorescence depends on the molar ratio of DNA and dye and shows a peak around 10 to 1. The spectral narrowing due to amplified spontaneous emission (ASE) is observed at the

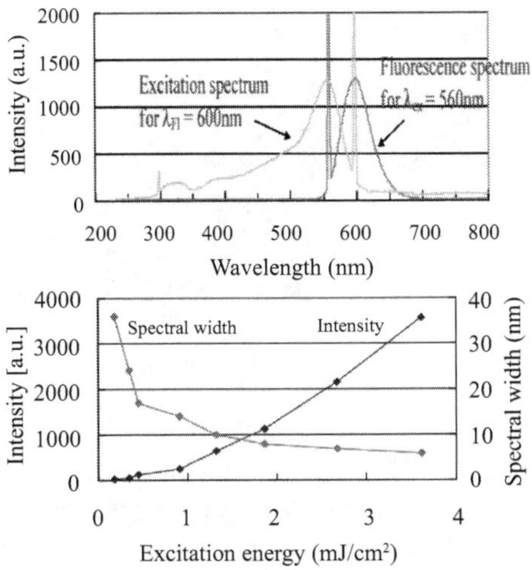

Figure 11.14 [Above] Typical fluorescence and excitation spectra of a DBASMPI-intercalated DNA–CTMA film. The sharp spikes at 560 nm and 600 nm are instrumental in origin. [Below] Peak intensity and spectral width of amplified spontaneous emission (ASE) spectra of DBASMPI-intercalated DNA–CTMA films (λ_{ex} = 532 nm). Reprinted with permission from Ref. [297], Copyright 2007, Society of Photo Optical Instrumentation Engineers.

excitation energy level higher than about 1 mJ/cm^2, as shown in Fig. 11.14 [Below] [297].

Figure 11.15 [Left] describes a fabricated DFB laser structure with an etchless grating fabrication procedure [297]. A DNA-CTMA solution was spin-coated or poured directly onto an etchless

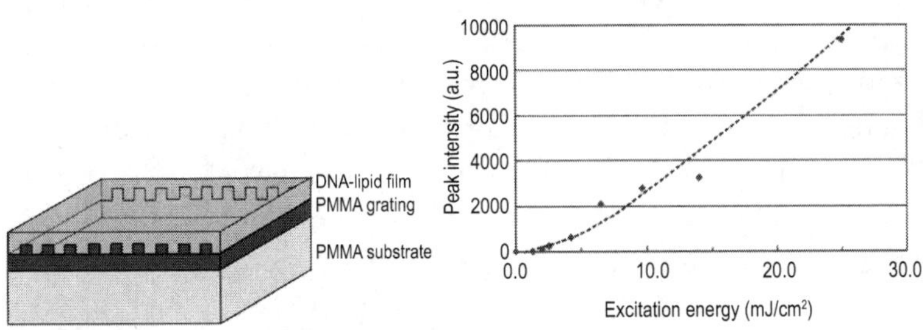

Figure 11.15 [Left] Experimental DFB laser structure. [Right] Input/Output characteristics of DBASMPI-intercalated DNA–CTMA DFB laser.

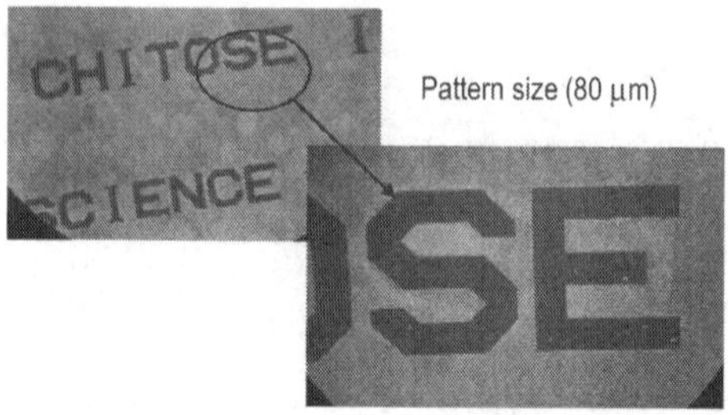

Figure 11.16 Long-distance transport in high resolution of 80 μm.

grating substrate of PMMA tuned to the lasing wavelength. The typical input–output characteristics of the DFB laser structure are shown in Fig. 11.15 [Right]. A single longitudinal mode operation is maintained in the whole range of excitation with somewhat large lasing threshold of about 1 mJ/cm^2. The DFB laser of DNA–CTMA–dye is applicable to a long-distance communication system, such as 10 km, and high resolution was confirmed, as shown in Fig. 11.16.

11.4 DNA as Physiological and Medical Materials

A biocompatible natural polymer of DNA is suitable for physiological and medical applications, in which insolubility in water is an inevitable property. Although it is confirmed that DNA–lipid is insoluble in water, the biocompatibility of DNA is spoilt by lipids, such as ammonium salt. The best way to make DNA films insoluble in water while preserving the biocompatibility and the double-helical structure of DNA is to form cross-linking in DNA films by UV light irradiation. A conventional UV lamp is utilized to generate cross-linking in DNA films. Thus, the obtained DNA films (UV-cured DNA film) are applied to physiological and medical purpose.

11.4.1 *UV-Cured DNA film for Skin Wound*

In the case of a UV-cured, cross-linked DNA film patch on a wound in pig skin, the recovery of the wound was accelerated as a result of guarding the wound well, as shown in Fig. 11.17. The acceleration resulted from the oxygen permeability of the DNA film.

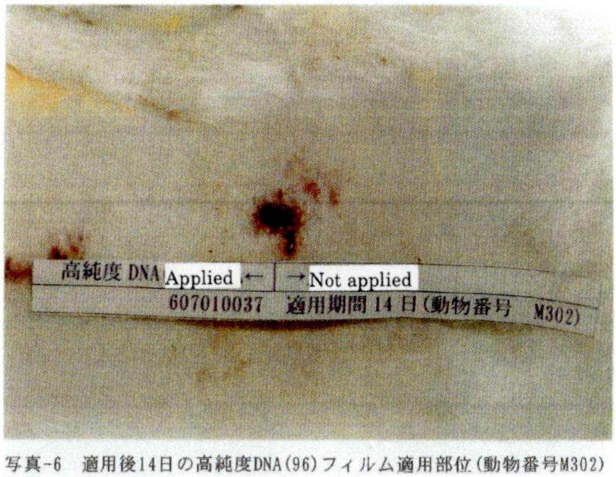

Figure 11.17 Application of a high-purity (98%) UV-cured DNA film on a wound in pig skin. Fourteen days after the application of DNA film, better recovery was noticed in the applied region in the form of yellowish furry skin.

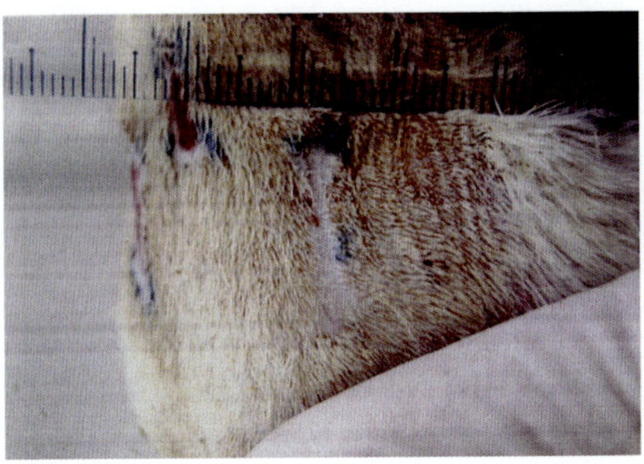

Figure 11.18 Application of a high-purity (98%) UV-cured DNA film on a wound in rat skin. The wound was almost perfectly healed in 14 days after being patched by a DNA film.

11.4.2 *UV-Cured DNA Film for Cell Cultivation*

As an application in the physiological field, the UV-cured DNA was applied to a substrate for cell cultivation. An aqueous 1% DNA solution was dried in a cell cultivation plate to form a substrate

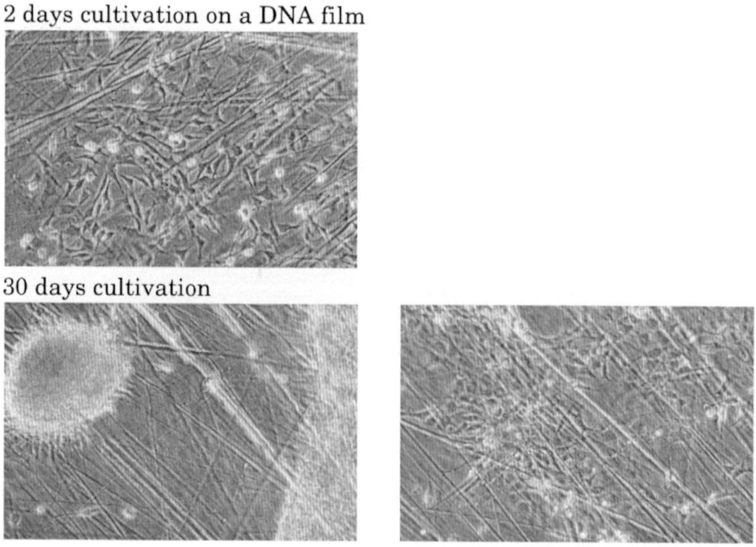

Figure 11.19 Affinity of a UV-cured DNA film with ATDC5 cells. Both living and dead cells were observed after 4 days, but the cells were well cultivated and formed massive multilayered structures after 18 days. New cells grew after 28 days in the space formed by the massive structures.

coated with a DNA film. A 10 μm thick DNA film coated on the cultivation plate was cured for 10 min at room temperature by irradiation with a 1 kW UV lamp and then made insoluble in water by cross-linking of the DNA film. Cartilage cells of a rat were cultivated on the DNA-coated cultivation plate, as described in Fig. 11.19. The cartilage cells adhered well to the DNA substrate and the cultivation speed was enhanced compared to the usual cases.

The cell cultivation study is undertaken as follows.

The affinity of several UV-cured DNA films to unspecialized mesenchymal tissue cells of rat bone marrow (C3H10T1/2) and rat cartilage cells (ATDC5) was investigated. The UV-cured DNA film coated on a Petri dish is immersed in a culture solution with a composition of

- Dulbecco modified eagle medium (DMEM): 500 ml
- Fetal bovine serum qualified (FBS): 55 ml
- Penicillin–streptomycin (AB): 6 ml.

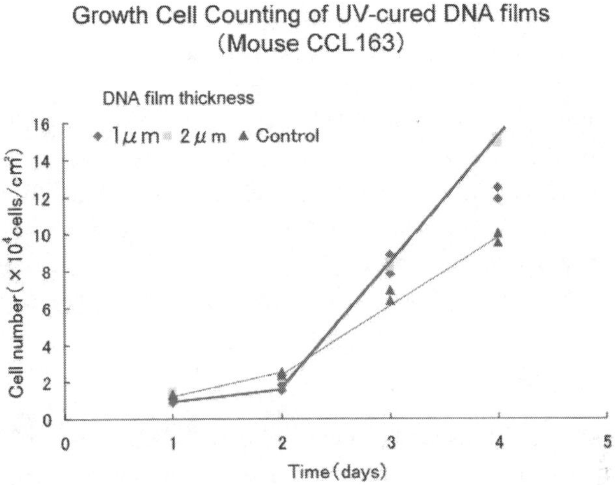

Figure 11.20 Growth cell counting of cartilage cells of rat (Mouse CCL163) on the UV-cured DNA films.

An example of the result is demonstrated in Fig. 11.19.

The growth rate of rat cartilage cells in a cultivation plate coated with the UV-cured DNA film insoluble in water is much faster than that in the conventional cultivation plate under well-controlled conditions, as shown in Fig. 11.20.

11.5 Summary of DNA as a Material

In this chapter, salmon-milt has been primarily used for producing high-purity DNA because a large number of salmons are caught in the Hokkaido Island of Japan. Moreover, since the black stomach sack of scallops contains DNA up to 10%, high-purity DNA is also produced from scallops. A substantial part of DNA application relies on the double-helical structure, but the sequences of base pairs have no role. Thus, DNAs produced from both salmon-milt and black stomach sacks of scallops have the same property as a material. High-purity DNA is also produced from onions and is utilized as optical materials, especially in radars in the United States. Clear images can be obtained by the application of a high-purity DNA film in the receiving/emitting system of a radar, which expands the practical usability of the radar. The issue to be resolved for actual use is to improve the durability of the system.

High-purity (more than 98%) DNA is inevitable for high-tech applications. On the other hand, since low-purity (70%) DNA has some merits, such as low price and relative ease of production, its application as a fire extinguisher fluid has been examined in Italy. DNA is a phosphorus compound but is biocompatible; thus, DNA extinguishers could reduce environmental contamination compared to other conventional extinguisher materials with phosphorus compounds.

Application of DNA as a material has become widespread all over the world. However, DNA is expensive, up to 1 million yen per kilogram, because it is separated and refined from natural resources. The range of use is limited now, and thus the market is limited too. The method to mass-produce DNA from bacteria, such as colon bacilli, is being explored, and it is expected that new technology would be developed for DNA as a material.

Chapter 12

Application of DNA to Electronic Devices

Norihisa Kobayashi and Kazuki Nakamura
Department of Image and Materials Science, Graduate School of Advanced Integration Science, Chiba University 1-33 Yayoi-cho, Inage-ku, Chiba, 263-8522, Japan
koban@faculty.chiba-u.jp

12.1 Introduction

This chapter describes the applications of DNA-functional molecule complexes for electronic and photonic devices studied by our group. The first section describes DNA-related organic transistors, and the successive sections review organic light-emitting diodes fabricated by DNA-based functional materials.

12.2 BiOTFT Memory with DNA Complex as Gate Dielectric

12.2.1 *OTFT using Bio-Related Materials and DNA-Surfactant Complexes*

Printing processes are attractive for the fabrication of electronic devices such as radio-frequency identification (RFID) tags [298],

DNA Engineering: Properties and Applications
Edited by Kenji Mizoguchi and Hirokazu Sakamoto

ISBN 978-981-4669-46-7 (Hardcover), 978-981-4669-47-4 (eBook)
www.panstanford.com

sensors [299] and active matrix displays [300] on a plastic substrate. Since organic compounds can be dissolved in a solvent, electric materials are promising candidates for use as inks in the printing process [301–303]. Further, functional molecules can be oriented along the highly ordered structure of biopolymers, such as DNA, proteins, and polysaccharides. On this basis, biopolymers have also attracted attention for their possible use in the fabrication of electronic devices; such device fabrication using the ordered structure of DNA has been reported, as mentioned earlier. Printable non-volatile memories also have high applicability in electronic devices, but research on promising materials using a low-temperature process has been little. Since an organic thin film transistor (OTFT) was fabricated using ferroelectric material as a gate dielectric and exhibited storage properties, the development of new ferroelectric materials has great significance for the potential fabrication of all printing devices.

Polyvinylidene fluoride and its derivatives are typical ferroelectric materials used as dielectric layers in OTFT memories because of their high ferroelectricity [304, 305]. It has also been reported that an OTFT memory was fabricated using biopolymers, and we reported the fabrication of a printable OTFT memory using a-helical poly(g-methyl-L-glutamate) (PMLG) as a ferroelectric layer in a previous study [306]. PMLG exhibited ferroelectric properties without any special treatment, such as poling, annealing, and stretching, possibly because of its rod-like, a-helical structure. DNA also has a rod-like secondary structure. Hence, OTFTs fabricated using DNA are also expected to exhibit novel storage properties. However, an OTFT memory typically has a low on/off ratio and short memory-retention time.

We reported that the BiOTFT memory device prepared by natural DNA as the gate dielectric showed poor device performance, such as high OFF current because the presence of sodium ions probably contributed to ion conduction [307]. Okahata *et al.* reported the preparation of the DNA–cetyltrimethylammonium (CTMA) complex, which is soluble in organic solvents and is effective for reducing the mobility of counter ions [286]. CTMA is a quaternary ammonium compound that results in a positive charge on the nitrogen (cation) due to the delocalization of the lone pair of electrons, with chlorine

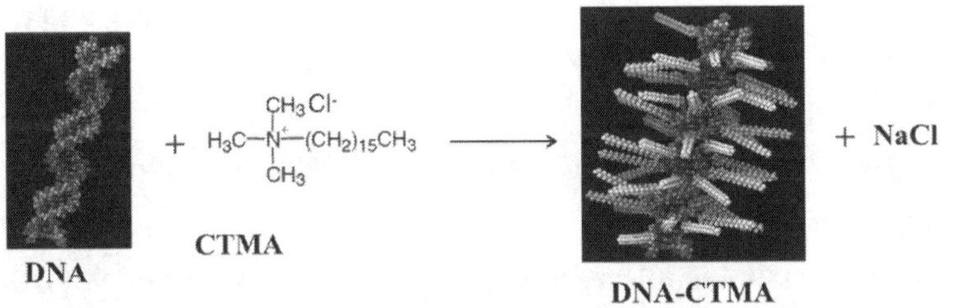

Figure 12.1 Schematic illustration of DNA–CTMA complex.

as the counter anion. The negatively charged phosphate groups in the DNA backbone and corresponding sodium cations in the aqueous solution form electrostatic charge pairs. The positive charge on the nitrogen of the CTMA replaces the sodium cation and ionically binds to the DNA molecule. The displaced sodium cation then ionically binds with the chlorine anion of CTMA, forming NaCl as the byproduct, as schematically shown in Fig. 12.1.

The cationic surfactant complexation of DNA with CTMA in the laboratory setting is a simple processing step. Since both materials are water soluble, they are dissolved in deionized water, and the CTMA solution is titrated into the DNA solution. The DNA–CTMA material is precipitated out, while the NaCl byproduct remains in solution. The mixture is filtered to obtain the DNA–CTMA, which is thoroughly washed with water to rinse out any unreacted CTMA and the remaining NaCl. DNA–CTMA is water insoluble but is soluble in organic solvents such as methanol, ethanol, isopropyl alcohol, and butanol. This makes the material usable in solvent fabrication techniques for thin films such as casting, dip coating, and spin coating. After this novel finding, to exclude the influence of DNA sodium ions and to prepare high-quality thin films, researchers utilized DNA surfactants for the application of electronic devices, including OTFT.

In this section, to investigate the influence of surfactant structure on BiOTFT device performance, the physicochemical properties of various DNA-surfactant complexes in solution and film state have been studied. Further, BiOTFT device performance, as inferred by aspects such as transfer properties, has also been examined, and the possible explanation of device performance is discussed.

12.2.2 *Photo-Electrical Properties of DNA-Surfactant Complexes*

DNA-surfactant complexes were prepared by adding 10 mM of DNA (the concentration of the phosphate group) aqueous solution to 10 mM of the surfactant [CTMA, octadecyltrimethylammonium (OTMA) chloride, Lau (lauroylcholine chloride), Fig. 12.2] aqueous solution, and then the precipitate was filtered, washed with ultrapure water, and dried in vacuo.

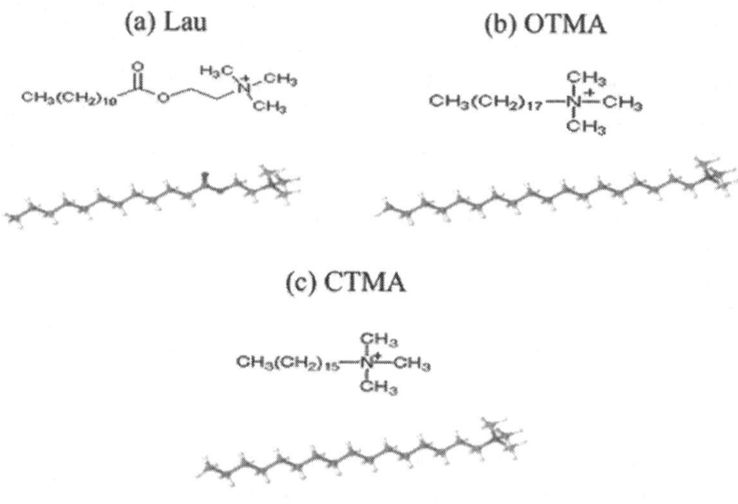

Figure 12.2 Chemical structures of (a) Lau, (b) OTMA, (c) CTMA.

The DNA-surfactant complexes were obtained as a white powder, and the yield was over 95%. Then, to obtain DNA-surfactant films, the 100 mM DNA-surfactant solution was spin coated on ITO glass substrates. The thicknesses of the DNA and DNA-surfactant complex films were approximately 6 μm and 2 μm, respectively.

The photophysical property of UV absorption by DNA is an important property for determining whether π–π stacking of nucleobases occurs. The DNA double helix in aqueous solution had a specific absorption band from 220 nm to 300 nm, with λ_{max} at 260 nm [287]. To compare the effect of different DNA molecular weights with different cationic surfactants on structural regularity, the UV-visible spectra were measured. Figures 12.3 and 12.4 show the UV-visible spectra of DNA–Lau, DNA–OTMA, and DNA–CTMA in butanol solutions.

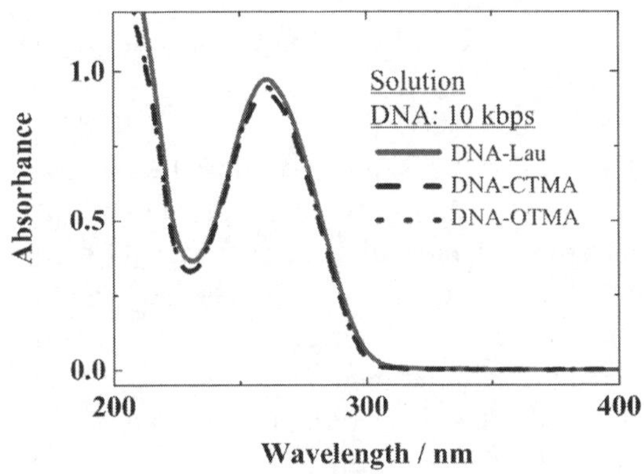

Figure 12.3 UV-vis spectra of DNA complex solution with high molecular weight.

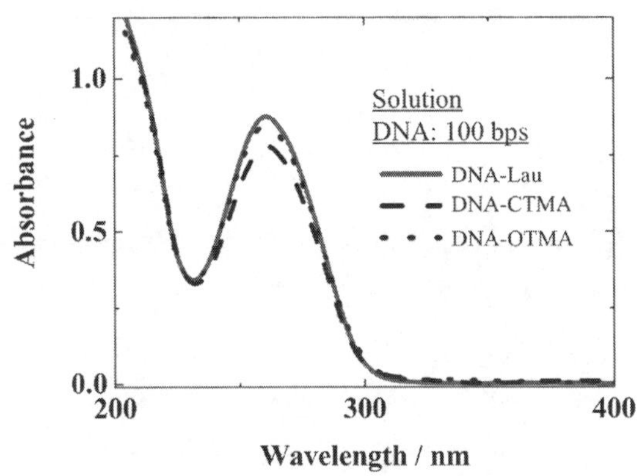

Figure 12.4 UV-vis spectra of DNA complex solution with low molecular weight.

It was clear that both high (10 kbps) and low (100 bps) molecular weights of different DNA complexes have the same peak absorption wavelength at 260 nm, which was assigned as the characteristic absorption of DNA nucleobases. The molecular weight had no relationship with the UV spectra. The π–π stacking of DNA nucleobases did not change when the molecular weight was altered, indicating that both the high and low DNA molecular weights retained the double-helix structure after an ion exchange reaction in the butanol solution.

Circular dichroism (CD) analysis is one of the most useful techniques for probing the conformation of DNA complexes in many kinds of solutions, including gels, films, and fibers [308]. In order to compare different DNA molecular weights with different surfactants and to analyze its effect on the double-helix structure, both the solution and film have been investigated. Figures 12.5–12.7 show CD spectra of the solution and film.

All the complexes in either butanol solution or film state showed positive and negative CD signals. In the case of the solution, a positive Cotton effect at about 280 nm and a negative Cotton effect at about 225 nm and 245 nm have been observed, which are similar in shape to natural DNA. Meanwhile, the A form of the DNA complexes in the butanol solution appeared to transform into the B form in the

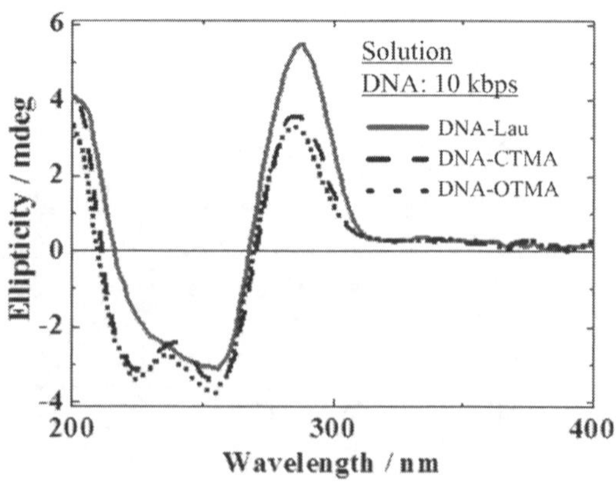

Figure 12.5 CD spectra of DNA complex solution with high molecular weight.

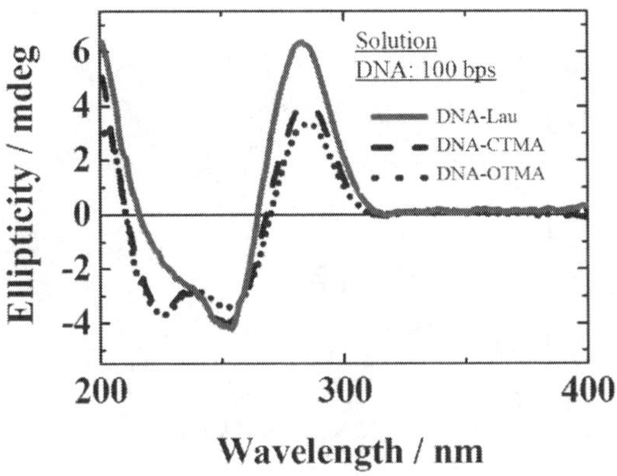

Figure 12.6 CD spectra of DNA complex solution with low molecular weight.

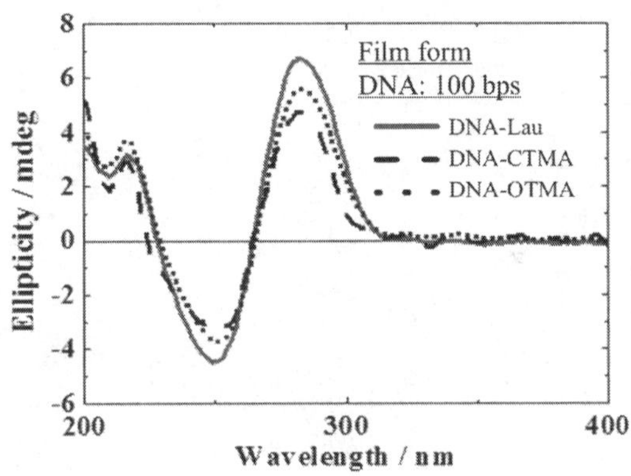

Figure 12.7 CD spectra of DNA complexes in film form.

film state. Additionally, with the alkyl chain increase, the CD signal decreased. This indicates the decrease in structure regularity, which would be caused by the hydrophobic properties of the alkyl chain and steric hindrance between the long alkyl chains.

We also evaluated the DNA complex resistivity because it plays an important role in the BiOTFT memory when used as the insulator layer. For comparison, the current–voltage (I–V) characteristics of DNA alone with different chain lengths are shown in Fig. 12.8.

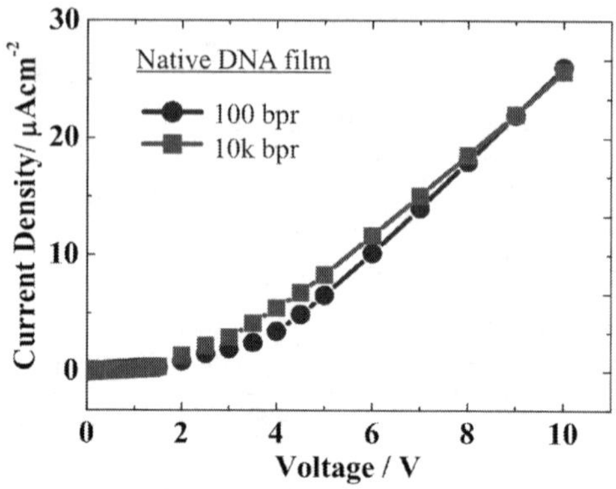

Figure 12.8 I–V curves of native DNA films with different chain lengths.

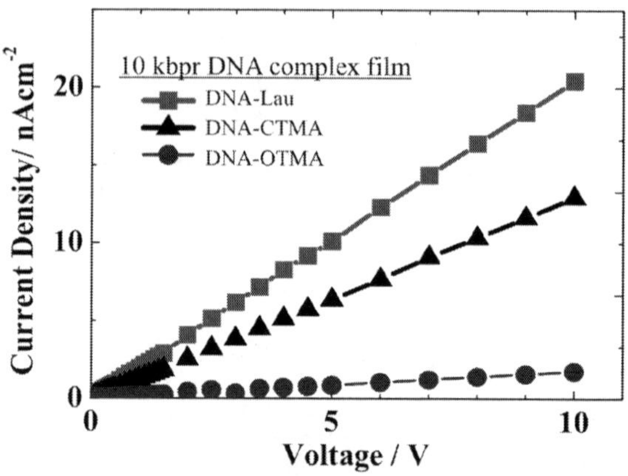

Figure 12.9 I–V characteristics of ITO/high DNA/Au cells.

Figures 12.9 and 12.10 show that the DNA complex resistivity increased significantly compared to DNA alone (Fig. 12.8) with both long and short chain lengths.

This is caused by the DNA, which is one of the poly-anions that could interact with cationic surfactants through the ion-exchange reaction, leading to the decrease in movable ions, such as sodium ions, and to improved resistivity. Additionally, Figs. 12.9 and 12.10 also show that resistivity increased as the alkyl chain

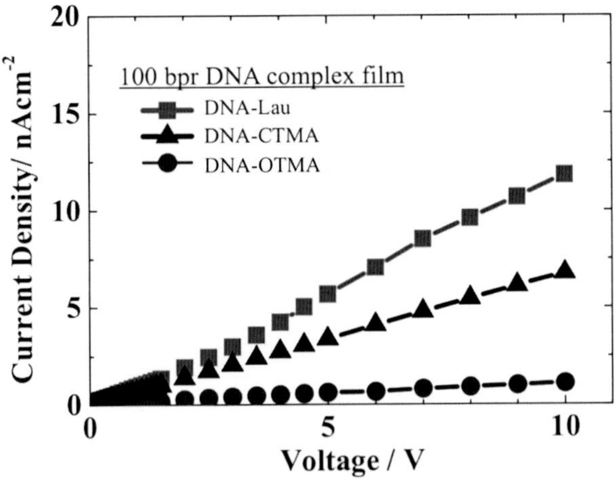

Figure 12.10 $I-V$ characteristics of ITO/low DNA/Au cells.

length increased, because the longer alkyl chains provided a higher proportion of the insulating alkyl chains in the DNA complex film, which would improve the insulating properties.

Furthermore, compared to the high-molecular weight DNA, low-molecular weight DNA showed better resistivity when the same surfactants were applied. This was due to the decrease in the carrier conductive pathway along the DNA chains when the DNA chain length was short. In conclusion, the kinds of surfactant and DNA chain length have substantial effects on film resistivity. The resistivities of the low molecular weights of DNA–Lau, DNA–CTMA, and DNA–OTMA were 1.41×10^{13} Ωcm, 8.2×10^{13} Ωcm, and 1.51×10^{14} Ωcm, respectively. Thus, it is reasonable to use them as dielectric material in the OTFT memory.

In general, thin film surface roughness and structure play crucial roles in the use of any insulating dielectric film for OTFT memory. This stems from the fact that charge transport takes place at the interface between dielectric and semiconductor films [309]. Figure 12.11 shows the 2 μm × 2 μm surface morphology of different DNA complexes with high and low molecular weights. The RMS of the high-molecular weight DNA complex was estimated to be 0.8–1.0 nm.

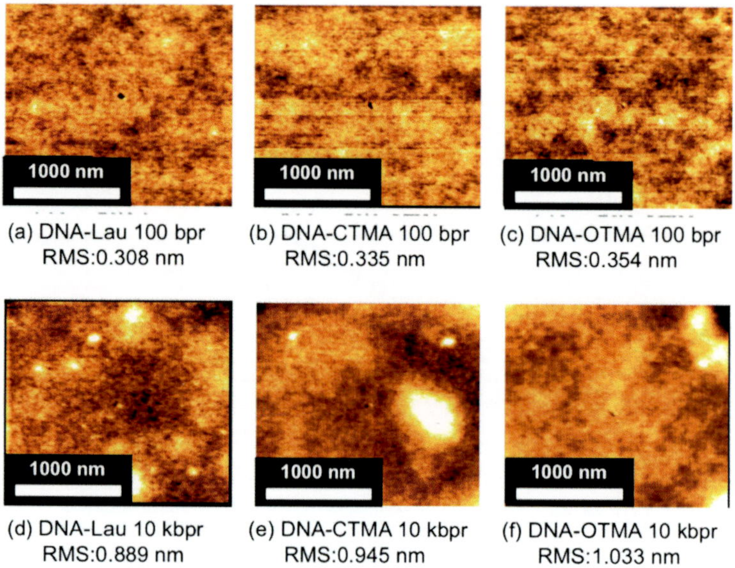

Figure 12.11 AFM topographical images of (a–d) DNA–Lau, (b–e) DNA–CTMA, and (c–f) DNA–OTMA films.

In contrast, the RMS of the low-molecular weight DNA complex was lower, at about 0.3 nm. Additionally, in comparison with the morphology of the low-molecular weight DNA complex, some bump structures were found in the high-molecular weight DNA complex. It is possible that this occurred because the longer chain length would lead to the entanglement of the DNA complex. Therefore, smoothness and film formability of the shorter DNA complex were substantially higher than that of the longer DNA complex. This indicates that low-molecular weight DNA is favorable for use in the dielectric layer of the OTFT memory.

12.2.3 *Fabrication of BiOTFT Devices using DNA-Surfactant Complexes*

Next we fabricated the OTFT memories using the following method. On the prepared ITO/DNA or ITO/DNA complex films, a pentacene layer (film thickness of 50 nm) was an active layer at the pressure

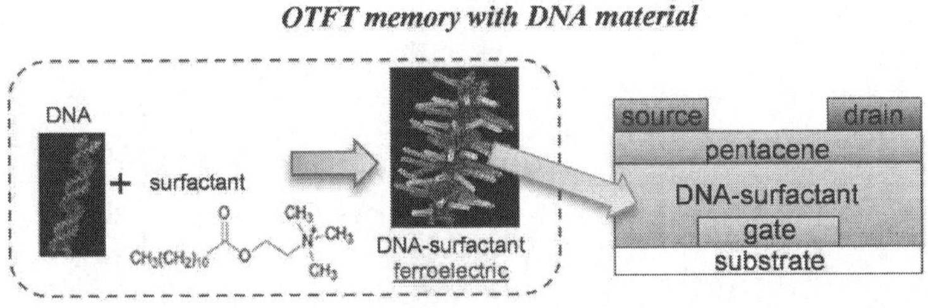

Figure 12.12 Schematic figure of the BiOTFT memory device configuration.

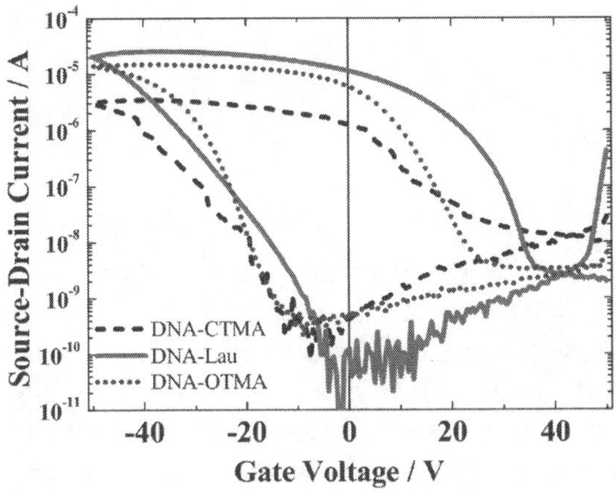

Figure 12.13 Transfer characteristics of BiOTFTs with DNA complex films as the gate dielectric.

of 2 × 10^{-3} Pa and the evaporation rate of 0.2–0.4 Ås^{-1}. The Au layer was the source, and drain electrodes (W/L = 5 mm/20 μm) were deposited by vacuum evaporation on the pentacene film. The BiOTFT structure using a top contact and gate bottom geometry is schematically depicted in Fig. 12.12.

Figure 12.13 shows the transfer characteristics, which represent the transistor current I_{ds}, plotted as a function of the gate voltage at constant V_d of the BiOTFTs fabricated by using DNA–OTMA, DNA–CTMA, and DNA–Lau films as a gate dielectric layer. Figure 12.14 also shows the transfer characteristics when the DNA film was used as the dielectric layer.

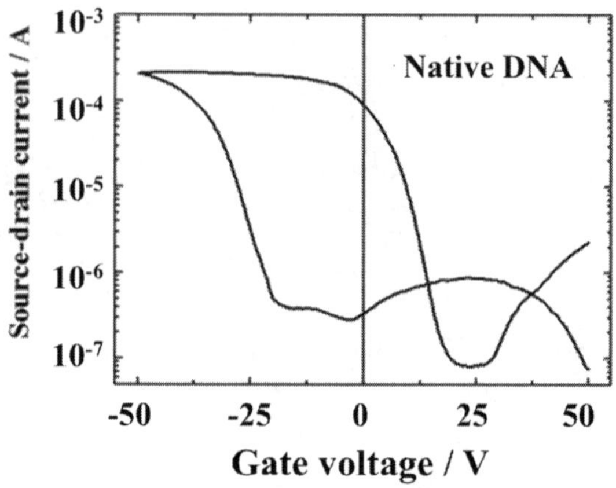

Figure 12.14 Transfer characteristics of BiOTFTs with native DNA film as the gate dielectric.

In all these BiOTFTs, hysteresis behaviors were observed, which indicated that both the DNA and DNA complex films work as memory layers. Because of the poor resistivity of the DNA film, the OFF current, which equals 3.8×10^{-7} A, was too high to use as the insulator in the OTFT memory. In contrast, when the DNA complex was applied as the dielectric layer, the OFF current decreased remarkably to about 10^{-9} A because of the resistivity improvement. At the same time, hysteresis of the BiOTFT memory with the DNA complex significantly increased, and the ON/OFF current ratio at $V_G = 0$ V was as high as 10^4.

Moreover, compared to DNA–OTMA and DNA–CTMA, when DNA–Lau was used as the dielectric layer, higher ON current and restrain of the OFF current were observed. This improvement in device performance might be effected by varying the composition and structure of the DNA complex, because the DNA–Lau film roughness was the lowest and had the best structure regularity when compared to the DNA–CTMA and DNA–OTMA complex films.

The appearance of hysteresis in the DNA-based BiOTFT was thought to be due to the accumulation of mobile ions at the interface. However, more hysteresis was observed in the DNA complex, indicating that the accumulation of mobile ions is not the dominant factor, but instead the ferroelectricity of DNA complex and/or the interface are thought to be possible explanations. Research is now under way to provide a better understanding of this memory performance.

12.2.4 *Summary*

To summarize, DNA and various DNA complex films were used as gate dielectric materials of the BiOTFT memories. In order to exclude the influence of sodium ions and other impurities, it is necessary to make the DNA complex with cationic surfactant through the ion-exchange reaction. At the same time, the molecular weight of DNA also plays an important role in the DNA complex film resistivity and morphology. In this study, the most favorable surfactant for preparing the BiOTFT is DNA–Lau, as compared to DNA–OTMA and DNA–CTMA. The ON current and restrain of the OFF current improve extensively by using DNA–Lau as the gate dielectric.

12.3 BiOLED with DNA Complexes

12.3.1 *Application of DNA Complexes as Charge Conductive Material*

In the previous section, we described the application of DNA-functional molecule complexes in organic electronics, such as BiOTFT memory. The DNA-surfactant complexes acted as ferroelectric materials in OTFT devices. On the other hand, DNA has sparked interest as a functional material since the stacked aromatic bases of DNA may act as a π-way for efficient charge transfer [310]. This is possibly because of application in novel electronic devices, such as highly integrated circuits with nano-structures. The electron transfer and conductivity of DNA have been extensively debated. While the actual magnitude of these conductive properties is unclear because of different measurement conditions and apparatus [3, 4, 6, 311], attempts to improve DNA conductivity have been carried out by modifying DNA with other functional materials [17, 312, 313]. In this section, we demonstrate the opt-electronic application of DNA complexes, which contain conductive polymers and luminescent metal complexes in the structure.

The applicability of conducting polymers such as poly(acetylene) in electronic devices has been gaining interest. These polymers are generally prepared by oxidative polymerization. If conducting polymers are prepared by the photoinduced electron-transfer process, which enables vectorial electron transfer (redox reaction) between photocatalysts and other molecules by illumination, patterning with conducting polymers is possible at any place and on any geometry. This process would allow the possibility of fabricating molecular electronic and/or optical devices, and for microprocessing. On the basis of this information, photopolymerization of pyrrole using $Ru(bpy)_3^{2+}$ as a photocatalyst has already been performed, resulting in conductive polypyrrole.

We have already reported photopolymerization of dimeric aniline by the photoinduced electron transfer between $Ru(bpy)_3^{2+}$ and MV^{2+} [314–316], because polyaniline (PAn) is one of the most promising conductive polymers due to its high environmental stability in air. We also demonstrated its application in imaging and micropatterning [317, 318]. However, for the development of electronic devices, the obtained PAn does not seem to be sufficient in its conjugation length, because the PAn photopolymerized in homogeneous systems involves a branched and/or compact coil structure. The use of a polyelectrolyte as a polymerization template has been reported for the polymerization of aniline by an enzyme in the presence of hydrogen peroxide in order to minimize branching [319]. We also carried out polymerization in the presence of clay minerals [320] and micelles [312], in order to utilize its specific structure as a template and to improve physicochemical properties of the photopolymerized PAn. Since template polymerization is expected to result in a characteristic structure reflecting template structure, template variations are advantageous for different applications.

In this section, structurally ordered DNA was employed as a template to prepare the DNA/PAn complex. The relevance of polymerization and its structure for electronic material have been discussed because DNA works as a rigid, straightforward template, due to its rod-like and double-helical structure. Further, the

DNA/PAn complex should include $Ru(bpy)_3^{2+}$ in its matrix even after purification, because $Ru(bpy)_3^{2+}$ could be electrostatically bound on the free anionic sites of the DNA and remained after complexation with PAn.

$Ru(bpy)_3^{2+}$ works as a light-emitting molecule as well as photosensitizer. In the complex, PAn should be *p*-type conductor. If materials such as $Ru(bpy)_3^{2+}$ showed electronic conduction, the DNA/PAn complex containing $Ru(bpy)_3^{2+}$ was expected to be a photovoltaic and light-emitting molecular system, where electrons and holes were injected from electrodes through DNA and PAn, respectively, and combined at $Ru(bpy)_3^{2+}$ to emit red light. We, therefore, analyzed the light-emitting properties of the device, which was fabricated from a novel, processable, and water-soluble DNA/photopolymerized PAn complex containing $Ru(bpy)_3^{2+}$. Further, a green light–emitting Alq_3 complex layer was deposited on the DNA/PAn complex. The device showed voltage-controlled emission color tunability. The mechanism of multi-emission was also investigated.

12.3.2 *Structure of Template Photopolymerized PAn/DNA Complex*

We first analyzed the structure of DNA associated with phenylene-diamine (PPD, dimeric aniline), and $Ru(bpy)_3^{2+}$ (photocatalyst) in order to obtain a highly ordered complex. The kinds of structures that can be obtained by the association of DNA with these chemicals are important for template photopolymerization. The DNA-based functional complex binding the PAn chain in its groove, DNA/PAn, was prepared as follows. The aerated hydrochloric acid (HCl) aqueous solution (pH 3.0–6.5) containing 6×10^{-5} M $Ru(bpy)_3^{2+}$, 1.0 mM PPD, and a given concentration of DNA (the concentration of phosphate group) was illuminated with a 500 W xenon lamp through a 420–600 nm filter. The light intensity was adjusted to 15 mW cm^{-2} at 450 nm.

Then we investigated the interaction between the $Ru(bpy)_3^{2+}$ photocatalyst and DNA in aqueous solution by analyzing the emission behavior of $Ru(bpy)_3^{2+}$. Since the polymerization of aniline derivatives is commonly carried out in acidic aqueous solutions, we used aqueous solutions with pH 3.1. The emission behavior of $Ru(bpy)_3^{2+}$ in a polyanionic environment is different from that in solutions without polyanions.

The emission intensity at about 600 nm from excited $Ru(bpy)_3^{2+}$ (6×10^{-5} M) increased with DNA concentration and then reached constant emission intensity at a DNA concentration of 1×10^{-3} M. Further, the increase in DNA melting temperature was also observed in the solution at pH 3.1 in the presence of $Ru(bpy)_3^{2+}$. The pH-titration measurements indicated that the pKa of the DNA was 4 to 4.5. We suggest that electrostatic interaction between $Ru(bpy)_3^{2+}$ and DNA minimally takes place at pH 3.1. Rather, $Ru(bpy)_3^{2+}$ is associated with duplex DNA by specific interaction, such as by intercalation at pH 3.1.

The absorption of PPD at 280 nm in the HCl aqueous solution of pH 3.1 containing DNA underwent a slight red shift and exhibited 25% hypochromicity by increasing the concentration of DNA from 7.5×10^{-5} M to 1.25×10^{-4} M. Further, the increase in the melting temperature of DNA was also observed in the solution at pH 3.1 in the presence of PPD, as shown in Fig. 12.15(a).

However, the absorption of PPD at pH 6.5 did not show hypochromicity. These results suggest that PPD, protonated by the primary amino group at pH 3.0 (pKa1 = −0.1, pKa2 = 5.72), is associated with duplex DNA by intercalation and/or electrostatic interaction. Considering the aforementioned results, DNA is associated with PPD and the $Ru(bpy)_3^{2+}$ photocatalyst, with a structure as schematically represented in Fig. 12.15(b). This kind of architecture, a so-called supramolecular polymer, can be easily obtained by dissolving only DNA, PPD, and $Ru(bpy)_3^{2+}$ in the aqueous solution.

When an aqueous solution (pH 6.5) containing 6×10^{-5} M $Ru(bpy)_3^{2+}$, 1×10^{-3} M PPD, and 1×10^{-3} M DNA was illuminated with a xenon lamp through a 420–600 nm filter, absorption peaks at 400 and 800 nm emerged, which are assignable to the polaron band of PAn [321]. The mechanism of photocatalytic polymerization has already been discussed [315]: protonated PPD is oxidized by excited $Ru(bpy)_3^{2+}$, and the formed $Ru(bpy)_3^{3+}$ is oxidized back to $Ru(bpy)_3^{2+}$ by an acceptor (e.g., oxygen in an aerated solution). Polymerization proceeds with the reaction of the protonated form

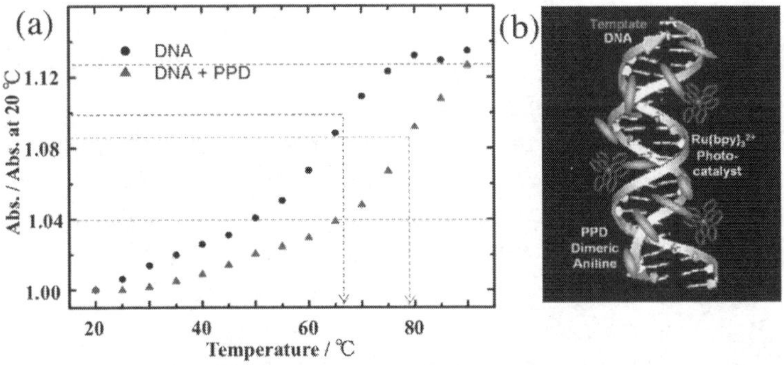

Figure 12.15 (a) The helix melting curves of 1 $\times 10^{-5}$ M DNA in the presence and absence of 1 $\times$ 10^{-5} M PPD (pH 3.1). (b) Schematic representation of the structure of DNA associated with PPD and $Ru(bpy)_3^{2+}$.

of doubly oxidized PPD with unoxidized PPD, followed by the successive reaction at the end of the chain via photoinduced electron transfer. When polymerization was carried out at pH 6.5 in the absence of DNA, the absorption peaks were observed at 300 and 600 nm, which are assignable to π–π^* transition of the benzenoid ring and exciton absorption of the PAn quinoid ring, respectively. These results strongly suggest that the photopolymerized PAn is acid-doped (i.e., protonated) by DNA, and that DNA phosphate groups provide a local, lower pH environment.

Figure 12.16 shows the time evolution of the absorption spectrum in the HCl aqueous solution containing 6 $\times$ 10^{-5} M $Ru(bpy)_3^{2+}$, 1 $\times$ 10^{-3} M PPD, and 1 $\times$ 10^{-3} M DNA at (a) pH 3.0 and (b) pH 4.0 upon visible light illumination. The polaron absorption also emerged from 600 to 800 nm at pH 3.0 in the absence of DNA. Its absorption maxima (λ_{max}), however, significantly shifted to the longer wavelength side (i.e., ca. 200 nm), in the presence of DNA. The longer-wavelength shift indicates the formation of PAn with a longer π-electron conjugation. MacDiarmid *et al.* reported that the shift is attributable to the delocalization of electrons in the polaron band of PAn due to the conformational change in PAn from compact coil to expanded coil. The shift, therefore, is also explained by the expanded coil formation of PAn photopolymerized at pH 3.0 in the presence of DNA.

When polymerization was carried out at pH 3.0 in the presence of DNA, a larger amount of PPD electrostatically interacted with DNA than at pH 6.5 because of PPD protonation. In other words, while DNA at pH 6.5 provides the local, lower-pH environment required for photopolymerization, the template effect of DNA on polymerization was insufficient. Differential CD spectra of the solution containing DNA and PPD and that containing only DNA at pH 3.0 showed a negative CD band at 280 nm assignable to π–π^* PPD transition. However, the negative CD band was not found at pH 6.5.

It is well known that the helical structure of DNA is deeply affected by the pH of aqueous solutions, and these results indicate that PPD in the DNA solution at pH 3.0 is aligned along the main chain of DNA through intercalation and/or electrostatic interaction, leading to the formation of an expanded-coil PAn structure in the PAn/DNA complex. DNA studied in these experiments was about 60% unwounded at pH 3.0. We carried out the same experiments at pH 4.0, where more than 90% of the helical structure remained. As can be seen in Fig. 12.16(b), a similar absorption spectral change was obtained at pH 4.0. This indicates that DNA works well at pH 4.0 as a template for the PAn photopolymerization.

Figure 12.17 shows CD spectra of the PAn/DNA complex in an aqueous solution at pH 4.0. CD signal emerged after photopolymerization at around 440 nm and 1100 nm, which were assignable to the

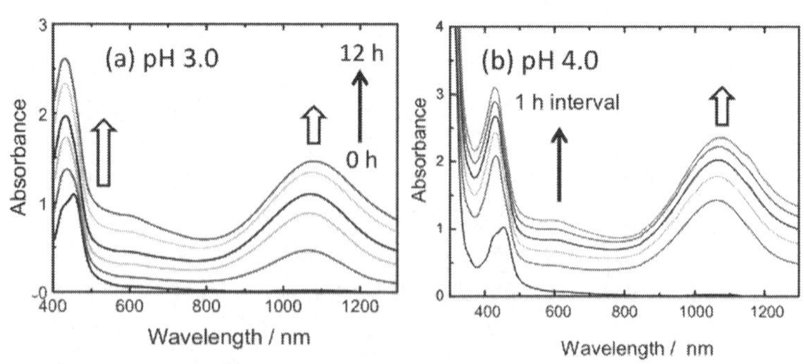

Figure 12.16 Change in absorption spectra of the aqueous solution containing 1 mM PPD, 6 × 10^{-5} M $Ru(bpy)_3^{2+}$, and 1 mM DNA at (a) pH 3.0 and at (b) pH 4.0 upon visible light illumination.

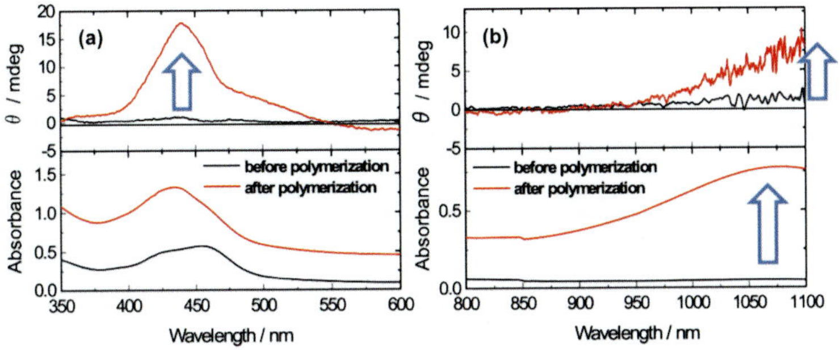

Figure 12.17 UV-vis and CD spectra of the aqueous solution (pH 4.0) containing DNA, PPD, and $Ru(bpy)_3^{2+}$ before and after photopolymerization.

absorption bands of PAn in the PAn/DNA complex. However, PAn in the absence of DNA did not show a CD signal between 350 and 1100 nm. Similar spectra were obtained in the solution at pH 3.0, but the intensity of the CD signal at pH 4.0 was larger than that at pH 3.0 ($\theta = 3$ mdeg at 440 nm and 2 mdeg at 1100 nm). This is due to the higher double-helical structure content of DNA at pH 4.0. This clearly indicates that PAn was associated with DNA, and PAn in the complex had a helical structure along the DNA main chain. Taking the aforementioned results into account, schematic representation of the structure of the PAn/DNA complex is depicted in Fig. 12.18.

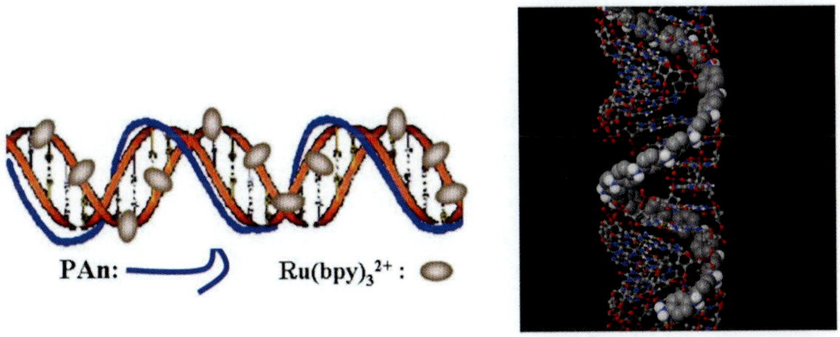

Figure 12.18 Schematic representation (left) and calculated structure using the molecular dynamics method (right) of the PAn/DNA complex.

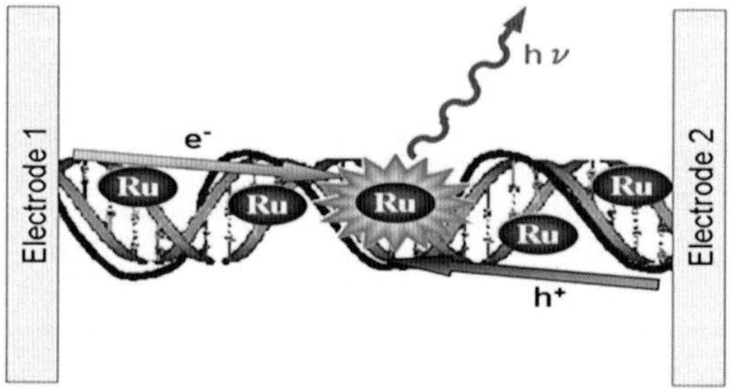

Figure 12.19 Schematic representation of the PAn/DNA light-emitting molecular system structure.

12.3.3 *EL Properties of $Ru(bpy)_3^{2+}$-Based DNA BiOLED*

As mentioned in Section 12.1, the DNA/PAn complex included $Ru(bpy)_3^{2+}$ in its matrix even after purification, because $Ru(bpy)_3^{2+}$ could be electrostatically bound to the free anionic sites of DNA, remaining after complexation with PAn. $Ru(bpy)_3^{2+}$ works as a light-emitting molecule as well as photocatalyst (photosensitizer). In this complex, PAn should be a *p*-type conductor. If such materials as $Ru(bpy)_3^{2+}$ attached regularly to the PAn/DNA complex were enabled to be conductive, the PAn/DNA complex containing $Ru(bpy)_3^{2+}$ would be expected to be a photovoltaic and light-emitting molecular system. The electrons and holes were injected from the electrodes through DNA and PAn, respectively, and combined at $Ru(bpy)_3^{2+}$ to emit red light, as schematically shown in Fig. 12.19.

The DNA/PAn complex was prepared by template photopolymerization of dimeric aniline, as mentioned earlier, and was purified repeatedly by precipitation with acetone. Since $Ru(bpy)_3^{2+}$ should work as an emitter and the concentration of $Ru(bpy)_3^{2+}$ on the DNA/PAn complex was not enough to fabricate an organic light-emitting diode (OLED), a solution containing 4.2×10^{-3} g/mL of DNA/PAn complex and 3.2×10^{-4} g/mL of $Ru(bpy)_3^{2+}$ was prepared by adding an appropriate amount of $Ru(bpy)_3^{2+}$to the solution. An indium-tin oxide (ITO) coated glass electrode was dipped in the solution to prepare the DNA/PAn complex containing $Ru(bpy)_3^{2+}$ on the electrode. The film thickness was estimated to be 30 to 50 nm.

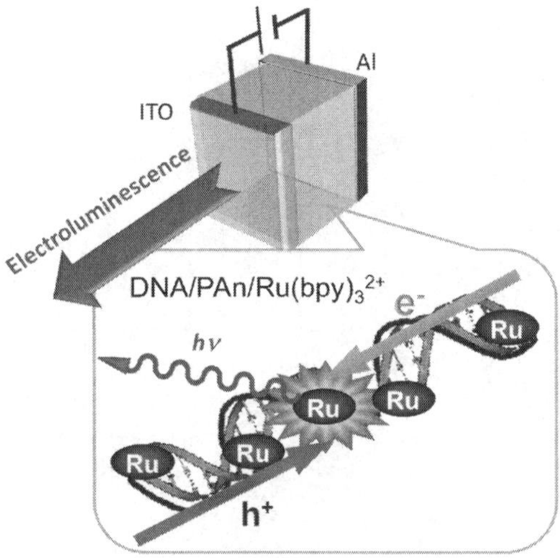

Figure 12.20 Schematic representation of the cell structure of PAn/DNA-based OLED.

The precipitate and crystal corresponding to $Ru(bpy)_3^{2+}$ itself were not found in the film micrograph, indicating that the incorporated $Ru(bpy)_3^{2+}$ was homogeneously dispersed in the film, possibly due to electrostatic interaction between the DNA phosphoric group and $Ru(bpy)_3^{2+}$. The Al top electrode was finally deposited at 5.3×10^{-4} Pa with a thickness of 100 nm. The emitting area was 0.2×0.2 cm^2. We then analyzed the light-emitting properties of the cell, which was fabricated from a novel, processable, and water-soluble DNA/photopolymerized PAn complex containing $Ru(bpy)_3^{2+}$. The Al electrode DNA/PAn complex/ITO cell with a DNA/PAn thickness from 30 to 50 nm was fabricated and the voltage was applied as schematically shown in Fig. 12.20 [322].

The cell I–V curve is shown in Fig. 12.21(a). The current abruptly increased above a bias voltage of 7 V, and a large current of 1600 mA/cm^2 was obtained at 11 V. The current over 7 V seems to obey the space-charge-limited current (SCLC) mechanism. The current at 11 V is relatively higher than that of devices commonly fabricated by vacuum evaporation of low-molar mass organic molecules. This is due to the employment of conductive PAn in the complex. With

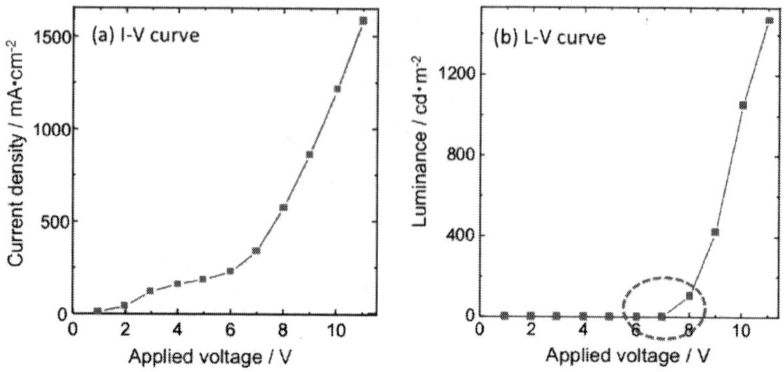

Figure 12.21 (a) Current–voltage characteristics and (b) luminance–voltage characteristics of a device fabricated with DNA/PAn complex containing Ru(bpy)$_3^{2+}$.

regard to the $I-V$ characteristics, light was emitted from the cell above the bias voltage of 7 V, and the luminance of the device increased with bias voltage as shown in Fig. 12.21(b).

A maximum luminance of 1500 cd/m^2 was observed at 11 V. As can be seen by comparison between Fig. 12.21(a) and (b), the luminance seems to increase linearly with injected charges. Turn-on response of the cell luminance was very fast for our device. Although we could not measure its quantitative response time, we do not doubt that the turn-on response was much faster than 1 s. We confirmed flicker due to emission-on and -off from the cell when an AC bias voltage of 9 V was applied to the cell at a frequency of 30 Hz. This indicates that the turn-on and -off response should be faster than 30 ms. We fabricated a cell composed of DNA/Ru(bpy)$_3^{2+}$ without PAn, and the emission-time response of the cell was measured with a photomultimeter by applying 0 to 10 V of rectangular wave to the cell at 10 Hz [323]. The luminance increased rapidly just after turning on the bias and reached a constant value within 70 μs. This behavior seems to be different from that of previously reported Ru complex-based devices. This indicates that the emitting mechanism is not electrochemical but is commonly observed in an electroluminescence (EL) device fabricated by the vacuum evaporation of organic molecules.

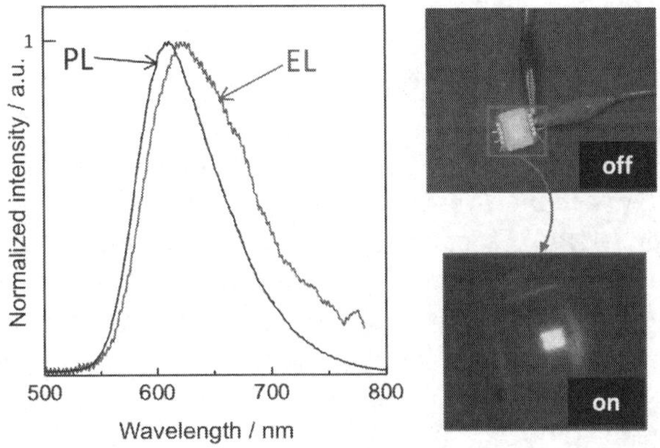

Figure 12.22 PL spectrum of $Ru(bpy)_3^{2+}$ in an aqueous solution and EL spectrum of the cell (left), and digital images of cell emission (right).

The EL spectrum of the cell and photoluminescence (PL) spectrum of $Ru(bpy)_3^{2+}$ are shown in Fig. 12.22 (left), and the photo images of the emission from the cell are shown in Fig. 12.22 (right).

The PL spectrum was measured for $Ru(bpy)_3^{2+}$ in an aqueous solution. The EL spectrum was found to be almost identical to PL spectrum, indicating that the red light was emitted from $Ru(bpy)_3^{2+}$ in the DNA/PAn matrix. However, the emission peaks were found to red-shift from 610 nm for PL in the solution to 620 nm for the OLED cell, and the EL spectrum was slightly broader than the PL spectrum. This is due to the electronic interaction between $Ru(bpy)_3^{2+}$ molecules, which increases as the separation between $Ru(bpy)_3^{2+}$ molecules decreases.

As described earlier, the role of DNA in this cell is interesting to examine. Whether DNA is an electronic conductor or not is unclear because we do not have any quantitative evidence that electrons or holes pass through DNA chains. Taking into account the presence of electroactive $Ru(bpy)_3^{2+}$ in the complex and the very low conductivity ($< 10^{-10}$ S/cm) of DNA film, electron transfer through $Ru(bpy)_3^{2+}$ from the cathode is plausible in the EL cell.

When poly(aniline sulfonic acid) (SPAn) was employed instead of the DNA/PAn complex to fabricate the EL device, no emission was observed from the device, even at the bias voltage of 11 V. This is due to the increase in ineffective current for emission, resulting in the decrease in recombination probability between holes and electrons at $Ru(bpy)_3^{2+}$ in the complex by employing more conductive SPAn as a matrix.

When poly(vinyl sulfate) (PVS) was employed instead of DNA, the PVS/PAn complex was also obtained by the template photopolymerization of PPD. A solution similar to the PAn/DNA complex containing $Ru(bpy)_3^{2+}$ was prepared for the PVS/PAn complex for EL device fabrication. The ITO glass electrode, however, repelled the PVS/PAn solution differently from the DNA/PAn complex solution, and it was difficult for us to prepare a thin, smooth, and homogeneous PVS/PAn film on the electrode. The film surface was qualitatively confirmed by the AFM technique, indicating that the DNA/PAn complex film on the ITO glass electrode was smooth and homogeneous.

Further, when DNA/$Ru(bpy)_3^{2+}$ without PAn was employed to fabricate the Al/DNA/$Ru(bpy)_3^{2+}$/ITO-based cell, a maximum luminance of 50 cd/m^2 was observed at the bias voltage of 15 V. The cell also showed the fast turn-on response, as mentioned earlier. In this experiment, a brand new cell was prepared in each measurement to avoid potential formation from previous electrochemical reactions. These results clearly suggest that DNA plays an important role for at least OLED cell fabrication.

12.3.4 *Color Tunable OLED Based on the DNA/PAn/$Ru(bpy)_3^{2+}$ Complex*

OLEDs have a wide spectral range, require low driving voltage, and are lightweight. Therefore, they are suitable for use in portable display devices such as full-color flat panel displays, as described earlier. Color-tunable OLEDs are of particular interest because they enable control of picture elements in large-screen displays and improve resolution and color quality. These devices can generate two or more colors depending on device parameters such as driving voltage, current, and local temperature [324, 325]. Voltage-controlled multicolor OLEDs are widely used in the fabrication of color-variable devices. These OLEDs are classified into several types based on their components: a dye-dispersed polymer emissive layer, a single quantum-well layer inserted in the emissive layer, two emissive layers separated by a carrier blocking layer, or two emissive parts stacked on a transparent electrode.

In this section, the color tunability of a DNA-based OLED was successfully demonstrated for the first time. The OLED consists of a DNA/PAn/Ru(bpy)$_3^{2+}$ complex layer (hole transport layer, red light–emitting material) and a tris(8-hydroxyquinolinate) aluminum layer (Alq$_3$, electron transport layer, green light–emitting material). Owing to this structure, the OLED exhibited multicolor emission, ranging from green to yellow to red, upon application of different voltages.

We first prepared the DNA/PAn complex with a simple complexation procedure as follows. Five mM of the DNA/PAn complex solution ([DNA]/[PAn] = 10/1) was prepared by mixing suitable amounts of a DNA aqueous solution, a PAn/NMP solution, and a dilute HCl (pH 3.8) solution. This solution was purified by reprecipitation with acetone to remove NMP, and the precipitated DNA/PAn complex was obtained by filtering the solution. The vis-NIR absorption spectra of the DNA/PAn complex measured in aqueous solution are shown in Fig. 12.23.

A large absorption band was observed around 750 nm, which was assignable to the localized polaron structure of PAn [326]. While PAn cannot dissolve in water by itself, the DNA/PAn complex can. This result indicates that the doped PAn interacted with the

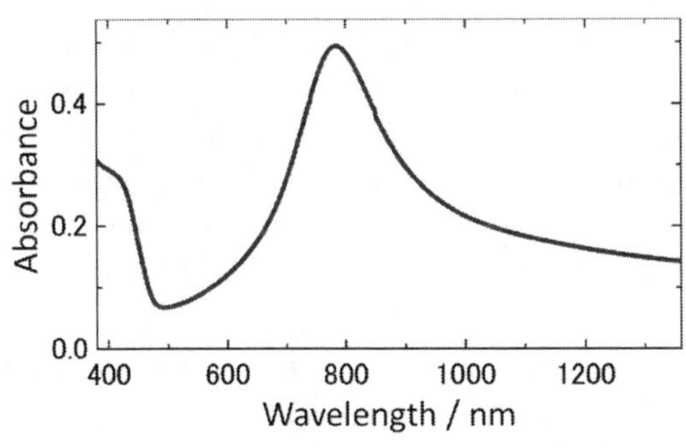

Figure 12.23 Vis-NIR absorption spectrum of the DNA/PAn complex in an aqueous solution.

DNA phosphate group in the aqueous solution. Further, a CD signal corresponding to the absorption band of PAn was observed at a wavelength of around 600 to 800 nm, suggesting formation of the DNA/PAn complex. The absorption peak wavelength was different from that of the aforementioned photopolymerized PAn/DNA complex (Figs. 12.16 and 12.17). This is possibly due to difference in the interaction between DNA and PAn.

We subsequently synthesized the DNA/PAn/Ru(bpy)$_3^{2+}$ complex by adding 5 mM of Ru(bpy)$_3$Cl$_2$ aqueous solution to 50 mM of DNA/PAn aqueous solution. The interaction between Ru(bpy)$_3^{2+}$ and the DNA/PAn complex was also studied by measuring their emission spectra. The luminescence spectra of Ru(bpy)$_3^{2+}$ varied with DNA concentration in the DNA/PAn/Ru(bpy)$_3^{2+}$ complex. The emission intensity of Ru(bpy)$_3^{2+}$ increased in conjunction with increasing DNA concentration. This enhancement in emission intensity indicates formation of the DNA/PAn/Ru(bpy)$_3^{2+}$ complex, because the complexation of DNA and Ru(bpy)$_3^{2+}$ suppressed the vibrational deactivation of the excited states of Ru(bpy)$_3^{2+}$ [327]. The detailed DNA complex structure and properties will be published in another paper in the near future.

We then fabricated the electroluminescent devices consisting of the DNA/PAn/Ru(bpy)$_3^{2+}$ complex as follows. A thin layer with 30 to 40 nm of the DNA/PAn/Ru(bpy)$_3^{2+}$ complex was obtained directly on the sufficiently washed ITO glass electrode by spin coating (1000 rpm, 30 s), and the layer was dried at 50°C for 12 h in vacuo. Subsequently, a thin layer of Alq$_3$ (thickness: 50 nm) was formed by vacuum deposition under reduced pressure ($< 4.0 \times 10^{-4}$ Pa). For the top electrode, an active diode area of 0.09 cm^2 was formed by sequentially depositing an aluminum layer (thickness: 110 nm) in vacuo through a shadow mask on top of the Alq$_3$ thin layer without breaking the vacuum. For spectral measurements of EL, the device was then encapsulated in a glass cover and sealed with UV-cured epoxy glue.

The current density–voltage characteristics of the ITO/(DNA/PAn/Ru(bpy)$_3^{2+}$)/Alq$_3$/Al OLED are shown in Fig. 12.24.

The thicknesses of the DNA/PAn/Ru(bpy)$_3^{2+}$ and Alq$_3$ layers were 30 nm and 50 nm, respectively. The current density increased abruptly above a bias voltage of 5 V, and at 18 V, a current density of

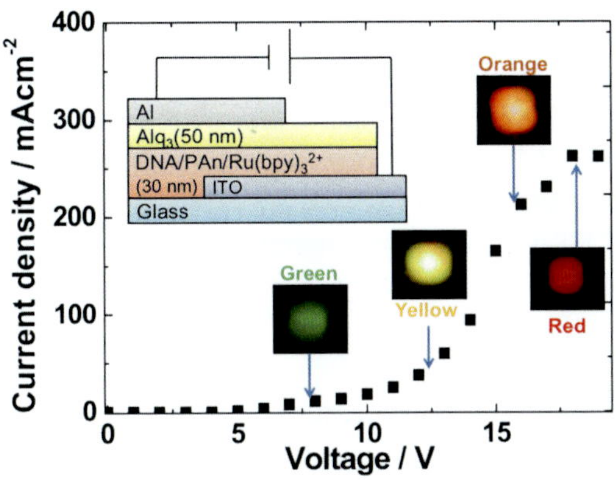

Figure 12.24 *I–V* characteristics and photographs of OLED (the inset shows a schematic of the OLED and digital images of the emission). Reprinted from Ref. [328] with the permission of AIP Publishing, Copyright 2010.

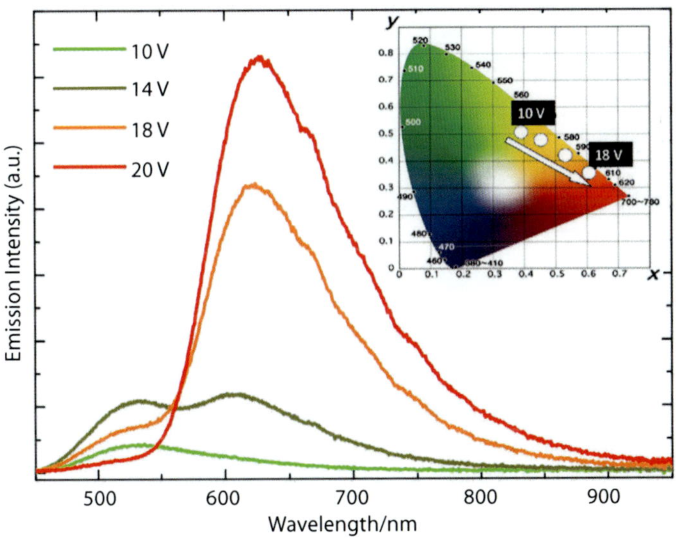

Figure 12.25 EL spectra of OLED at various bias voltages (the inset shows changes in the CIE emission coordinates of the OLED for various bias voltages). Reprinted from Ref. [328] with the permission of AIP Publishing, Copyright 2010.

300 mA/cm^2 was obtained. The current above 5 V obeyed Child's law for SCLC. Upon reflection of the $I-V$ characteristics, green luminescence was observed from the device above a bias voltage of 5 V. Interestingly, the emission color of the device changed with increasing applied voltage, from green to yellow (14 V) to orange (16 V) and finally to red (18 V), as shown in the inset photo of Fig. 12.24. The maximum luminance of the OLED was 10 cd/m^2, which was lower than that of the previously reported DNA-based OLED [322, 323].

These results are caused by the insufficient amounts of PAn in the DNA complex compared with that of the previous studies [312, 322, 323]. In the previous report, the DNA/PAn/Ru(bpy)$_3^{2+}$ complex was prepared by photopolymerization of the aniline dimer in the presence of DNA and Ru(bpy)$_3^{2+}$. Additionally, the concentration of the PAn monomer was equal to that of the DNA phosphate group. On the other hand, the [DNA]/[PAn] ratio was 10/1 in this study. The present method facilitates a very easy way to introduce PAn into DNA, but it is difficult to increase the concentration of PAn in DNA. As a result, the low concentration of PAn is responsible for the luminance decrease.

To confirm multicolor emission from the device, we measured its EL spectra at various voltages (Fig. 12.25).

At a bias voltage of 10 V, a green emission band was observed around 540 nm. This band was found to be similar to that observed in the PL spectrum of the Alq$_3$ layer [327], suggesting that the green emission originated from this layer. In addition, a red emission band with a maximum at 610 nm appeared above a bias voltage of 14 V. This band was found to be almost identical to that observed in the PL spectrum of the Ru(bpy)$_3^{2+}$ solution, indicating that the red emission originates from the Ru(bpy)$_3^{2+}$ layer. The height of the green emission band of Alq$_3$ decreased gradually when the applied voltage exceeded 14 V, and it decreased to almost zero at a bias voltage of 18 V. In contrast, the red emission band of Ru(bpy)$_3^{2+}$ (610 nm) increased when the applied voltage exceeded 14 V. The inset in Fig. 12.25 shows changes in the collisional ionization equilibrium (CIE) emission coordinates of the device at various applied voltages; the coordinates changed from (0.39, 0.51) at 10 V to (0.61, 0.37) at 18 V. In this manner, the applied voltage dependence of the Alq_3 emission intensity was found to be different from that of $Ru(bpy)_3^{2+}$, and multicolor emission from one device was confirmed.

To discuss the multicolor emission mechanism, OLEDs with configurations different from the one mentioned earlier were used [328]. The current density–voltage characteristics of an OLED modified by increasing the thickness of the Alq_3 layer (130 nm, previous: 50 nm) were measured. A comparison with the I–V characteristics of the original OLED structure (50 nm) revealed that the green emission occurred from the Alq_3 layer at a bias voltage of 10 V, and its intensity increased above 10 V. However, no red emission was observed from the $Ru(bpy)_3^{2+}$ layer in the entire range of applied voltages. This result suggests that the recombination of the holes and the electrons occurred in the Alq_3 layer, because the thickness increase of the Alq_3 layer caused the degradation of its electron transporting ability. We also analyzed the current–voltage characteristics of another modified OLED, in which the PAn component (hole transport layer) was removed from the DNA complex layer. As a result, only the red emission from $Ru(bpy)_3^{2+}$ was observed, and its intensity increased above a bias voltage of 11 V. The removal of PAn resulted in a decrease in hole mobility; therefore, carrier recombination occurred only in the DNA/$Ru(bpy)_3^{2+}$ layer.

Next, the carrier mobility of the DNA complex was measured by the time of flight (TOF) technique. Figure 12.26 shows the dependence of the hole mobility on the applied electric field for a 1:5 mixture of the DNA/PAn complex and native DNA. The hole mobility of this mixture layer was found to be 1×10^{-6} to 4×10^{-6} cm^2/Vs, and it showed negative field dependence. The negative–electric field dependence of the hole mobility in semiconducting or molecularly doped polymers has been previously reported [329, 330], in which the negative dependence has been attributed to the diffusion of the charge carriers against the direction of the applied electric field as the result of a large positional disorder. In the case of our device, the negative dependence was a result of the one-dimensional structure of the DNA/PAn complex. In the spin-coated DNA complex layer, the main chains of the DNA complex were expected to lie on the substrate, resulting in the formation of detour-routes perpendicular to the electric field (Fig. 12.27).

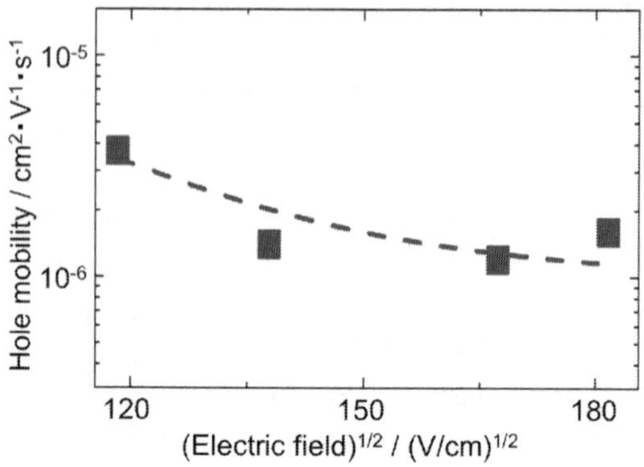

Figure 12.26 Applied-electric-field dependence of hole mobility for a 1:5 mixture layer of the DNA/PAn complex and native DNA. Reprinted from Ref. [328] with the permission of AIP Publishing, Copyright 2010.

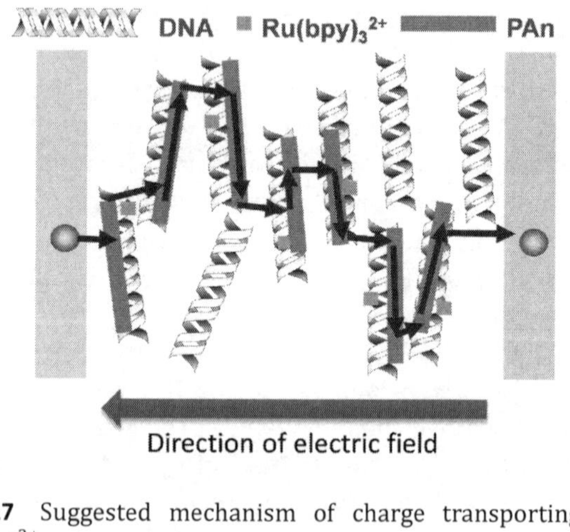

Figure 12.27 Suggested mechanism of charge transporting in DNA/PAn/Ru(bpy)$_3^{2+}$.

In contrast, the electron mobility of the Alq_3 layer is known to show the positive field dependence over a wide range of electric fields [331]. Based on these differences in the electric-field dependence of the carriers, the recombination distribution of the OLED studied herein can be explained as follows. At low voltages, since the hole mobility of the DNA complex layer was superior to the electron mobility of the Alq3 layer, the recombination occurred mainly in the Alq3 layer. As the voltage was increased, the electron mobility of the Alq3 layer increased rapidly with the electric field, and the electrons could move through the interface and recombine with the holes in the DNA complex layer. This resulted in a continuous shift of the recombination region from the Alq3 layer to the DNA/PAn/Ru(bpy)$_3^{2+}$ complex layer, as studied theoretically in the previous literature [332].

Thus, the multicolor emission mechanism of the OLED studied herein can be summarized as follows. At a bias voltage of 5–10 V, the green emission was observed to occur from the Alq3 layer because the carrier recombination occurred in this layer. As the applied voltage increased, the carrier recombination region shifted from the Alq3 layer to the DNA/PAn/Ru(bpy)$_3^{2+}$ complex layer, and the recombination occurred in both the Alq3 and DNA/PAn/Ru(bpy)$_3^{2+}$ complex layers. As a result, both the green emission from Alq3 and the red emission from Ru(bpy)$_3^{2+}$ were observed. Finally, only the red emission from Ru(bpy)$_3^{2+}$ was observed above the applied voltages of 18 V.

12.3.5 *Summary*

A PAn/DNA complex was successfully prepared by the template photopolymerization of dimeric aniline (PPD) via photocatalytic reaction of Ru(bpy)$_3^{2+}$ in the presence of DNA. The aqueous solution of PPD, Ru(bpy)$_3^{2+}$, and DNA was first prepared to analyze the structure of the resulting complex and to carry out photopolymerization. PPD and Ru(bpy)$_3^{2+}$ were bound to duplex DNA regularly in the solution through the electrostatic and/or hydrophobic interaction with DNA.

Photopolymerization was carried out by illuminating the solution containing DNA/PPD/Ru(bpy)$_3^{2+}$ with visible light. The photopolymerization reaction occurred even in solutions at pH from 3.0 to 6.0, owing to the specific local, lower-pH environment provided by DNA. The absorption and CD spectra indicate that the PAn/DNA complex has an ordered structure in which PAn is associated with DNA and winds around the duplex DNA, just like a triple helix. It was revealed that DNA works as the template for the photopolymerization of PPD with Ru(bpy)$_3^{2+}$ to prepare a highly ordered, novel PAn/DNA complex. The complex contains the photocatalyst, $Ru(bpy)_3^{2+}$, even after purification. $Ru(bpy)_3^{2+}$ also works as emitting material.

An Al/DNA/PAn containing $Ru(bpy)_3^{2+}$/ITO-based OLED cell was fabricated to analyze EL properties. It was revealed that the novel, processable, and water-soluble PAn/DNA complex worked as material for a $Ru(bpy)_3^{2+}$ complex-based red-emitting diode with a fast turn-on response. Further, we demonstrated green-to-red color-tunable emission from an OLED consisting of a DNA/PAn/$Ru(bpy)_3^{2+}$ complex (hole transport layer) and Alq_3 (electron transport layer). The DNA/PAn/$Ru(bpy)_3^{2+}$ complex was prepared with a simple complexation procedure between DNA and PAn, which is different from the template photopolymerization method. The emission colors of the Alq_3-stacked OLED could be tuned by varying the applied voltage. The color tunability of this emission was principally a result of shifts in the carrier recombination region that are attributable to increasing applied voltage.

Work is in progress to improve the luminance and the stability of the OLED. We expect that the results of our research will contribute to the development of organic electronic devices, such as BiOLED, using biomaterials characterized by DNA-hybrid materials.

Chapter 13

Fabrication and Characteristics of DNA-Biofuel Cell

Yasumitsu Matsuo[a] and Seiichiro Ikehata[b]

[a]*Department of Life Science, Setsunan University, 17-8 Ikeda-Nakamachi, Neyagawashi, Osaka, 572-8508, Japan*
[b]*Department of Applied physics, Tokyo University of Science, 1-3 Kagurazaka, Shinjyuku-ku, Tokyo, 162-8601, Japan*
ymatsuo@lif.setsunan.ac.jp

13.1 Introduction

As shown in Chapter 4, DNA shows relatively high ionic conductivity under humidified conditions, which originates from the formation of water bridges between DNA and water molecules. This means DNA has the potential to be used as sensors and batteries, applying ion or proton conduction. Especially, we can fabricate biofuel cells based on the DNA electrolyte (DNA fuel cell) with proton conductivity. In this section, we show the characteristics of fuel cells with DNA films.

DNA Engineering: Properties and Applications
Edited by Kenji Mizoguchi and Hirokazu Sakamoto

ISBN 978-981-4669-46-7 (Hardcover), 978-981-4669-47-4 (eBook)
www.panstanford.com

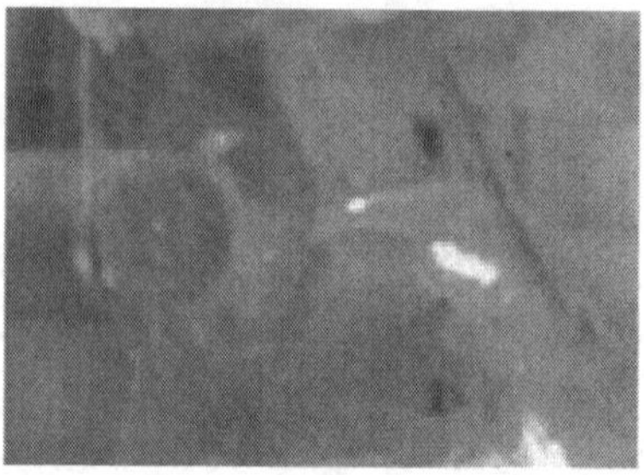

Figure 13.1 A DNA–CTAB film.

13.1.1 *Fuel Cell Construction*

We show the experimental results of fuel cells based on DNA and DNA–CTAB films. DNA–CTAB is synthesized with DNA and CTAB $(C_{16}H_{33})N(CH_3)_3Br$, which is known as a surface-active agent. Figure 13.1 shows a DNA–CTAB film, which is very stable compared with a DNA film.

Figure 13.2 shows the schematic structure of a fuel cell. The Pt-C electrodes attached to both sides of a DNA membrane as the electrolyte by Pt-C paste are sandwiched between the SUS mesh electrodes (current collector), and H_2 and O_2 are supplied. The

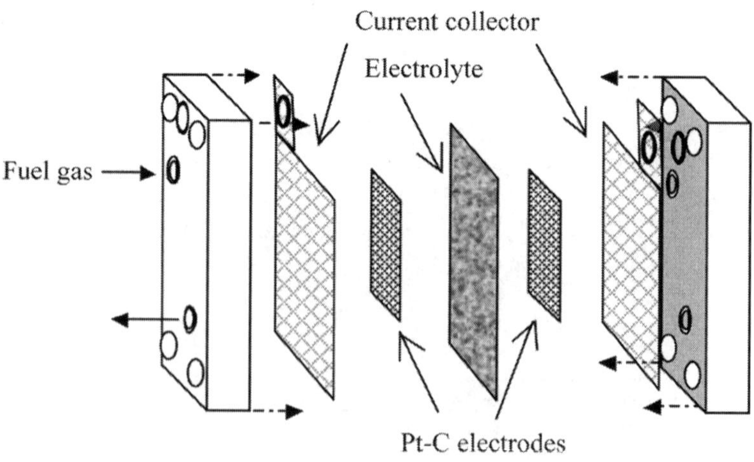

Figure 13.2 Schematic structure of a fuel cell.

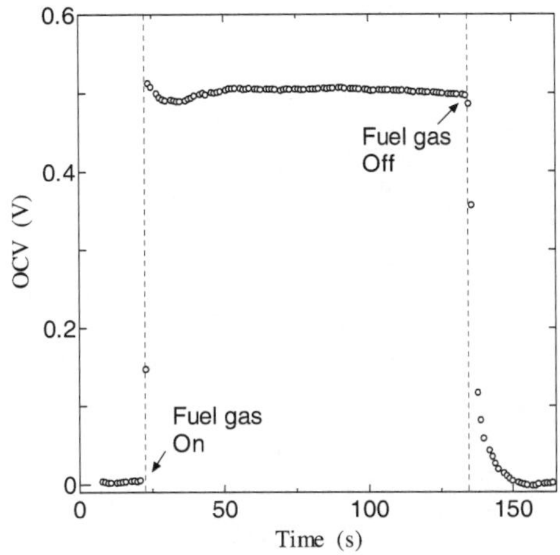

Figure 13.3 Time dependence of OCV in a DNA fuel cell.

sealing is set between the DNA film and SUS electrodes, and gas leaks are attached.

13.1.2 *Results and Discussion*

Figure 13.3 shows the time dependence of the open-circuit voltage (OCV) in a DNA-based fuel cell. OCV increases with the amount of the supplied fuel gas and then saturates at about 0.5 V. On the other hand, when the flow of the fuel gas is stopped, OCV decreases soon, as shown in Fig. 13.3. In this way, in a DNA fuel cell, voltage is generated by the supplied fuel gas. Figure 13.4 shows the I–V characteristic of a fuel cell, which is typical of the I–V curve of the batteries. These results indicate that the DNA film works as the electrolyte of the fuel cell. However, the current density is very small in this fuel cell, which derives from the stability of the DNA electrolyte. The DNA film is water soluble and becomes soft under high humidity conditions and then cannot operate as the electrolyte of the fuel cell. In order to use DNA films as the electrolyte for a fuel cell, it is necessary for them to have hydrophobicity. Therefore, we have prepared the DNA–CTAB synthetic film and have fabricated a fuel cell based on the DNA–CTAB electrolyte (DNA–CTAB fuel cell).

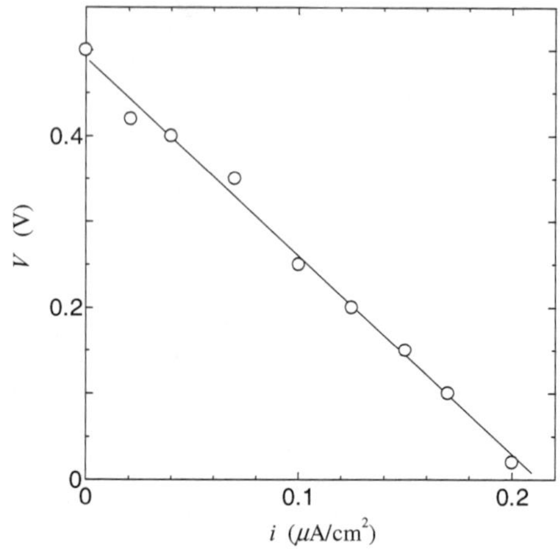

Figure 13.4 $I-V$ characteristic of a DNA fuel cell.

Figure 13.5 shows the humidity dependence of the OCV in a DNA–CTAB fuel cell along with that of the OCV in a DNA fuel cell for comparison. It is evident that the OCV monotonically increases with humidity above about 20% RH and tends to saturate at around 0.6 V thereafter. This result indicates that the DNA–CTAB film works as the electrolyte of the fuel cell. The $I-V$ characteristic of a DNA–CTAB

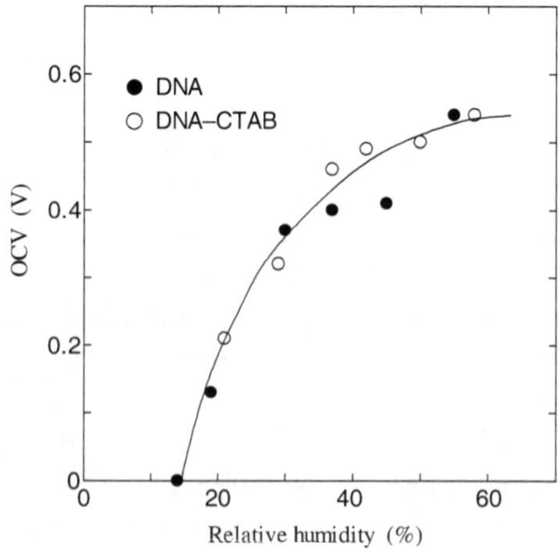

Figure 13.5 Humidity dependence of OCV in a DNA–CTAB fuel cell.

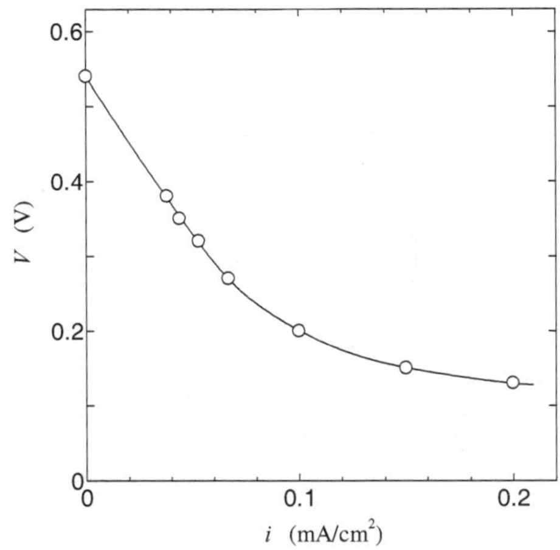

Figure 13.6 I–V characteristic of a DNA–CTAB fuel cell.

fuel cell is shown in Fig. 13.6, which is typical of the fuel cell with electrode overpotential [333]. Moreover, it is noted that the current density becomes around 600 times larger than that of a fuel cell based on DNA as electrolyte. This result indicates that the stability of the DNA–CTAB electrolyte yields a fuel cell with higher current density. In this way, by improving the stability, DNA can be used as the electrolyte in fuel cells. In the near future, fuel cells or sensors employing the proton conductivity of DNA will appear.

13.2 Conclusion

In order to study the application of DNA in electrical devices, fuel cells based on DNA have been fabricated and their properties have been investigated. It has been found that DNA works as the electrolyte of a fuel cell. Biofuel cells based on natural DNA operate below 55% RH, while DNA film becomes soft above 60% RH and cannot operate as electrolyte. On the other hand, DNA–CTAB, which is a cross-linked DNA with CTAB, is stable and can operate as the electrolyte of a fuel cell even under higher humidity conditions. These results indicate that cross-linking of DNA molecules is effective for the application of biofuel cells based on DNA, and by controlling the cross-linking, biofuel cells based on DNA will become fit for practical use in the future.

Acknowledgments

We would like to thank the collaborators Mr. Go Kumasaka, Mr. Kazuya Inoue, Dr. Yukihiko Yoshida, and Mr. Hitoki Semizo for carrying out various measurements in fuel cells based on DNA. Every result described in this chapter was accomplished with the help and support of the above collaborators.

Bibliography

1. Goeddel, D. V., Kleid, D. G., Bolivar, F., Heyneker, H. L., Yansura, D. G., Crea, R., Hirose, T., Kraszewski, A., Itakura, K., and Riggs, A. D. (1979). Expression in Escherichia coli of chemically synthesized genes for human insulin, *Proc. Natl. Acad. Sci. USA* **76**, 1, pp. 106–110.
2. Watson, J. D. and Crick, F. H. (1953). Molecular structure of nucleic acid, *Nature* **171**, pp. 737–738.
3. Fink, H.-W. and Schenenberger, C. (1999). Electrical conduction through DNA molecules, *Nature* **398**, pp. 407–410.
4. Porath, D., Bezryadin, A., Vries, S. D., and Dekker, C. (2000). Direct measurement of electrical transport through DNA molecules, *Nature* **403**, pp. 635–638.
5. Tran, P., Alavi, B., and Grüner, G. (2000). Charge transport along the λ-DNA double Helix, *Phys. Rev. Lett.* **85**, 7, pp. 1564–1567.
6. Kasumov, A. Y., Kociak, M., Gueron, S., Reulet, B., Volkov, V. T., Klinov, D. V., and Bouchiat, H. (2001). Proximity-induced superconductivity in DNA, *Science* **291**, pp. 280–282.
7. Yoo, K.-H., Ha, D. H., Lee, J.-O., Park, J. W., Kim, J., Kim, J. J., Lee, H.-Y., Kawai, T., and Choi, H. Y. (2001). Electrical conduction through poly(dA)-poly(dT) and poly(dG)-poly(dC) DNA molecules, *Phys. Rev. Lett.* **87**, 19, pp. 198102.
8. Zhang, Y., Austin, R. H., Kraeft, J., Cox, E. C., and Ong, N. P. (2002). Insulating behavior of λ-DNA on the micron scale, *Phys. Rev. Lett.* **89**, 19, pp. 198102 (1–4).
9. Heim, T., Mélin, T., Deresmes, D., and Vuillaume, D. (2004). Localization and delocalization of charges injected in DNA, *Appl. Phys. Lett.* **85**, 13, pp. 2637–2639.
10. Lee, H.-Y., Tanaka, H., Otsuka, Y., Yoo, K.-H., Lee, J.-O., and Kawai, T. (2002). Control of electrical conduction in DNA using oxygen hole doping, *Appl. Phys. Lett.* **80**, pp. 1670–1672.

11. Kutnjak, Z., Filipic, C., Podgornik, R., Nordenskiold, L., and Korolev, N. (2003). Electrical conduction in native deoxyribonucleic acid: Hole hopping transfer mechanism?, *Phys. Rev. Lett.* **90**, 9, pp. 098101 (1–4).
12. Conwell, E. M. and Basko, D. M. (2003). Polarons and conduction in DNA, *Synth. Met.* **137**, pp. 1381–1383.
13. Starikov, E. B. (2004). Why DNA electrical properties change on molecular oxygen doping: a quantum-chemical study, *Mod. Phys. Lett. B* **18**, 16, pp. 785–790.
14. Taniguchi, M., Lee, H., Tanaka, H., and Kawai, T. (2003). Electrical properties of poly(dA)-poly(dT) and poly(dG)-poly(dC) DNA doped with iodine molecules, *Jpn. J. App. Phys.* **42**, pp. L215–L216.
15. Taniguchi, M., Otsuka, Y., Tabata, H., and Kawai, T. (2003). Humidity dependence of electrical resistivity in poly(dG)·poly(dC) DNA thin film, *Jpn. J. Appl. Phys.* **42**, pp. 6629–6630.
16. Lee, J. S., Latimer, L. J. P., and Reid, R. S. (1993). A cooperative conformational change in duplex DNA induced by Zn^{2+} and other divalent metal ions, *Biochem. Cell Biol.* **71**, pp. 162–168.
17. Rakitin, A., Aich, P., Papadopoulos, C., Kobzar, Y., Vedeneev, A. S., Lee, J. S., and Xu, J. M. (2001). Metallic conduction through engineered DNA: DNA nanoelectronic building blocks, *Phys. Rev. Lett.* **86**, 16, pp. 3670–3673.
18. Tanaka, K., Tengeiji, A., Kato, T., Toyama, N., and Shionoya, M. (2003). A discrete self-assembled metal array in artificial DNA, *Science* **299**, 5610, pp. 1212–1213.
19. Wettig, S. D., Li, C.-Z., Long, Y.-T., Kraatz, H.-B., and Lee J. S. (2003). M-DNA: a self-assembling molecular wire for nanoelectronics and biosensing, *Analytical Sciences* **19**, pp. 23–26.
20. Kino, H., Tateno, M., Boero, M., Torres, J. A., Ohno, T., Terakura, K., and Fukuyama, H. (2004). A possible origin of carrier doping into DNA, *J. Phys. Soc. Jpn.* **73**, 8, pp. 2089–2092.
21. Alexandre, S. S., Soler, J. M., Seijo, L., and Zamora, F. (2006). Geometry and electronic structure of M-DNA (M $= Zn^{2+}$, Co^{2+}, and Fe^{2+}), *Phys. Rev. B* **73**, pp. 205112 (1–5).
22. Fuentes-Cabrera, M., Sumpter, B. G., Šponer, J. E., Šponer, Jiří, Petit, L., and Wells, J. C. (2007). Theoretical study on the structure, stability, and electronic properties of the guanine-Zn-cytosine base pair in *M*-DNA, *J. Phys. Chem. B* **111**, pp. 870–879.
23. Mallajosyula, S. S. and Pati, S. K. (2007). Structure and transport characteristics of modified DNA with magnetic ions, *Phys. Rev. Lett.* **98**, pp. 136601 (1–4).

24. Omerzu, A., Anželak, B., Turel, I., Strancar, J., Potocnik, A., Arcon, D., Arcon, I., Mihailović, D., and Matsui, H. (2010). Strong correlations in highly electron-doped Zn(II)-DNA complexes, *Phys. Rev. Lett.* **104**, pp. 156804 (1–4).

25. Mizoguchi, K. (2011). Metal incorporated M-DNA: structure, magnetism, optical absorption, *Proc. SPIE* **8103**, pp. 810307 (1–10).

26. Tseng, S.-H., Jangjian, P.-C., Tsai, C.-M., Cheng, T.-M., Chu, H.-L., Chang, Y.-C., Chung, W.-H., and Chang, C.-C. (2011). Ni^{2+}-enhanced charge transport via π–π stacking corridor in metallic DNA, *Biophys. J.* **100**, pp. 1042–1048.

27. Tsuburaya, M., Sakamoto, H., and Mizoguchi, K. (2014). Electronic states of DNA and M-DNA studied by optical absorption, *Phys. Rev. E* **89**, pp. 022719 (1–7).

28. Kumeta, T., Sakamoto, H., and Mizoguchi, K. (2014). Freeze-dried Zn-DNA: Magnetism dominated by water molecules, *J. Phys. Soc. Jpn.* **83**, 8, pp. 084801 (1–6).

29. Taniguchi, M. and Kawai, T. (2006). DNA electronics, *Physica E.* **33**, pp. 1–12.

30. Kwon, Y.-W., Lee, C. H., Choi, D.-H., and Jin, J.-I. (2009). Materials science of DNA, *J. Mater. Chem.* **19**, pp. 1353–1380.

31. Endres, R. G., Cox, D. L., and Singh, R. R. P. (2004). Colloquium: The quest for high-conductance DNA, *Rev. Mod. Phys.* **76**, pp. 195–214.

32. Nakamae, S., Cazayous, M., Sacuto, A., Monod, P., and Bouchiat, H. (2005). Intrinsic low temperature paramagnetism in B-DNA, *Phys. Rev. Lett.* **94**, pp. 248102 (1–4).

33. Mizoguchi, K., Tanaka, S., and Sakamoto, H. (2006). Comment on "intrinsic low temperature paramagnetism in B-DNA", *Phys. Rev. Lett.* **96**, 8, pp. 089801.

34. Lee, C. H., Kwon, Y.-W., Do, E.-D., Choi, D.-H., Jin, J.-I., Oh, D.-K., and Kim, J. (2006). Electron magnetic resonance and SQUID measurement study of natural A-DNA in dry state, *Phys. Rev. B* **73**, 9, pp. 224417 (1–7).

35. Rothemund, P. W. K. (2006). Folding DNA to create nanoscale shapes and patterns, *Nature* **440**, pp. 297–302.

36. Elson, D. and Chargaff, E. (1952). On the deoxyribonucleic acid content of sea urchin gametes, *Experientia* **8**, 4, pp. 143–145.

37. Wilkins, M. H. F., Stokes, A. R., and Wilson, H. R. (1953). Molecular structure of deoxypentose nucleic acids, *Nature* **171**, pp. 738–740.

38. Franklin, R. E. and Gosling R. G. (1953). Molecular configuration in sodium thymonucleate, *Nature* **171**, pp. 740–741.

39. Bowater, R. P. (2003). DNA structure, in *Nature Encyclopedia of the Human Genome*, Vol. 1 (ed. Cooper, D. M.) (Macmillan), pp. 1–9.
40. Sinden, R. R. (1994). *DNA Structure and Function* (1st ed.) (Academic Press, London).
41. Rich, A., Norheim, A., and Wang, A. H. J. (1984). The chemistry and biology of left-handed Z-DNA, *Annual Review of Biochemistry* **53**, pp. 791–846.
42. Ho, P. S. (1994). The non-B-DNA structure of d(CA/TG)n does not differ from that of Z-DNA, *Proc. Natl. Acad. Sci. (USA)* **91**, 20, pp. 9549–9553.
43. Abdurakhman, H., Tajiri, K., Yokoi, H., Kuroda, N., Matsui, H., Yanagimachi, T., Taniguchi, M., Kawai, T., and Toyota, N. (2007). Infrared spectroscopy on poly(dG)-poly(dC) DNA at low hydration, *J. Phys. Soc. Jpn* **76**, No. 2, pp. 024009 (1–6).
44. De Rosa, M., De Sanctis, D., Rosario, A. L., Archer, M., Rich, A., Athanasiadis, A., and Carrondo, M. A. (2010). Crystal structure of a junction between two Z-DNA helices, *Proc. Natl. Acad. Sci. USA* **107**, 20, pp. 9088–9092.
45. Helgren, E., Omerzu, A., Grüner, G., Mihailović, D., Podgornik, R., and Grimm, H. (2001). Electrons on the double helix: optical experiments on native DNA, *e-print* cond-mat/0111299.
46. Omerzu, A., Mihailović, D., Anželak, B., and Turel, I. (2007). Optical spectra of wet and dry M-DNA, *Phys. Rev. B* **75**, pp. 121103R (1–4).
47. Eley, D. D. and Spivey, D. I. (1962). First study on the nature of charge migration in DNA, *Trans. Faraday Soc.* **58**, pp. 411–415.
48. Roth, S. R. (1995). *One-Dimensional Metals* (VCG Verlagsgesellschaft, Weinheim, Germany).
49. Chen, J. H. and Seeman, N. C. (1991). Synthesis from DNA of a molecule with the connectivity of a cube, *Nature* **350**, pp. 631–633.
50. Zhang, Y. and Seeman, N. (1994). The construction of a DNA truncated octahedron, *J. Am. Chem. Soc.* **116**, 5, pp. 1661–1669.
51. Shih, W. M., Quispe, J. D., and Joyce, G. F. (2004). A 1.7-kilobase single-stranded DNA that folds into a nanoscale octahedron, *Nature* **427**, pp. 618–621.
52. Labean, T. H., Park, S., Ahn, S. J., and Reif, J. H. (2005). Stepwise DNA self-assembly of fixed-size nanostructures, in *Foundations of Nanoscience (FNANO11), Self-Assembled Architectures and Devices* (Snowbird, Utah), pp. 179–181.

53. Douglas, S. M., Dietz, H., Liedl, T., Hogberg, B., Graf, F., and Shih, W. M. (2009). Self-assembly of DNA into nanoscale three-dimensional shapes, *Nature* **459**, 7245, pp. 414–418.

54. Dietz, H., Douglas, S. M., and Shih, W. M. (2009). Folding DNA into twisted and curved nanoscale shapes, *Science* **325**, 5941, pp. 725–730.

55. Douglas, S. M., Bachelet, I., and Church, G. M. (2012). A logic-gated nanorobot for targeted transport of molecular payloads, *Science* **335**, pp. 831–834.

56. Nagatori, M., Ojima, M., Ibuki, Y., Sakamoto, H., and Mizoguchi, K. (2011). Electronic states of metal ions incorporated in Mn-DNA, *J. Phys. Soc. Jpn.* **80**, 11, pp. 114803 (1–5).

57. Aich, P., Labuik, S. L., Tari, L. W., Delbaere, L. J. T., Roesler, W. J., Falk, K. J., Steer, R. P., and Lee, J. S. (1999). *M*-DNA: a complex between divalent metal ions and DNA which behaves as a molecular wire, *J. Mol. Biol.* **294**, pp. 477–485.

58. Voet, D., Gratzer, W. B., Cox, R. A., and Doty, P. (1963). Absorption spectra of nucleotides, polynucleotides, and nucleic acids in the far ultraviolet, *Biopolymers* **1**, 193–208.

59. Clark, L. B., Peschel, G. G., and Tinoco, I., Jr. (1965). Vapor spectra and heats of vaporization of some purine and pyrimidine bases, *J. Phys. Chem.* **69**, 3615–3618.

60. Clark, L. B. and Tinoco, I. (1965). Correlations in the ultraviolet spectra of the purine and pyrimidine bases, *J. Am. Chem. Soc.* **87**, 11–15.

61. Yamada, T. and Fukutome, H. (1968). Vacuum ultraviolet absorption spectra of sublimed films of nucleic acid bases, *Biopolymers* **6**, 43–54.

62. Voelter, W., Records, R., Bunnenberg, E., and Djerassi, C. (1968). Magnetic circular dichroism studies. VI. Investigation of some purines, pyrimidines, and nucleosides, *J. Am. Chem. Soc.* **90**, 6163–6170.

63. Sutherland, J. C. and Griffin, K. (1984). Magnetic circular dichroism of adenine, hypoxanthine, and guanosine 5′-diphosphate to 180 nm, *Biopolymers* **23**, 2715–2724.

64. Broo, A. and Holmén, A. (1997). Calculations and characterization of the electronic spectra of DNA bases based on ab initio MP2 geometries of different tautomeric forms, *J. Phys. Chem. A* **101**, 19, pp. 3589–3600.

65. Bouvier, B., Gustavsson, T., Markovitsi, D., and Millié, P. (2002). Dipolar coupling between electronic transitions of the DNA bases and its

relevance to exciton states in double helices, *Chem. Phys.* **275**, 1–3, pp. 75–92.

66. Fiebig, T. (2009). Exciting DNA, *J. Phys. Chem. B* **113**, pp. 9348–9349.
67. Stewart, R. F. and Davidson, N. (1963). Polarized absorption spectra of purines and pyrimidines, *J. Chem. Phys.* **39**, 255–266.
68. Chen, H. H. and Clark, L. B. (1973). On the electronic spectrum of protonated adenine: A single crystal study of adenine hydrochloride, *J. Chem. Phys.* **58**, 2593–2603.
69. Zaloudek, F., Novros, J. S., and Leigh, L. B. (1985). The electronic spectrum of cytosine, *J. Am. Chem. Soc.* **107**, 7344–7351.
70. Clark, L. B. (1990). Electronic spectrum of the adenine chromophore, *J. Phys. Chem.* **94**, 2873–2879.
71. Clark, L. B. (1994). Electronic spectra of crystalline guanosine: transition moment directions of the guanine chromophore, *J. Am. Chem. Soc.* **116**, 5265–5270.
72. Hug, W. and Tinoco, I. (1973). Electronic spectra of nucleic acid bases. I. Interpretation of the in-plane spectra with the aid of all valence electron MO-CIA [configuration interaction] calculations, *J. Am. Chem. Soc.* **95**, 2803–2813.
73. Hug, W. and Tinoco, I. (1974). Electronic spectra of nucleic acid bases. II. Out-of-plane transitions and the structure of the nonbonding orbitals, *J. Am. Chem. Soc.* **96**, 665–673.
74. Lorentzon, J., Fulscher, M. P., and Roos, B. O. (1995). Theoretical study of the electronic spectra of uracil and thymine, *J. Am. Chem. Soc.* **117**, 9265–9273.
75. Fülscher, M. P. and Roos, B. O. (1995). Theoretical study of the electronic spectrum of cytosine, *J. Am. Chem. Soc.* **117**, 2089–2095.
76. Fülscher, M. P., Serrano-Andres, L., and Roos, B. O. (1997). A theoretical study of the electronic spectra of adenine and guanine, *J. Am. Chem. Soc.* **119**, 6168–6176.
77. Iguchi, K. (2004). π electrons in a single strand of DNA: a phenomenological approach, *Int. J. Mod. Phys. B* **18**, 13, pp. 1845–1910.
78. Volosov, A. (1989). Role of doubly excited states in calculations of optical activity, *Int. J. Quant. Chem.* **36**, 4, pp. 473–486.
79. Petke, J. D., Maggiora, G. M., and Christoffersen, R. E. (1990). Ab initio configuration interaction and random phase approximation calculations of the excited singlet and triplet states of adenine and guanine, *J. Am. Chem. Soc.* **112**, 14, pp. 5452–5460.

80. Roos, B. O., Fülscher, M. P., Malmqvist, P. A., Merchán, M., and Serrano-Andrés, L. (ed. Langhoff, S. R.) (1995). *Quantum Mechanical Electronic Structure Calculations with Chemical Accuracy* (Kluwer, Dordrecht).

81. Theiste, D., Callis, P. R., and Woody, R. W. (1991). Effects of the crystal field on transition moments in 9-ethylguanine, *J. Am. Chem. Soc.* **113**, 9, pp. 3260–3267.

82. Eisinger, J. and Shulman, R. G. (1968). Excited electronic states of DNA, *Science* **161**, pp. 1311–1319.

83. Weiss, S. B. and Nakamoto, T. (1961). On the participation of DNA in RNA biosynthesis, *Proc. Nat. Acad. Sci. USA* **47**, 5, pp. 694–697.

84. Blois Jr., M. S., Maling, J. E., and Taskovich, L. T. (1963). On the electron spin resonances in DNA, *Biophys. J.* **3**, 4, pp. 275–297.

85. Shields, H. and Gordy, W. (1959). Electron-spin resonance studies of radiation damage to the nucleic acids and their constituents, *Proc. Nat. Acad. Sci.* **45**, pp. 269–281.

86. Shen, P. K., Bliumenfeld, L. A., Kalmanson, A. E., and Pasynskii, N. G. (1959). Electrical paramagnetic resonance spectra of biological objects. III. Effect of ionizing radiations on nucleic compounds, *Biofizika* **4**, pp. 263–274.

87. Bliumenfeld, L. A. (1959). Anomalous magnetic properties of nucleic acids, *Biofizika* **4**, pp. 515–520.

88. Müller, A., Hotz, G., and Zimmer, K. G. (1961). Electron spin resonances in bacteriophage: alive, dead and irradiated, *Biochem. Biophys. Res. Commun.* **4**, pp. 214–217.

89. Blios Jr., M. S., and Maling, J. E. (1961). Electron spin resonance in nucleic acids, nucleotides, and the nitorgenous bases of DNA, *Biochem. Biophys. Res. Commun.* **4**, pp. 252–257.

90. Shulman, R. G., Walsh Jr. W. M., Williams, H. J., and Wright, J. P. (1961). Ferromagnetic resonance in DNA samples, *Biochem. Biophys. Res. Commun.* **5**, pp. 52–56.

91. Walsh Jr. W. M., Shulman, R. G., and Heidenriech, R. D. (1961). Ferromagnetic inclusions in Nucleic Acid samples, *Nature* **16**, pp. 1041–1043.

92. Malino, J. E., Taskovich, L., and Blois, Jr. M. S., (1963). Electron spin resonance in ATP and RNA, *Biophysic. J.* **3**, pp. 79–95.

93. Mizoguchi, K. (2010). Electronic states of M-DNA incorporated with divalent metal ions, *Proc. SPIE* **7765**, pp. 77650R (1–12).

94. Mizoguchi, K., Tanaka, S., Ojima, M., Sano, S., Nagatori, M., Sakamoto, H., Yonezawa, Y., Aoki, Y., Sato, H., Furukawa, K., and Nakamura, T. (2007).

AF-like ground state of Mn-DNA and charge transfer from Fe to base-π-band in Fe-DNA, *J. Phys. Soc. Jpn.* **76**, 4, pp. 043801 (1–4).

95. Mizoguchi, K. (2008). Physical properties of natural DNA and metal ion inserted M-DNA, *Proc. SPIE* **7040**, pp. 70400Q (1–9).
96. Altshuler, B. L., Gefen, Y., and Imry, Y. (1991). Persistent differences between canonical and grand canonical averages in mesoscopic ensembles: Large paramagnetic orbital susceptibilities, *Phys. Rev. Lett.* **66**, 7 January, pp. 88–91.
97. Ambegaokar, V. and Eckern, U. (1990). Coherence and persistent currents in mesoscopic rings, *Phys. Rev. Lett.* **65**, pp. 381–384.
98. Socrates, G. (2001). *Infrared and Raman Characteristic Group Frequencies Third Edition*, (John Wiley & Sons, Chichester).
99. Larkin, P. J. (2011). *IR and Raman Spectroscopy*, (Elsevier, Amsterdam).
100. Stuart, B. (1997). *Biological Application of Infrared Spectroscopy*, (John Wiley and Sons, New York).
101. Saenger, W. (1984). *Principles of Nucleic Acid Structure*, (Springer-Verlag, New York).
102. Drew, H. R. and Dickerson, R. E. (1981). Structure of a B-DNA Dodecamer III, *J. Mol. Biol.* **151**, pp. 535–556.
103. Fuller, W., Forsyth, T., and Machendrasingam, A. (2004). Water-DNA interactions as studied by X-ray and neutron fibre diffraction, *Phil. Trans. R. Soc. Lond. B* **359**, pp. 1237–1248.
104. Arai, S., Chatake, T. Ohhara, T., Kurihara, K., Tanaka, I., Suzuki, N., Fujimoto, Z., Mizuno, H., and Niimura, N. (2005). Complicated water orientations in the minor groove of the B-DNA decamer $d(CCATTAATGG)_2$ observed by neutron diffraction measurements, *Nucl. Acids Res.* **33**, pp. 3017–3024.
105. Stuhrmann H. B. (2004). Unique aspects of neutron scattering for the study of biological systems, *Rep. Prog. Phys.* **67**, pp. 1073–1115.
106. Langridge, R., Marvin, D. A., Seeds, W. E., Wilson, H. R., Hooper, C. W., Wilkins, M. H. F., and Hamilton, L. D. (1960). The molecular configuration of deoxyribonucleic acid: molecular models and their Fourier transforms, *J. Mol. Biol.* **2**, pp. 38–64.
107. Fuller, W., Wilkins, M. H., Wilson, H. R., and Hamilton, L. D. (1965). The molecular configuration of deoxyribonucleic acid. IV. X-ray diffraction study of the A form, *J. Mol. Biol.* **12**, pp. 60–76.
108. Watson, J. D., Baker, T. A., Bell, S. P., Gann, A., Levine, M., and Losick, R. (2004). *Molecular Biology of the Gene* (Pearson Education, San Francisco).

109. Tsuboi, M. (1969). Application of infrared spectroscopy to structure studies of nucleic acids, *Appl. Spec. Rev.* **3**, pp. 45–90.

110. Liquier, J. and Taillandier, E. (1996). In Mantsch, H. and Chapman, D. (eds.) *Infrared Spectroscopy of Nucleic Acids, Infrared Spectroscopy of Biomolecules*, (Wiley-Liss, New York), pp. 131–158.

111. Kornilova, S. V., Miskovsky, P., Tomkova, A., Kapinos, L. E., Hackl, E. V., Andruschenko, V. V., Grigoriev, D. N., and Blagoi, Y. P. (1997). Vibrational spectroscopic studies of the divalent metal ion effect on DNA structural transitions, *J. Mol. Struct.* **408/409**, pp. 219–223.

112. Weidlich, T., Lindsay, S. M., and Ruprecht, A. (1988). Counterion effects on the structure and dynamics of solid DNA, *Phys. Rev. Lett.* **61**, pp. 1674–1677.

113. Lavalle, N., Lee, S. A., and Ruprecht, A. (1990). Counterion effects on the physical properties and the A to B transition of calf-thymus DNA films, *Biopolymers* **30**, pp. 877–887.

114. Keller, P. B. and Hartman, K. A. (1986). Structural forms, stabilities and transitions in double-helical poly(dG-dC) as a function of hydration and NaCl content. An infrared spectroscopic study, *Nucleic Acids Res.* **14**, pp. 8167–8182.

115. Wolf, B. and Hanlon, S. (1975). Structural transitions of deoxyribonucleic acid in aqueous electrolyte solutions. II. Role of hydration, *Biochemistry* **14**, pp. 1661–1670.

116. Albiser, G., Lamiri, A., and Premilat, S. (2001). The A-B transition: Temperature and base composition effects on hydration of DNA, *Int. J. Biol. Macromol.* **28**, pp. 199–203.

117. Ghosh, A. and Bansal, M. (2003). A glossary of DNA structures from A to Z, *Acta Cryst.* **D59** pp. 620–626.

118. Andruschenko, V. V., Kornilova, S. V., Kapinos, L. E., Hakle, E. V., Galkin, V. L., Grigoriev, D. N., and Blagoi, Y. P. (1997). IR-spectroscopic studies of divalent metal ion effects on DNA hydration, *J. Mol. Struct.* **408–409**, pp. 225–228.

119. Taboury, J. A., Liquier, J., and Taillandier, E. (1985). Characterization of DNA structures by infrared spectroscopy: double helical forms of poly(dG-dC)·poly(dG-dC), poly(dD_8G-dC)·poly(dD_8G-dC), and poly(dG-dm_5C)·poly(dG-dm_5C), *Can. J. Chem.* **63**, pp. 1904–1909.

120. Lee, S. L., Debenedeetti, P. G., Errington, J. R., Pethica, B. A., and Moore, D. J. (2004). A calorimetric and spectroscopic study of DNA at low hydration, *J. Phys. Chem. B* **108**, pp. 3098–3106.

121. Falk, M., Hartman Jr, K. A., and Lord, R. C. (1962). Hydration of deoxyribonucleic acid. I. gravimetric study, *J. Am. Chem. Soc.* **84**, pp. 3843–3846.

122. Falk, M., Hartman Jr, K. A., and Lord, R. C. (1963). Hydration of deoxyribonucleic acid. II. An infrared study, *J. Am. Chem. Soc.* **85**, pp. 387–391.

123. Falk, M., Poole, A. G., and Goymour, C. G. (1970). Infrared study of the water in the hydration shell of DNA, *Can. J. Chem.* **48**, pp. 1536–1542.

124. Keller, P. B. and Hartman, K. A. (1986). The effect of ionic environment and mercury(II) binding on the alternative structures of DNA. An infrared spectroscopic study, *Spectrochim. Acta A* **42**, pp. 299–306.

125. Eisenberg, D. and Kauzmann, W. (1969) *The Structure and Properties of Water*, (Clarendon, Oxford).

126. Finney, J. L. (2004). Water? What's so special about it?, *Phil. Trans. R. Soc. Lond. B* **359**, pp. 1145–1165.

127. Marechal, Y. (2007). *The Hydrogen Bond and the Water Molecule*, (Elsevier, Amsterdam).

128. Conner, B. N., Takano, T., Tanaka, S., Itakura, K., and Dickerson, R. E. (1982). The molecular structure of d(ICpCpGpG), a fragment of right-handed double helical A-DNA, *Nature* **295**, pp. 294–299.

129. Tereshko, V., Minasov, G., and Egli, M. (1999). A "hydrat-ion" spine in a B-DNA minor groove, *J. Am. Chem. Soc.* **121**, pp. 3590–3595.

130. Minasov, G., Tereshko, V., and Egli, M. (1999). Atomic-resolution crystal structures of B-DNA reveal specific influences of divalent metal ions on conformation and packing, *J. Mol. Biol.* **291**, pp. 83–99.

131. Shotton, M. W., Pope, L. H., Forsyth, T., Langan, P., Denny, R. C., Giesen, U., Dauvergne, M. T., and Fuller, W. (1997). A high-angle neutron fibre diffraction study of the hydration of deuterated A-DNA, *Biophys. Chem.* **69**, pp. 85–96.

132. Matsui, H., Ohhta, Y., Iida, C., Horii, M., and Tadokoro, M. (2010). Observation of quasi-one dimensional proton conductions in molecular porous crystal $[Co^{III}(H_2bim)_3](TMA)\cdot 20H_2O$, *J. Phys. Soc. Jpn.* **79**, pp. 103601 (1–4).

133. Matsui, H. and Tadokoro, M. (2012). Eigen-like hydrated protons traveling with a local distortion through the water nanotube in new molecular porous crystals $[M^{III}(H_2bim)_3](TMA)\cdot 20H_2O_n$ ($M =$ Co, Rh, Ru), *J. Chem. Phys.* **137**, pp. 144503 (1–11).

134. Matsui, H., Suzuki, Y., Fukumochi, H., and Tadokoro, M. (2014). Defect dynamics of the dipole ordered water chain in a polar nanochannel, *J. Phys. Soc. Jpn.* **83**, pp. 054708 (1–10).

135. Agre, P. (2004). Aquaporin water channels (Nobel Lecture), *Angew Chem. Int. Ed.* **43**, pp.4278–4290.

136. Angell, C. L. (1961). An infrared spectroscopic investigation of nucleic acid constituents, *J. Chem. Soc. (London)*, pp. 504–515.

137. Susi, H. and Ard, J. S. (1971). Vibrational spectra of nucleic acid constituents I, *Spectrochimica Acta* **27A**, pp. 1549–1562.

138. Susi, H., Ard, J. S., and Purcell, J. M. (1973). Vibrational spectra of nucleic acid constituents II, *Spectrochimica Acta* **29A**, pp. 725–733.

139. Susi, H. and Ard, J. S. (1973). Planar valence force constants and assignments for pyrimidine derivatives, *Spectrochimica Acta* **30A**, pp. 1843–1853.

140. Rosenberg, M., Shoham, G., Reva, I., and Hausto, R. (2004). Low temperature FTIR spectroscopy and hydrogen bonding in cytosine polycrystals, *Spectrochimica Acta* **60A**, pp. 463–470.

141. Blout, E. R. and Fields, M. (1950). Absorption spectra. VIII. The infrared spectra of some purines and pyrimidines, *J. Am. Chem. Soc.* **72**, pp. 479–484.

142. Abdurakhman, H., Tajiri, K., Yokoi, H., Kuroda, N., Matsui, H., Yanagimachi, T., Taniguchi, M., Kawai, T., and Toyota, N. (2007). Infrared spectroscopy on poly(dG)-poly(dC) DNA at low hydration, *J. Phys. Soc. Jpn.* **76**, pp. 024009.

143. Hojo, A., Matsui, H., Iwamoto, K., Yanagimachi, T., Abrurakhan, H., Taniguchi, M., Kawai, T., and Toyota, N. (2008). Hydration effects on the microwave dielectricity in dry poly(dA)-poly(dT) DNA, *J. Phys. Soc. Jpn.* **77**, pp. 044802 (1–7).

144. Matsui, H., Nagatori, M., Sakamoto, H., Mizoguchi, K., and Toyota, N. (2010). Hydration states and metallic intercalation effects in dry DNA, *J. Phys. Chem. Solids* **71**, pp. 440–443.

145. Semenov, M., Bolbukh, T., and Maleev, V. (1997). Infrared study of the influence of water on DNA stability in the dependence on AT/GC composition, *J. Mol. Struct.* **408/409**, pp. 213–217.

146. Sclavi, B., Peticolas, W. L., and Powell, J. W. (1994). Fractal-like patterns in DNA films, B form at 0% relative humidity, and antiheteronomous DNA: An IR study, *Biopolymers* **34**, pp. 1105–1113.

147. Peppas, N. A. and Langer, R. (1994). New challenges in biomaterials, *Science* **263**, pp. 1715–1720.

148. Bettinger, C. J. and Bao, Z. (2010). Biomaterials-based organic electronic devices, *Polym. Int.* **59**, pp. 563–567.

149. Meller, A., Nivon, L., Brandin, E., Golovchenko, J., and Branton, D. (2000). Rapid nanopore discrimination between single polynucleotide molecules, *Proc. Natl. Acad. Sci.* **97**, pp. 1079–1084.

150. Li, J. L., Gershow, M., Stein, D., Brandin, E., and Golovchenko, J. A. (2003). DNA molecules and configurations in a solid-state nanopore microscope, *Nat. Mater.* **2**, pp. 611–615.

151. Liu, Y., Garcia, C., and Henry, C. (2003). Recent progress in the development of TAS for clinical analysis, *Analyst* **128**, pp. 1002–1008.

152. Malsch, I. (2002). Biomedical applications of nanotechnology, *Ind. Phys.* pp. 15–17.

153. Matsuo, Y. and Hatori, J. (2011). Fuel cell based on natural sausage casing, *J. Biobased Mater. Bioenergy* **5**, pp. 562–564.

154. Matsuo, Y., Hatori, J., and Oyama, H. (2012). Proton conductivity and impedance analysis in submucosa membrane, in *Solid State Ionics: Ionics Sustainable World (the 13th Asian Conference)* (Sendai, Japan), pp. 334–339.

155. Kasumov, A. Y., Klinov, D. V., Roche, P.-E, Gueron, S., and Bouchiat, H. (2004), Thickness and low temperature conductivity of DNA molecules, *Appl. Phys. Lett.* **84**, pp. 1007–1009.

156. Kutnjak, Z., Filipic, C., Podgornik, R., Nordenskiold, L., and Korolev, N. (2003). Electrical conduction in native deoxyribonucleic acid: Hole hopping transfer mechanism, *Phys. Rev. Lett.* **90**, pp. 098101 (1–4).

157. Mizoguchi, K., Tanaka, S., Ogawa, T., Shiobara, N., and Sakamoto, H. (2005). Magnetic study of the electronic states of B-DNA and M-DNA doped with metal ions, *Phys. Rev. B* **72**, 033106 (1–4).

158. Watanabe, H., Manabe, C., Shigematsu, T., Shimotani, K., and Shimizu, M. (2001). Single molecule DNA device measured with triple-probe atomic force microscope, *Appl. Phys. Lett.* **79**, pp. 2462–2464.

159. Yang, C., Moses, D., and Heeger, A. J. (2003). Base-pair stacking in oriented films of DNA, *Adv. Mater.* **15**, pp. 1364–1367.

160. Lei, C. H., Das, A., Elliott, M., and Mcdonald, J. E. (2003). Conductivity of macromolecular networks measured by electrostatic force microscopy, *Appl. Phys. Lett.* **83**, pp. 482–484.

161. Cole, K. S. and Cole, R. H. (1941). Dispersion and absorption in dielectrics I. Alternating current characteristics, *J. Chem. Phys.* **9**, pp. 341–351.

162. Matsuo, Y., Hatori, J., Yoshida, Y., and Ikehata, S. (2010). Humidity dependence of proton conductivity in DNA film studied by NMR and AC conductivity, *J. Phys. Soc. Jpn.* **79**, Suppl. A, pp. 12–14.

163. Matsuo, Y., Sugita, K., and Ikehata, S. (2005). Doping effect for ionic conductivity in DNA film, *Synth. Met.* **154**, pp. 13–16.

164. Matsuo, Y., Kumasaka, G., Hatori, J., Saito, K., and Ikehata, S. (2006). Electrical properties of fuel cell based on DNA film, *Curr. Appl. Phys.* **6**, pp. 340–343.

165. Davis, R. D., Bur, A. J., McBrearty, M., Lee, Y.-H., Gilman J. W., and Start, P. R. (2004). Dielectric spectroscopy during extrusion processing of polymer nanocomposites: a high throughput processing/characterization method to measure layered silicate content and exfoliation, *Polymer* **45**, pp. 6487–6493.

166. Wadati, H., Okazaki, K., Niimi, Y., Fujimori, A., Tabata, H., Pikus, J., and Lewis, J. P. (2005). Photoemission study of poly(dA)-poly(dT) DNA: Experimental and theoretical approach to the electronic density of states, *Appl. Phys. Lett.* **86**, pp. 023901 (1–3).

167. Hieda, K. (1994). DNA damage induced by vacuum and soft X-ray photons from synchrotron radiation, *Int J. Radiat. Biol.* **66**, 5, pp. 561–567.

168. Purkayastha, S., Milligan, J. R., and Bernhard, W. A. (2007). On the chemical yield of base lesions, strand breaks, and clustered damage generated in plasmid DNA by the direct effect of X rays, *Radiat. Res.* **168**, pp. 357–366.

169. Yokoya, A. and Akamatsu, K. (2001). EPR spectrometer installed in a soft X-ray beamline at SPring-8 for biophysical studies, *Nucl. Instrum. Meth. A* **1333**, pp. 467–468.

170. Oka, T., Yokoya, A., and Fujii, K. (2011). Electron paramagnetic resonance study of unpaired electron species in thin films of pyrimidine bases induced by nitrogen and oxygen K-shell photoabsorption, *Appl. Phys. Lett.* **98**, 10, pp. 103701 (1–3).

171. Oka, T., Yokoya, A., and Fujii, K. (2012). Lifetime of the unpaired electron species in calf thymus DNA thin films induced by nitrogen and oxygen K-shell photoabsorption, *Int. J. Radiat. Biol.* **88**, pp. 884–887.

172. Oka, T., Yokoya, A., Fujii, K., Fukuda, Y., and Ukai, M. (2012). Unpaired electron species in thin films of calf-thymus DNA molecules induced by nitrogen and oxygen K-shell photoabsorption, *Phys. Rev. Lett.* **109**, 21, pp. 213001 (1–5).

173. Schmidt, V. (1997). *Electron Spectrometry of Atoms using Synchrotron Radiation* (Cambridge University Press, Cambridge, England).

174. Boudaïffa, B., Cloutier, P., Hunting, D., Huels, M. A., and Sanche, L. (2000). Resonant formation of DNA strand breaks by low-energy (3 to 20 eV) electrons, *Science* **287**, 5458, pp. 1658–1660.

175. Macnaughton, J. B., Kurmaev, E. Z., Finkelstein, L. D., Lee, J. S., Wettig, S. D., and Moewes, A. (2006). Electronic structure and charge carriers in metallic DNA investigated by soft x-ray spectroscopy, *Phys. Rev. B* **73**, pp. 205114 (1–7).

176. Macnaughton, J. B., Yablonskikh, M. V., Hunt, A. H., Kurmaev, E. Z., Lee, J. S., Wettig, S. D., and Moewes, A. (2006). Solid versus solution: Examining the electronic structure of metallic DNA with soft x-ray spectroscopy, *Phys. Rev. B* **74**, pp. 125101 (1–5).

177. Dekker, C. and Ratner, M. A. (2001). Electronic properties of DNA, *Physics World* pp. 29–33.

178. Heath, J. R. and Ratner, M. A. (2003). Molecular electronics, *Physics Today* pp. 43–50.

179. Hjort, M. and Stafström, S. (2001). Band resonant tunneling in DNA molecules, *Phys. Rev. Lett.* **87**, pp. 228101 (1–4).

180. Sugiyama, H. and Saito, I. (1996). Theoretical studies of GG-specific photocleavage of DNA via electron transfer: significant lowering of ionization potential and 5′-localization of HOMO of stacked GG bases in B-Form DNA, *J. Amer. Chem. Soc.* **118**, (30), pp. 7063–7068.

181. Hutter, J., Carloni, P., and Parrinello, M. (1996). Nonempirical calculations of a hydrated RNA duplex, *J. Amer. Chem. Soc.* **118** (38), pp. 8710–8712.

182. Carloni, P. and Andreoni, W. (1996). Platinum-modified nucleobase pairs in the solid state: a theoretical study, *J. Phys. Chem.* **100** (45), pp. 17797–17800.

183. Lewis, J. P., Ordejón, P., and Sankey, O. F. (1997). Electronic-structure based molecular-dynamics method for large biological systems: application to the 10 basepair poly(dG)-poly(dC) DNA double helix, *Phys. Rev. B* **55**, pp. 6880–6887.

184. Machado, M., Ordejón, P., Artacho, E., Sánchez-Portal, D., and Soler, J. M. (1999). Density functional calculations of planar DNA base-pairs, *e-print* cond-mat/Physics/9908022, pp. 1–12.

185. Kurita, N. and Kobayashi, K. (2000). Density functional MO calculation for stacked DNA base-pairs with backbones, *Comp. Chem.* **24**, pp. 351–357.

186. Hobza, P. and Šponer, J. (1999). Structure, energetics, and dynamics of the nucleic acid base pairs: nonempirical ab initio calculations, *Chem. Rev.* **99** (11), pp. 3247–3276.
187. Di Felice, R., Calzolari, A., Molinari, E., and Garbesi, A. (2001). Ab initio study of model guanine assemblies: the role of π-π coupling and band transport, *Phys. Rev.* **B55**, pp. 045104 (1–10).
188. Starikov, E. B. (2002). Mechanisms of charge carrier generation in polycrystalline DNA fibers, *Phys. Chem. Chem. Phys.* **4**, pp. 4523–4527; Starikov, E. B. (2002). Quantum chemistry of nucleic acids: how it could help and when it is necessary, *J. Photochem. Photobiol. C: Photochem. Rev.* **3**, pp. 147–164. References therein.
189. Gervasio, L. F., Carloni, P., and Parrinello, M. (2002). Electronic structure of wet DNA, *Phys. Rev. Lett.* **89**, pp. 108102 (1–4).
190. Hückel, E. (1932). Quantum theoretical contributions to the problem of aromatic and non-saturated compounds, *Z. Phys.* **76**, pp. 628–648.
191. Coulson, H. C. C. A. and Longuet-Huggins, H. C. (1947). The electronic structure of conjugated systems. I. General theory, *Proc. R. Soc. London* **A191**, pp. 39–60; Coulson, H. C. C. A. and Longuet-Huggins, H. C. (1947). The electronic structure of conjugated systems. II. Unsaturated hydrocarbons and their hetero-derivatives, *Proc. R. Soc. London* **A192**, pp. 16–32.
192. Coulson, H. C. C. A. and Longuet-Huggins, H. C. (1948). The electronic structure of conjugated systems. III. Bond orders in unsaturated molecules; IV. Force constants and interaction constants in unsaturated hydrocarbons, *Proc. R. Soc. London* **A193**, pp. 447–464; Coulson, H. C. C. A. and Longuet-Huggins, H. C. (1948). The electronic structure of conjugated systems. V. The interaction of two conjugated systems, *Proc. R. Soc. London* **A195**, pp. 188–197.
193. Coulson, H. C. C. A. and Longuet-Huggins, H. C. (1950). The electronic structure of conjugated systems. VI, *Proc. R. Soc. London* **A201**, pp. 196–209.
194. Hoffman, R. (1963). An extended Hückel theory. I. Hydrocarbons, *J. Chem. Phys.* **39**, pp. 1397–1412.
195. Ladik, J. J. (1999). Polymers as solids: A quantum mechanical treatment, *Phys. Rep.* **313**, pp. 171–235.
196. Ashcroft, N. W. and Mermin, N. D. (1976). *Solid State Physics*, (Holt-Saunders, New York).

197. Yomosa, S. (1964). Physical methods for treating the π-electron systems of biopolymers such as polypeptides and polynucleotides in the Hückel approximation, *J. Phys. Soc. Jpn.* **19**, pp. 1718–1730.

198. Bakhshi, A. K. (1994). Investigation of electronic conduction in proteins and DNA, *Prog. Biophys. molec. Biol.* **61**, pp. 187–253. References in 1990's therein.

199. Ladik, J. and Hoffman, T. A. (1963). Quantum mechanical calculation of the electronic structure of DNA, *Biopolymers Symposia* **13**, pp. 117–127.

200. Hoffman, T. A. and Ladik, J. (1964). Quantum mechanical considerations on some properties of DNA, *Adv. Chem. Phys.* **VII**, pp. 84–158.

201. Ladik, J. and Biczó, G. (1965). Energy-band calculations for periodic DNA models in the Hückel approximation, *J. Chem. Phys.* **42**, pp. 1658–1662. References in 1960's therein. This is about the time of birth of quantum chemistry.

202. Iguchi, K. (1997). Tight-binding model for DNA double chains: Metal-insulator transition due to the formation of a double strand of DNA, *Int. J. Mod. Phys.* **11**, pp. 2405–2423.

203. Iguchi, K. (2001). Semiconductivity and band gap of a double strand of DNA, *J. Phys. Soc. Jpn.* **70**, pp. 593–597.

204. Iguchi, K. (2003). Extrinsic semiconductor character of a double strand of DNA: Holstein's polarons as donors and acceptors, *Int. J. Mod. Phys.* **17**, pp. 2565–2578.

205. Roche, S. (2003). Sequence dependent DNA-mediated conduction, *Phys. Rev. Lett.* **91**, pp. 108101 (1–4); Roche, S., Bicout, D., Maciá, E., and Kats, E. (2003). Long range correlations in DNA: Scaling properties and charge transfer efficiency, *Phys. Rev. Lett.* **91**, 228101 (1–4).

206. Maciá Barber, E. (2008). *Aperiodic Structures in Condensed Matter: Fundamentals and Applications*, (Taylor & Francis, New York). References therein.

207. Yamada, H. (2004). Electronic localization properties of a double strand of DNA: A simple model with long-range correlated hopping disorder, *Int. J. Mod. Phys.* **B18**, pp. 1697–1716; Yamada, H. (2004). Localization of electronic states in chain models based on real DNA sequence, *Phys. Lett. A* **332**, pp. 65–73.

208. Yamada, H. and Iguchi, K. (2010). Some effective tight-binding models for electrons in DNA conduction: A review, *Adv. in Cond. Mat. Phys.* **2010**, pp. 380710 (1–28).

209. Klotsa, D., Römer, R. A., and Turner, M. S., (2005). Electronic transport in DNA, *Biophys. J.* **89**, pp. 2187–2198.

210. Wang, H., Marsh, R., Lewis, J. P., and Römer, R. A. (2006). Electronic transport and localization in short and long DNA, in *Modern Methods for Theoretical Physical Chemistry of Biopolymers*, edited by E. B. Starikov, J. P. Lewis, S. Tanaka, (Elsevier B. V., Amsterdam), Chapter 21, pp. 407–428.

211. Cuniberti, G., Maciá, E., Rodríguez, A., and Römer, R. A. (2007). Tight-binding modeling of charge migration in DNA devices, in *Charge Migration in DNA*, edited by Prof. Dr. T. Chakraborty, (Springer Berlin Heidelberg) pp. 1–20.

212. Fukui, K. (1976). *Chemical Reactions and Electronic Orbitals*, (Maruzen, Tokyo), in Japanese.

213. Burdett, J. K. (1984). From bonds to bands and molecules to solids, *Prog. Solid St. Chem.* **15**, pp. 173–255.

214. Nagata, Chikayoshi (2002). Private communications through postal mails (Jan. 25). He was one of graduate students from Ken-Ichi Fukui who got Nobel Prize for the discovery of frontier electrons.

215. Stryer, L. (1988). *Biochemistry*, Third Ed., (Freeman and Company, New York).

216. Yomosa, S. (1965). Electronic structures of proteins, in *Quantum Chemistry II*, Lecture Series in Biophysics, (Yoshioka Shoten, Kyoto), pp. 3–74 (in Japanese).

217. Nagata, C. (1965). Electronic structures of DNA, in *Quantum Chemistry II*, Lecture Series in Biophysics, (Yoshioka Shoten, Kyoto), pp. 75–112 (in Japanese).

218. Nagata, C. (1975). *Introduction to Quantum Biology*, (Gakkai Shuppan Center, Tokyo), (in Japanese).

219. Pullman, B. and Pullman, A. (1960). Some electronic aspects of biochemistry, *Rev. Mod. Phys.* **32**, pp. 428–436.

220. Chiang, C. K., Fincher, Jr., C. R., Park, Y. W., Heeger, A. J., Shirakawa, H., Louis, E. J., Gau, S. C., and MacDiarmid, A. G. (1977). Electrical conductivity in doped polyacetylene, *Phys. Rev. Lett.* **39**, pp. 1098–1101.

221. Heeger, A. J., Kivelson, S., Schrieffer, J. R., and Su, W.-P. (1988). Solitons in conducting polymers, *Rev. Mod. Phys.* **60**, pp. 781–850.

222. Fukui, K., Morokuma, K., and Nagata, C. (1960). A molecular orbital treatment of phosphate bonds of biochemical interest. I. Simple LCAO MO treatment, *Bull. Chem. Soc. Jpn.* **33**, pp. 1214–1219.

223. Fukui, K., Imamura, A., and Nagata, C. (1963). A molecular orbital treatment of phosphate bonds of biochemical interest. II. Metal chelates of adenosine triphosphate, *Bull. Chem. Soc. Jpn.* **33**, pp. 1450–1453.

224. Boyd, D. B. and Lipscomb, W. N. (1969). Electronic structures for energy-rich phosphates, *J. Theor. Biol.* **25**, pp. 403–420.

225. Mulliken, R. S. (1955). Electronic population analysis on LCAO-MO molecular wave functions. II. Overlap populations, bond orders, and covalent bond energies, *J. Chem. Phys.* **23**, pp. 1841–1846.

226. Wolfsberg, M. and Helmholtz, L. (1952). The spectra and electronic structure of the tetrahedral ions MnO_4^-, CrO_4^{--}, and ClO_4^-, *J. Chem. Phys.* **20**, pp. 837–843.

227. Sandorfy, C. (1949). Etude sur le role des substitutants et des heteroatomes dans le noyau benzenique, *Bull. Soci. Chim. de France* **16**, pp. 615–626.

228. Streiwieser, Jr., A. (1961). *Molecular Orbital Theory for Organic Chemists*, (John Willey & Sons, Inc., New York), Chapter 5.

229. Mulliken, R. S., Rieke, C. A., Orloff, D., and Orloff, H. (1949). Formulas and numerical tables for overlap integrals, *J. Chem. Phys.* **17**, pp. 1248–1267.

230. Ballhausen, C. J. and Gray, H. B. (1964). *Molecular Orbital Theory*, (Benjamin, New York), p. 158.

231. Prichard, H. O. and Skinner, H. A. (1955). The concept of electronegativity, *Chem. Rev.* **1955**, 55 (4), pp. 745–786.

232. Vela, A. and Gázquez, J. L. (1988). Extended Hückel parameters from density functional theory, *J. Phys. Chem.* **92**, pp. 5688–5693.

233. Zhao, J. and Lu, J. P. (2003). A nonorthogonal tight-binding total energy model for molecular simulations, *Phys. Lett. A* **319**, pp. 523–529.

234. Arnold, V. I. (1989). *Mathematical Methods of Classical Mechanics*, Second Edition, (Springer-Verlag, Berlin), p. 225.

235. Meade, T. J. and Kayyem, J. F. (1995). Electron transfer through DNA: Site-specific modification of duplex DNA with ruthenium donors and acceptors, *Angew. Chem. Int. Ed. Engl.* **34**, 352–354.

236. White, J. H. and Bauer, W. R. (1986). Calculation of the twist and the writhe for representative models of DNA, *J. Mol. Biol.* **189**, pp. 329–341.

237. Mila, F., Stafford, C. A., and Capponi, S. (1998). Persistent currents in a Möbius ladder: A test of interchain coherence of interacting electrons, *Phys. Rev. B* **57**, pp. 1457–1460.

238. Hayashi, M. and Ebisawa, H. (2001). Little-parks oscillation of superconducting Möbius strip, *J. Phys. Soc. Jpn.* **70**, pp. 3495–3498.

239. Schrödinger, E. (1979). *What is life? & Mind and Matter*, (Cambridge University, Cambridge, 1967).

240. Shechtman, D., Blech, I., Gratias, D., and Cahn, J. W. (1984). Metallic phase with long-range orientational order and no translational symmetry, *Phys. Rev. Lett.* **53**, pp. 1951–1953.

241. Kohmoto, M., Kadanoff, L. P., and Tang, C. (1983). Localization problem in one dimension: Mapping and escape, *Phys. Rev. Lett.* **50**, pp. 1870–1872.

242. Hofstadter, D. R. (1979). *Gödel, Esher, Bach: An Eternal Golden Braid*, (Vintage Books, New York); translated into Japanese (Hakuyosha, Tokyo, 1985).

243. Kohmoto, M., Sutherland, B., and Iguchi, K. (1987). Localization of optics: Quasiperiodic media, *Phys. Rev. Lett.* **58**, pp. 2436–2438.

244. Sutherland, B. (1991). Quasiperiodic physics in one dimension, in *Disorder and Nonlinearity*, edited by Bishop, A. B., Campbell, D. K., and Pnevmatikos, S. (Springer-Verlag, New York), pp. 62–69.

245. Iguchi, K. (1990). *Theory of quasiperiodic lattices*, Ph. D. Thesis, Department of Physics, The University of Utah, Salt Lake City, Utah, USA.

246. Iguchi, K. (1991). Theory of quasiperiodic lattices. I. Scaling transformation for a quasiperiodic lattice, *Phys. Rev. B* **43**, pp. 5915–5918; Theory of quasiperiodic lattices. II. Generic trace map and invariant surface, *Phys. Rev. B* **43**, pp. 5919–5923.

247. Iguchi, K. (1992). Exact wave functions of an electron on a quasiperiodic lattice: Definition of an infinite dimensional Riemann theta function, *J. Math. Phys.* **33**, pp. 3938–3947.

248. Iguchi, K. (1994). Some remarks on the theory of an electron on binary quasiperiodic lattices, *Int. J. Mod. Phys.* **B 8**, pp. 1931–1964.

249. Iguchi, K. (1995). Electronic structure of one-dimensional quasiperiodic materials of $AB_{1-x}C_x$, *J. Phys. Soc. Jpn.* **64**, pp. 177–184.

250. Iguchi, K. (1994). The Schrödinger's dream: Electronic structure of DNAs and proteins: Quantum automata, *RIKEN Review* **6**, pp. 49–50.

251. Wood, D. O., Dinsmore, M. J., Bare, G. A., and Lee, J. S. (2002). M-DNA is stabilised in G·C tracts or by incorporation of 5-fluorouracil, *Nuc. Acids Res.* **30**, 10, pp. 2244–2250.

252. Mizoguchi, K. and Sakamoto, H. (2015). Role of water molecules to the electronic states of M-DNA, *Crystals* **5**, pp. 475–490.

253. Matsui, H., Toyota, N., Nagatori, M., Sakamoto, H., and Mizoguchi, K. (2009). Infrared spectroscopic studies on incorporating the effect of metallic ions into a M-DNA double helix, *Phys. Rev. B* **79**, pp. 235201 (1-8).

254. Berg, J. and Tymoczko, J. (2006). *Biochemistry 6th edition* (W. H. Freeman and Company).

255. Abragam, A. (1961). *The Principles of Nuclear Magnetism* (Oxford University Press, Oxford).

256. Slichter, C. P. (1989). *Principles of Magnetic Resonance* (Springer, Berlin, Heidelberg).

257. Anderson, P. W. and Weiss, P. R. (1953). Exchange narrowing in paramagnetic resonance, *Rev. Mod. Phys.* **25**, pp. 269–276.

258. Kittel, C. (1966). *Introduction to Solid State Physics* (Wiley, New York).

259. Hennessy, M. J., McElwee, C. D., and Richards, P. M. (1973). Effect of interchain coupling on electron-spin resonance in nearly one-dimensional systems, *Phys. Rev. B* **7**, pp. 930–947.

260. Nagata, K. and Tazuke, Y. (1972). Short range order effect on EPR frequencies in Heisenberg linear chain antiferromagnets, *J. Phys. Soc. Jpn.* **32**, pp. 337–345.

261. Abragam, A. and Bleany, B. (1970). *Electron Paramagnetic Resonance of Transition Ions* (Clarendon Press, Oxford).

262. Gordy, W. (1980). *Theory and Applications of Electron Spin Resonance* (John Wiley & Sons, New York).

263. Zhitomirskii, A. N. (1968). Relationship between the degree of ionic character of a bond and the EPR hyperfine coupling constant for monotypal binary crystals doped with manganese, *Journal of Structural Chemistry, translated from Zhurnal Strukturnoi Khimii, Vol. 9, pp. 612–615, 1968* **9**, 4, pp. 532–535.

264. Hackl, E. V., Kornilova, S. V., Kapinos, L. E., Andrushchenko, V. V., Galkin, V. L., Grigoriev, D. N., and Blagoi, Y. P. (1997). Study of Ca^{2+}, Mn^{2+} and Cu^{2+} binding to DNA in solution by means of IR spectroscopy, *J. Mol. Struct.* **408/409**, pp. 229–232.

265. Wittlin, A., Genzel, L., Kremer, F., Haseler, S., Poglitsch, A., and Rupprecht, A. (1986). Far-infrared spectroscopy on oriented films of dry and hydrated DNA, *Phys. Rev. A* **34**, pp. 493–500.

266. Starikov, E. B. (2003). Role of electron correlations in deoxyribonucleic acid duplexes: is an extended Hubbard Hamiltonian a good model in this case?, *Phil. Mag. Lett.* **83**, 11, pp. 699–708.

267. Gao, X.-T., Fu, X., Mei, L.-M., and Xie, S.-J. (2006). Electrical conductance of DNA molecules with varied density of itinerant electrons, *J. Chem. Phys.* **124**, pp. 234702 (1–6).

268. Krisch, M., Mermet, A., Grimm, H., Forsyth, V. T., and Rupprecht, A. (2006). Phonon dispersion of oriented DNA by inelastic X-ray scattering, *Phys. Rev. E* **73**, pp. 061909 (1–10).

269. Drew, H. R., Wing, R. M., Takano, T., Broka, C., Tanaka, S., Itakura, K., and Dickerson, R. E. (1981). Structure of a B-DNA dodecamer: conformation and dynamics, *Proc. Natl. Acad. Sci. USA* **78**, pp. 2179–2183.

270. Tanaka, K., Tengeiji, A., Kato, T., Toyama, N., Shiro, M., and Shionoya, M. (2002). Efficient incorporation of a copper hydroxypyridone base pair in DNA, *J. Am. Chem. Soc.* **124**, pp. 12494–12498.

271. Binnig, G. and Rohrer, H. (1982). Scanning tunneling microscopy, *Helv. Phys. Acta* **55**, pp. 726–735.

272. Tanaka, H. and Kawai, T. (2009). Partial sequencing of a single DNA molecule with a scanning tunnelling microscope, *Nature Nanotechnology* **4**, pp. 518–522.

273. Lee, G., Arscott, P. G., Bloomfield, V. A., and Evans, D. F. (1989). Scanning tunneling microscopy of nucleic acids, *Science* **244**, pp. 475–477.

274. Thomas P. Beebe, J., Wilson, T. E., Ogletree, D. F., Katz, J. E., Balhorn, R., Salmeron, M. B., and Siekhaus, W. J. (1989). Direct observation of native DNA structures with the scanning tunneling microscope, *Science* **243**, pp. 370–372.

275. Dunlap, D. D. and Bustamante, C. (1989). Images of single-stranded nucleic acids by scanning tunnelling microscopy, *Nature* **342**, pp. 204–206.

276. Eigler, D. M., Weiss, P. S., Schweizer, E. K., and Lang, N. D. (1991). Imaging Xe with a low-temperature scanning tunneling microscope, *Phys. Rev. Lett.* **66**, 9, pp. 1189–1192.

277. Clemmer, C. R. and Thomas P. Beebe, J. (1991). Graphite: A mimic for DNA and other biomolecules in scanning tunneling microscope studies, *Science* **251**, pp. 640–642.

278. Heckl, W. M. and Binnig, G. (1992). Domain walls on graphite mimic DNA, *Ultramicroscopy* **42**, pp. 1073–1078.

279. Guckenberger, R., Heim, M., Cevc, G., Knapp, H. F., Wiegrabe, W., and Hillebrand, A. (1994). Scanning tunneling microscopy of insulators and biological specimens based on lateral conductivity of ultrathin water films, *Science* **266**, pp. 1538–1540.

280. Cluzel, P., Lebrun, A., Heller, C., Lavery, R., Viovy, J.-L., Chatenay, D., and Caron, F. (1996). DNA: An extensible molecule, *Scinece* **271**, 5250, pp. 792–794.

281. Léger, J. F., Romano, G., Sarkar, A., Robert, J., Bourdieu, L., Chatenay, D., and Marko, J. F. (1999). Structural transitions of a twisted and stretched DNA molecule, *Phys. Rev. Lett.* **83**, 5, pp. 1066–1069.

282. Lebrun, A. and Lavery, R. (1996). Modelling extreme stretching of DNA, *Nuc. Acids Res.* **24**, 12, pp. 2260–2267.

283. Mori, E., Usui, H., Sakamoto, H., Mizoguchi, K., and Naito, T. (2011). Charge distribution in the surface BEDT-TTF layer of α-(BEDT-TTF)$_2$I$_3$ at room temperature with scanning tunneling microscopy, *J. Phys. Soc. Jpn.* **81**, 1, pp. 014707 (1–7).

284. Bardeen, J. (1961). Tunnelling from a many-particle point of view, *Phys. Rev. Lett.* **6**, pp. 57–59.

285. Tersoff, J. and Hamann, D. R. (1983). Theory and application for the scanning tunneling microscope, *Phys. Rev. Lett.* **50**, pp. 1998–2001.

286. Tanaka, K. and Okahata, Y. (1996). A DNA-lipid complex in organic media and formation of an aligned cast film, *J. Am. Chem. Soc.* **118**, pp. 10679–10683.

287. Wang, L., Yoshida, J., Ogata, N., Sasaki, S., and Kamiyama, T. (2001). Self-assembled supramolecular films derived from marine deoxyribonucleic acid (DNA) - cationic lipid complexes: large-scale preparation and optical and thermal properties, *Chem. Mater.* **13**, 4, pp. 1273–1281.

288. Kawabe, Y., Wang, L., Nakamura, T., and Ogata, N. (2002). Thin-film lasers based on dye-deoxyribonucleic acid-lipid complexes, *Appl. Phys. Lett.* **81**, pp. 1372–1374.

289. Ogata, N. (2003). Light and DNA: How to utilize DNA as photonic materials?, *Monograph on Photonic Science and Technology: DNA for Functional Materials*, (PWC Publishing, Chitose-shi, Japan), pp. 3–21.

290. Yoshida, J., Wang, L., Kobayashi, S., Zhang, G.-J., Ikeda, H., and Ogata, N. (2004). Optical properties of photochromic-compound-doped marine-biopolymer DNA-surfactant complex films for switching applications, in Proc. SPIE, *Organic Photonic Materials and Devices VI* (San Diego, Californina, USA) **5351**, pp. 260–267.

291. Heckman, W. M., Grote, J., Yaney, P. P., and Hopkins, F. K. (2004). DNA-based nonlinear photonic materials, in Proc. SPIE, *Nonlinear Optical Transmission and Multiphoton Processes in Organics II* **5516**, pp. 47–51.

292. Heckman, E. M., Hagen, J. A., Yaney, P. P., Grote, J. G., and Hopkins, F. K. (2005). Processing techniques for deoxyribonucleic acid: Biopolymer for photonics applications, *Appl. Phys. Lett.* **87**, pp. 211115 (1–3).

293. Yaney, P. P., Heckman, E. M., Diggs, D. E., Hopkins, F. K., and Grote, J. (2005). Development of chemical sensors using polymer optical waveguides fabricated with DNA, in Proc. SPIE, *Organic Photonic Materials and Devices VII* (San Diego, California, USA) **5724**, pp. 224–233.

294. Watanuki, A., Yoshida, J., Kobayashi, S., Ikeda, H., and Ogata, N. (2005). Optical and photochromic properties of spiropyran-doped marine-biopolymer DNA-lipid complex films for switching applications, in Proc. SPIE, *Organic Photonic Materials and Devices VII* **5724**, pp. 234–241.

295. Yoshida, J., Watanuki, A., Takano, H., Kobayashi, S., Ikeda, H., and Ogata, N. (2006). Optically controlled photonic switches based on spiropyran-doped marine-biopolymer DNA-lipid complex films, in Proc. SPIE, *Organic Photonic Materials and Devices VIII* **6117**, pp. 61170L (1–8).

296. Grote, J., Heckman, E., Hagen, J., Yaney, P., Diggs, D., Subramanyam, G., Nelson, R., Zetts, J., Zang, D., Singh, B., Saricifici, N., and Hoplins, K. (2006). DNA - new class of polymer, in Proc. SPIE, *Organic Photonic Materials and Devices VIII* **6117**, pp. 61170J (1–6).

297. Yoshida, J., Takano, H., Narisawa, S., Takenaka, S., Nakai, N., Fukuda, M., and Ogata, N. (2007). Thin film dye lasers based on DNA-lipid complex materials, in Proc. SPIE, *Nanobiotronics* (San Diego, California, USA) **6646**, pp. 664608 (1–8).

298. Baude, P. F., Ender, D. A., Haase, M. A., Kelley, T. W., Muyres, D. V., and Theiss, S. D. (2003). Pentacene-based radio-frequency identification circuitry, *Appl. Phys. Lett.* **82**, pp. 3964–3966.

299. Zhu, Z. T., Mason, J. T., Dieckmann, R., and Malliaras, G. G. (2002). Humidity sensors based on pentacene thin-film transistors, *Appl. Phys. Lett.* **81**, pp. 4643–4645.

300. Zhou, L., Wanga, A., Chu Wu, S., Sun, J., Park, S., and Jackson, T. N. (2006). All-organic active matrix flexible display, *Appl. Phys. Lett.* **88**, pp. 083502 (1–3).

301. Kawasaki, M., Imazeki, S., Hirota, S., Arai, T., Shiba, T., Ando, M., Natsume, Y., Minakata, T., Uemura, S., and Kamata, T. (2008). High-mobility solution-processed organic thin-film transistor array for active-matrix color liquid-crystal displays, *J. SID* **16**, pp. 161–167.

302. Zhou, L., Park, S., Bai, B., Sun, J., Wu, S. C., Jackson, T. N., Nelson, S., Freeman, D., and Hong, Y. (2005). Pentacene TFT Driven AM OLED Displays, *IEEE Electron Dev. Lett.* **26**, pp. 640–642.

303. Mizukami, M., Hirohata, N., Iseki, T., Ohtawara, K., Tada, T., Yagyu, S., Abe, T., Suzuki, T., Fujisaki, Y., Inoue, Y., Tokito, S., and Kurita, T. (2006). Flexible AM OLED panel driven by bottom-contact OTFTs, *IEEE Electron Dev. Lett.* **27**, pp. 249–251.

304. Naber, R. C. G., Tanase, C., Blom, P. W. M., Gelinck, G. H., Marsman, A. W., Touwslager, F. J., Setayesh, S., and de Leeuw, D. M. (2005). High-performance solution-processed polymer ferroelectric field-effect transistors, *Nat. Mater.* **4**, pp. 243–248.

305. Zhu, G., Zeng, Z., Zhang, L., and Yan, X. (2005). Anomalous low-voltage tunneling current characteristics of ultrathin gate oxide ($\approx$2 nm) after high-field stress, *Appl. Phys. Lett.* **89**, pp 102905 (1–3).

306. Hasegawa, M., Kobayashi, N., Uemura, S., and Kamata, T (2009). Memory mechanism of printable ferroelectric TFT memory with tertiary structured polypeptide as a dielectric layer, *Synth. Met.* **159**, pp. 961–964.

307. Grote, J. G. and Hopkins, F. K. (2006). Bio-organic-semiconductor-field-effect-transistor based on deoxyribonucleic acid gate dielectric, *J. Appl. Phys.* **100**, pp. 024514 (1–4).

308. Steckl, A. J., Spaeth, H., You, H., Gomez, E., and Grote, J. (2011). DNA as an optical material, *Optics and Photonics News* **22**, pp. 34–39.

309. Deman, A.-L., Erouel, M., Lallemand, D., Phaner-Goutorbe, M., Lang, P., and Tardy, J. (2008). Growth related properties of pentacene thin film transistors with different gate dielectrics, *J. Non-Cryst. Solids* **354**, pp. 1598–1607.

310. Murphy, C. J., Arkin, M. R., Jenkins, Y., Ghatlia, N. D., Bossmann, S. H., Turro, N. J., and Barton, J. K. (1993). Long-range photoinduced electron transfer through a DNA helix, *Science* **262**, pp. 1025–1029.

311. Arkin, M. R., Stemp, E. D. A., Holmlin, R. E., Barton, J. K., Hörmann, A., Olson, E. J. C., and Barbara, P. F. (1996). Rates of DNA-mediated electron transfer between metallointercalators, *Science* **273**, pp. 475–480.

312. Uemura, S., Shimakawa, T., Kusabuka, K., Nakahira, T., and Kobayashi, N. (2001). Template photopolymerization of dimeric aniline by photocatalytic reaction with $Ru(bpy)_3^{2+}$ in the presence of DNA, *J. Mater. Chem.* **11**, pp. 267–268.

313. Nagarajan, R., Liu, W., Kumar, J., Tripathy, S. K., Bruno, F. F., and Samuelson, L. A. (2001). Manipulating DNA conformation using

intertwined conducting polymer chains, *Macromolecules* **34**, pp. 3921–3927.

314. Teshima, K., Yamada, K., Kobayashi, N., and Hirohashi, R. (1996). Photopolymerization of aniline with a tris(2,2′-bipyridyl)ruthenium complex-methylviologen polymer bilayer electrode system, *Chem. Commun.*, No. 7, pp. 829–830.

315. Teshima, K., Uemura, S., Kobayashi, N., and Hirohashi, R. (1998). Effect of pH on photopolymerization reaction of aniline derivatives with the tris(2,2-bipyridyl)ruthenium complex and the methylviologen system, *Macromolecules* **31**, pp. 6783–6788.

316. Uemura, S., Teshima, K., Tokuda, S., Kobayashi, N., and Hirohashi, R. (1999). Photopolymerized conducting polyaniline micropattern and its application, *Synth. Met.* **101**, pp. 701–702.

317. Kim, Y., Kobayashi, N., Teshima, K., and Hirohashi, R. (1999). Photorewritable conducting polyaniline image formation with photoinduced electron transfer, *Synth. Met.* **101**, pp. 699–700.

318. Kim, Y., Teshima, K., and Kobayashi, N. (2000). Improvement of reversible photoelectrochromic reaction of polyaniline in polyelectrolyte composite film with the dichloroethane solution system, *Electrochimica Acta* **45**, pp. 1549–1553.

319. Liu, W., Cholli, A. L., Nagarajan, R., Kumar, J., Tripathy, S., Bruno, F. F., and Samuelson, L. (1999). The role of template in the enzymatic synthesis of conducting polyaniline, *J. Am. Chem. Soc.* **121**, pp. 11345–11355.

320. Uemura, S., Yoshie, M., Kobayashi, N., and Nakahira, T. (2000). Photopolymerization of aniline dimer by photocatalytic reaction of ruthenium trisbipyridyl in the interlayer of hectorite clay, *Polym. J.* **32**, pp. 987–990.

321. Furukawa, Y., Ueda, F., Hyodo, Y., Harada, I., Nakajima, T., and Kawagoe, T. (1988). Vibrational spectra and structure of polyaniline, *Macromolecules* **21**, pp. 1297–1305.

322. Kobayashi, N., Uemura, S., Kusabuka, K., Nakahira, T., and Takahashi, H. (2001). An organic red-emitting diode with a water-soluble DNA-polyaniline complex containing $Ru(bpy)_3^{2+}$, *J. Mater. Chem.* **11**, pp. 1766–1768.

323. Kobayashi, N., Hashimoto, M., and Kusabuka, K. (2004). Structure of DNA-$Ru(bpy)_3^{2+}$ complex and its application to orange color electroluminescence device, *Electrochemistry* **72**, pp. 440–442.

324. Berggren, M., Inganas, O., Gustafsson, G., Rasmusson, J., Andersson, M. R., Hjertberg, T., and Wennerstrom, O. (1994). Light emitting diodes with variable colours from polymer blends, *Nature* **372**, pp. 444–446.

325. Hamaguchi, M. and Yoshino, K., (1996), Color-variable electroluminescence from multilayer polymer films, *Appl. Phys. Lett.* **69**, pp. 143–145

326. Lu, F.-L., Wudl, F., Nowak, M., and Heeger, A. J. (1986). Phenyl-capped octaaniline (COA): An excellent model for polyaniline, *J. Am. Chem. Soc.* **108**, pp. 8311–8313.

327. Xu, Y. F., Zhang, H. J., Li, H. Y., Bao, S. N., and He, P. (2006). Photoluminescence spectroscopy study on tris(8-hydroxyquinoline) aluminum film, *Appl. Surf. Sci.* **252**, pp. 2328–2333.

328. Nakamura, K., Ishikawa, T., Nishioka, D., Ushikubo, T., and Kobayashi, N. (2010). Color-tunable multilayer organic light emitting diode composed of DNA complex and tris(8-hydroxyquinolinato)aluminum, *Appl. Phys. Lett.* **97**, pp. 193301 (1–3).

329. Mozer, A. J. and Sariciftci, N. S. (2004). Negative electric field dependence of charge carrier drift mobility in conjugated, semiconducting polymers, *Chem. Phys. Lett.* **389**, pp. 438–442.

330. Raj Mohan, S., Joshi, M. P., and Singh, M. P. (2009). Negative electric field dependence of mobility in TPD doped polystyrene, *Chem. Phys. Lett.* **470**, pp. 279–284.

331. Park, H., Shin, D.-S., Yu, H.-S., and Chae, H.-B. (2007). Electron mobility in tris(8-hydroxyquinoline)aluminum (Alq3) films by transient electroluminescence from single layer organic light emitting diodes, *Appl. Phys. Lett.* **90**, pp. 202103 (1–3).

332. Chen, C.-H. and Meng, H.-F. (2005). Recombination distribution and color tuning of multilayer organic light-emitting diode, *Appl. Phys. Lett.* **86**, pp. 201102 (1–3).

333. Zhang, J. (ed. Zhang, J.) (2008). *PEM Fuel Cell Electrocatalysts and Catalyst Layers* (Springer, London), Section 7.

Index